MW01633297

Lenk's Television Handbook

Operation and Troubleshooting

Other McGraw-Hill Books of Interest

Consumer Electronics Series
Books by John D. Lenk

Lenk's Video Handbook
Lenk's Laser Handbook
Lenk's Audio Handbook
Lenk's RF Handbook
Lenk's Digital Handbook

Other Books by John D. Lenk

McGraw-Hill Circuit Encyclopedia and Troubleshooting Guide, Vols. 1 & 2
McGraw-Hill Electronic Testing Handbook
McGraw-Hill Electronic Troubleshooting Handbook

Handbooks

BENSON • *Audio Engineering Handbook*
BENSON • *Television Engineering Handbook*
BENSON AND WHITAKER • *Television and Audio Handbook*
COOMBS • *Printed Circuits Handbook*
COOMBS • *Electronic Instrument Handbook*
CROFT AND SUMMERS • *American Electricians' Handbook*
FINK AND BEATY • *Standard Handbook for Electrical Engineers*
FINK AND CHRISTIANSEN • *Electronic Engineers' Handbook*
HICKS • *Standard Handbook of Engineering Calculations*
KAUFMAN AND SEIDMAN • *Handbook of Electronics Calculations*
MEE AND DANIEL • *Magnetic Recording Handbook*
TUMA • *Engineering Mathematics Handbook*
WILLIAMS AND TAYLOR • *Electronic Filter Design Handbook*

Other

BARTLETT • *Cable Television Technology and Operations*
DAVIDSON • *Troubleshooting and Repairing Solid State TVs*
DAVIDSON • *TV Repair for Beginners*
INGLIS • *Video Engineering*

Lenk's Television Handbook

Operation and Troubleshooting

John D. Lenk

McGraw-Hill, Inc.

New York San Francisco Washington, D.C. Auckland Bogotá
Caracas Lisbon London Madrid Mexico City Milan
Montreal New Delhi San Juan Singapore
Sydney Tokyo Toronto

Library of Congress Cataloging-in-Publication Data
Lenk, John D.
Lenk's television handbook : operation and troubleshooting / John D. Lenk.
p. cm.
Includes index.
ISBN 0-07-037517-8
1. Television—Receivers and reception—Maintenance and repair--Handbooks, manuals, etc. I. Title. II. Title: Television handbook.
TK6653.L417 1994
621.388'8—dc20 94-17678
CIP

1 2 3 4 5 6 7 8 9 0 DOC/DOC 9 9 8 7 6 5 4

ISBN 0-07-037517-8

The sponsoring editor for this book was Dan Gonneau. The executive editor was Joanne Slike. Andrew Yoder was the managing editor. Melanie D. Holsher was the manuscript editor. The director of production was Katherine G. Brown. This book was set in ITC Century Light. It was composed in Blue Ridge Summit, Pa.

Printed and bound by R.R. Donnelly & Sons, Crawfordsville.

Greetings from the Villa Buttercup!
To my wonderful wife Irene;
Thank you for being by my side all these years!
To my lovely family, Karen, Tom, Brandon, Justin, and Michael;
And to our Lambie and Suzzie, be happy wherever you are!

To my special readers, may good fortune find your doorways to good health and happy things.
Thank you for buying my books.
And a special thanks to Steve Chapman, Dan Gonneau, Stephen Fitzgerald, and Robert McGraw of McGraw-Hill/TAB for making me a best seller again!
This is book number 80.
Abundance!

Contents

Acknowledgments xi
Introduction xiii

Chapter 1. Introduction to Television 1

1.1 Basic Black-and-White TV Broadcast System 1
1.2 Basic Black-and-White TV Set Circuits 4
1.3 Basic Color TV Broadcast System 10
1.4 Basic Color Circuits 17

Chapter 2. Front-End Circuits 23

2.1 Front-End Basics 23
2.2 FS and PLL Basics 24
2.3 Tuner with Separate PLL 30
2.4 Tuner with Built-In PLL 35
2.5 Antenna Circuits 38

Chapter 3. VIF and SIF Circuits 41

3.1 VIF and SIF Basics 41
3.2 VIF and Audio 45
3.3 Phase Lock VIF 49
3.4 PIF and SIF 55

Chapter 4. Video and Chroma Processing Circuits 59

4.1 Video and Chroma Processing Basics 59
4.2 Color-Circuit Basics 64
4.3 Basic IC Video and Chroma Processing 68
4.4 Video and Chroma Processing with a Jungle Chip 76
4.5 Sony Trinitone Switching 84

Chapter 5. Vertical and Horizontal Sweep Circuits 89

5.1 Sync-Separator Basics 89
5.2 Vertical-Sweep Basics 93
5.3 Horizontal-Sweep Basics 97
5.4 High-Voltage and Horizontal-Output Basics 101
5.5 Basic IC Vertical and Horizontal Sweep 107
5.6 Vertical and Horizontal Sweep with a Jungle Chip 116
5.7 Horizontal and Vertical Sweep (25 Inch) 124

Chapter 6. RGB Interface Circuits 131

6.1 RGB Basics 131
6.2 RGB Digital and Analog 135
6.3 Recommended Troubleshooting Approach 137

Chapter 7. System-Control Circuits 139

7.1 System-Control Basics 139
7.2 System Control (19 Inch) 145
7.3 System Control (25 Inch) 148
7.4 Key Encoder 152

Chapter 8. Low-Voltage Power-Supply Circuits 155

8.1 Low-Voltage Power-Supply Basics 155
8.2 Low-Voltage Supply (13 Inch) 161
8.3 Low-Voltage Supply (19 Inch) 167
8.4 Low-Voltage Supply (25 Inch) 167

Chapter 9. On-Screen Display and Remote Circuits 175

9.1 OSD Basics 175
9.2 Remote-Control Basics 177
9.3 OSD and Remote Circuits (13 Inch) 180
9.4 OSD Circuits (19 Inch) 184

Chapter 10. Special TV Circuits 187

10.1 Stereo-TV Basics 187
10.2 Stereo-TV Circuits 198
10.3 Surround-Sound Circuits 205
10.4 External-Audio Circuits 210
10.5 Electronic Volume-Control Circuits 213
10.6 Digital-TV Circuits 217
10.7 Picture-in-Picture Circuits 226

Chapter 11. TV Test Equipment and Procedures 235

11.1 Safety Precautions 235
11.2 Signal Generators for TV Service 237
11.3 Oscilloscopes for TV Service 241
11.4 Color Generators 242
11.5 Miscellaneous Test Equipment 254
11.6 Analyzing the Composite-Video Waveform 256
11.7 Testing Video Transformers, Yokes, and Coils 257
11.8 Purity, Convergence, and Linearity Adjustments 261
11.9 Color Setup Using a Color Generator 266
11.10 Using Vectorscopes 270
11.11 Stereo-TV Generators 273
11.12 Stereo-TV Tests and Adjustments 273

Chapter 12. TV Troubleshooting Approach 279

12.1 Troubleshooting Sequence 280
12.2 Practical TV Troubleshooting Sequence 281
12.3 Trouble Symptoms 283
12.4 Trouble Localization 285
12.5 Trouble Isolation 289
12.6 Locating a Specific Trouble 292
12.7 Troubleshooting Notes 298
12.8 Signal Injection versus Signal Tracing 299
12.9 Basic Analyst/NTSC-Generator Troubleshooting 300

Index 305

Acknowledgments

Many professionals have contributed to this book. I gratefully acknowledge the tremendous effort needed to make this book. Such a comprehensive work is impossible for one person, and I thank all who contributed, both directly and indirectly.

I give special thanks to the following: Joe Cagle and Rinaldo Swayne of Alpine/Luxman; Bob Carlson and Martin Plude of B&K-Precision Dynascan Corporation, Tom Roscoe, Dennis Yuoka and Terrance Miller of Hitachi; Pat Wilson and Ray Krenzer of Philips Consumer Electronics; Thomas Lauterback of Quasar; Donald Woolhouse of Sanyo; Theodore Zrebiec of Sony; J.W. Phipps of Thomson Consumer Electronics (RCA); John Taylor and Matthew Mirapaul of Zenith.

I also wish to thank Joseph A. Labok of Los Angeles Valley College for help and encouragement throughout the years.

And a very special thanks to Steve Chapman, Dan Gonneau, Stephen Fitzgerald, Patrick Hansard, Fred Perkins, Robert McGraw, Barbara McCann, Kimberly Martin, Jane Schmidt, Tracy Baer, Judith Reiss, Charles Love, Betty Crawford, Jeanne Myers, Peggy Lamb, Thomas Kowalczyk, Suzanne Babeuf, Jaclyn Boone, Katherine Brown, Kathy Greene, Kriss Helman, Susan Wahlman, Bob Ostrander, Andrew Yoder, Stacy Spurlock, and Midge Haramis of the McGraw-Hill/TAB organization for having that much confidence in me.

And to Irene, my wife and Super Agent, I extend my thanks. Without her help, this book could not have been written.

Introduction

This is a "something for everyone" television handbook. No matter where you are in electronics, this book provides you with how-it-works and how-to-service-it information that can be put to immediate use on any TV set.

If you are a beginner or student with little knowledge of TV, test equipment, or troubleshooting, read chapters 1, 11, and 12, which describe the basics of these subjects, respectively.

If you are an experienced technician but want information on how a TV works and what to do if it doesn't work, chapters 2 through 10 provide data for every circuit in a TV set.

It is virtually impossible in one book to cover detailed troubleshooting for all TV sets. Similarly, it is impractical to attempt such coverage, because rapid technological advances soon make such a book's details obsolete. To overcome that problem, this book concentrates on basic approaches to TV service—approaches that can be applied to any TV (both those now in use and those to be found in the future).

The approaches (based on the techniques found in my best-selling troubleshooting books) involve breaking a TV down into several related circuits. For each circuit, you will find: a full description of the circuit function, how the function is accomplished, a summary of circuit inputs and outputs, a summary of test and adjustment points in the circuits, and, of greatest importance, a recommended troubleshooting approach.

All of the circuit descriptions are supplemented with simplified and/or block diagrams that show all points of importance to the troubleshooter (input, output, adjustment controls, test points), but without the often confusing details found in service-literature schematics. The simplified block diagrams given in this book become "universal" guides to understanding and using the full schematics for TV set.

The circuits covered are common to many present-day sets. Where there are substantial variations for a particular circuit, a cross-section of circuits are described. By studying the circuits in this book, you should have no difficulty understanding the schematics of similar TV sets.

Chapter 1 is devoted to the basics of television. While it is not intended that this introduction provide a complete course in TV broadcast and reception, the chapter does provide a thorough summary of all related subjects.

Chapter 2 is devoted to TV-set circuits between the antenna or cable input and the IF stages. These include the electronic-tuner, antenna-switching, and other front-end circuits, including LP.

Chapter 3 is devoted to TV-set circuits between the front-end tuner and audio and video processing circuits, generally known as the VIF (video IF) and SIF (sound IF) stages.

Chapter 4 is devoted to TV-set circuits between the VIF and SIF stages and the picture-tube or CRT, including comb filters and jungle-chip ICs.

Chapter 5 is devoted to TV-set circuits that provide the vertical and horizontal sweep for the CRT. Because such circuits are closely associated with the high-voltage and flyback-transformer stages, as well as the sync-separator, these circuits are also included in this chapter.

Chapter 6 is devoted to TV-set circuits that provide interconnection or interface between the video and audio circuits and the CRT and loudspeakers.

Chapter 7 is devoted to TV-set circuits that control such functions as keyboard (front-panel controls) remote control, channel tuning, and band selection.

Chapter 8 is devoted to TV-set circuits that provide low-voltage power to the remaining circuits of the set.

Chapter 9 is devoted to TV-set circuits that provide on-screen display (OSD) and remote-control operation. This includes circuits that provide for the presetting of channel on/off times, blocking of channels, setting and display of day/time, and so on.

Chapter 10 is devoted to special circuits found on some TV sets. These include stereo-TV, surround sound, external audio inputs, electronic volume controls, digital TV, and picture-in-picture circuits.

Chapter 11 is devoted to test equipment used in TV service and to general procedures for use of the equipment during service.

Chapter 12 provides a summary or refresher of my basic troubleshooting approach for TV sets of all types.

Chapter

1

Introduction to Television

This chapter is devoted to the basics of television. On the assumption that you might not know how television works or that you need a refresher, I'll start with descriptions of TV broadcast and reception for black-and-white sets, as well as color. These descriptions are kept to the block-diagram level, because you are interested in service (not theory) and so that the descriptions will apply to the greatest number of TV sets (new, old, discrete-component, IC, digital, picture-in-picture, etc.). This introduction is not intended to provide a complete course in TV broadcast and reception; it is merely a summary.

1.1 Basic Black-and-White TV Broadcast System

As shown in Figure 1.1, the basic TV broadcast system consists of a transmitter capable of producing both AM and FM signals, a television camera, and a microphone. TV program sound (or audio) is transmitted through a microphone, which modulates the FM portion of the transmitter. The picture (or video) portion of the program is broadcast by means of the AM transmitter.

The TV broadcast channels are about 6 MHz wide, with the sound (FM) carrier at a frequency of 4.5 MHz higher than the picture (AM) carrier. For example, on VHF channel 7, the picture (AM) is transmitted at a frequency of 175.25 MHz, with the sound (FM) transmitted at 179.75 MHz. Channel 7 occupies the band of frequencies between 174 and 180 MHz.

The sound portion of the TV signal uses conventional FM broadcast principles, which need not be described here. However, the picture portion of

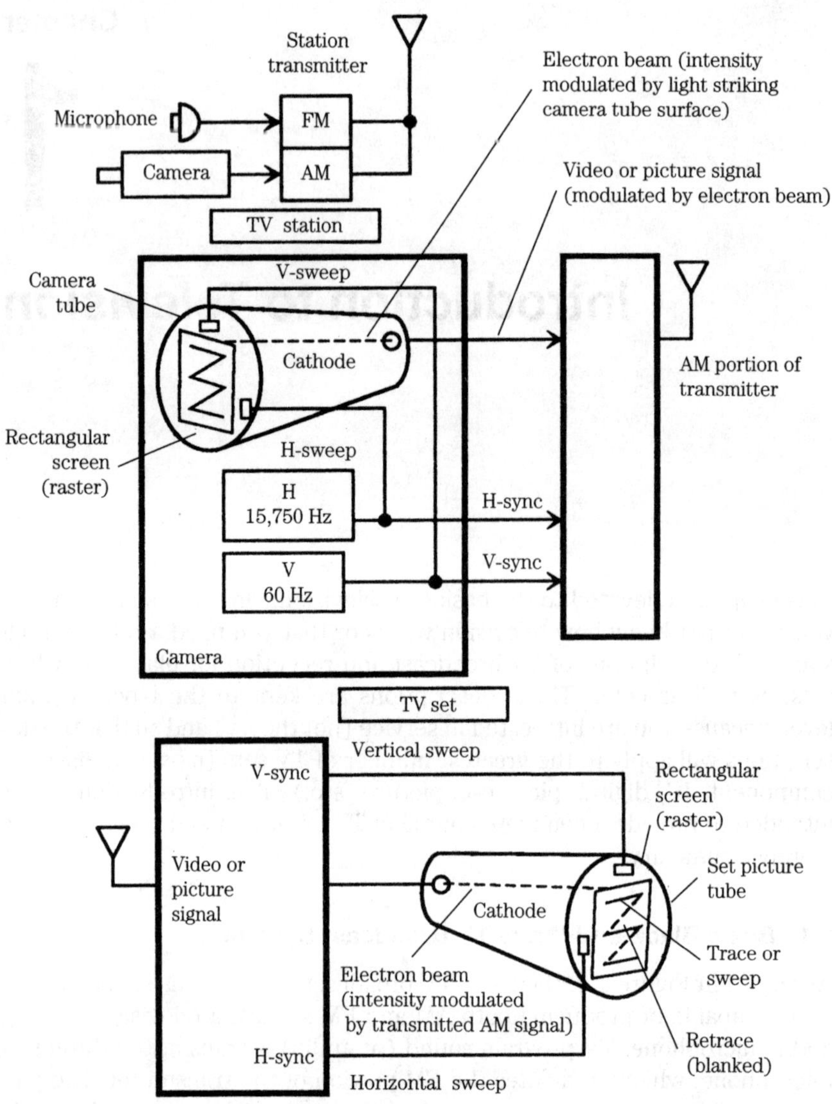

Figure 1.1 Basic TV broadcast system.

the signal is unique to TV in that both picture information and synchronizing pulses are transmitted on the AM carrier.

Figure 1.1 also shows the relationship between the picture tube in the TV set and the station-camera picture tube. Both tubes have an electron beam, which is emitted by the tube cathode and strikes the tube surface. Both tubes have horizontal and vertical sweep systems that deflect the beam to produce a rectangular screen (or raster) on the tube surface. The vertical sweep is at a rate of 60 Hz, with the horizontal sweep at 15,750 Hz.

The electron beam of the camera picture tube is modulated by the amount of light that strikes the camera tube screen. In turn, the transmitted AM signal is modulated by the electron beam. The amplitude of the transmitted AM signal is determined by the intensity of the light at any given instant. The position of the electron beam at a given instant is set by the horizontal and vertical sweep circuits, which are triggered by pulses in the camera. These same pulses are transmitted on the AM carrier and act as synchronizing pulses (sync pulses) for the TV set horizontal and vertical sweep circuits.

The electron beam of the TV set tube is modulated by the transmitted AM signals to "paint" a picture on the tube screen. The amplitude of the AM signal determines the intensity of the light produced on the TV screen at any instant. For example, if the camera sees an increase in light, the electron beam in both tubes (camera and receiver) is increased, and the TV set tube shows an increase in light.

The TV set tube horizontal and vertical deflection systems are triggered by the horizontal and vertical sync pulse transmitted on the AM portion of the TV broadcast signal. The electron beam of the TV set follows the beam in the camera tube. Both beams are at the same corresponding spot at the same instant.

Assume that the camera is focused on a white card with a black number at the center, as shown in Figure 1.2. As the camera-tube electron beam is swept across the surface, light is reflected from the card onto the camera-tube surface. When the beam passes across any portion of the light reflected from the white background, beam intensity is maximum. Beam intensity drops to minimum when the beam is at any portion of the light reflected from the black number. This varying electron beam modulates the transmitted AM carrier.

The electron beam starts at the top of the camera-tube screen, sweeps across to one side, and is blanked during the return to the other side (during retrace), until the beam finally reaches the bottom of the screen. The beam is then blanked and returned to the screen top. If the camera is focused on the numeral 3 pattern, the camera translates the pattern, line by line, into a picture signal (a voltage that varies in amplitude with intensity of the reflected light).

At the TV set, the picture-tube beam follows the camera beams increase during the sweep across any of the white background. Likewise, both beams decrease during the sweep across a portion of the numeral 3. The TV-set tube reproduces the numeral 3 in black and white (black or minimum beam intensity on the numeral portion of the sweep; white or maximum intensity on the card background).

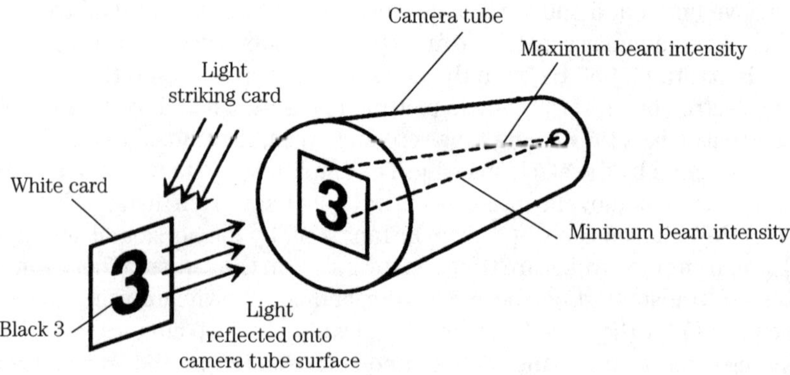

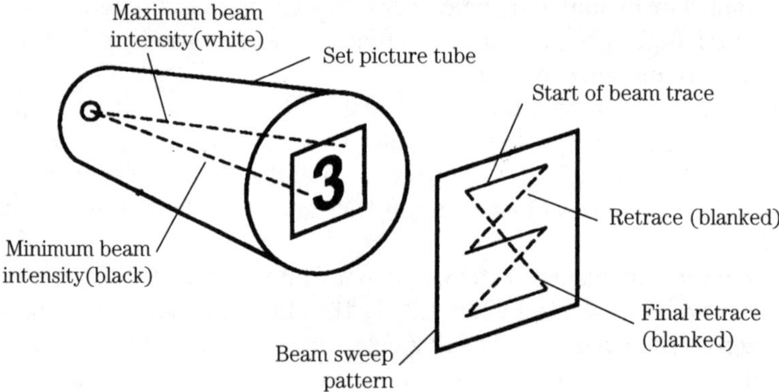

Figure 1.2 Electron beam in camera and receiver picture tubes.

1.2 Basic Black-and-White TV Set Circuits

Figure 1.3 shows the block diagram of a black-and-white TV set. The diagram shown is a composite of several types of sets and is presented as a point of reference for understanding TV circuit operation. Many of the discrete-component circuit groups of Figure 1.3 are found as ICs in present-day sets. However, the circuit groups generally have essentially the same functions, as well as inputs and outputs.

As an example, the amplifier, mixer, and oscillator shown as part of the tuner group in Figure 1.1 are usually part of a replaceable package in present-day IC sets. Individual tuner components or circuits are not available. However, the tuner still has an input (from the antenna), an output (sound and video in IF form), and an AGC input. These input-output points must be checked during troubleshooting, whether the set has discrete components, ICs, or solder-in modules.

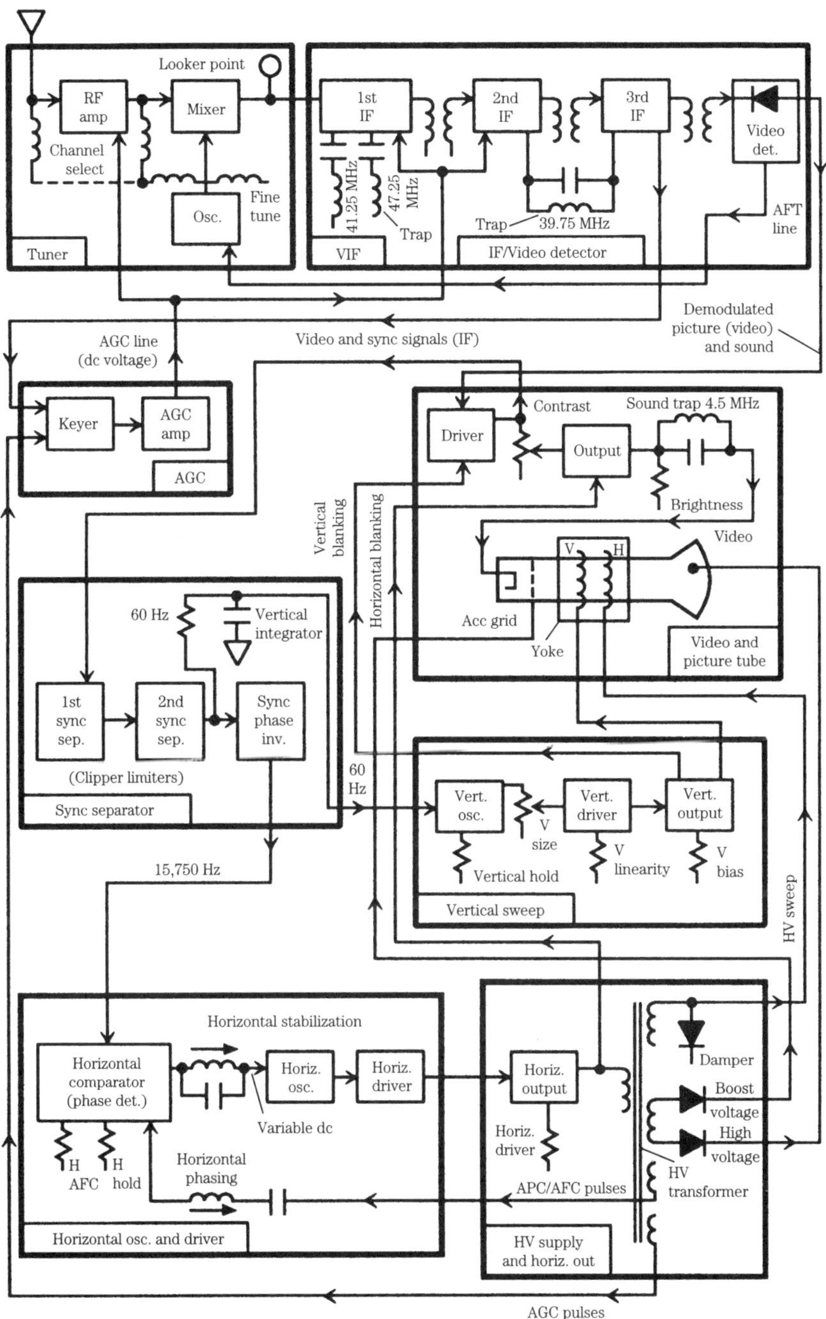

Figure 1.3 Composite discrete-component TV set.

The following paragraphs of this section describe the principles of all black-and-white TV sets and monitors. The detailed circuit descriptions (and troubleshooting approaches) for present-day TV sets are given in the related chapter.

1.2.1 Low-voltage power supply

The basic functions of the low-voltage power supply (not shown) is to provide direct current to all circuits of the set, except the high voltages required by the picture tube. The high voltages are provided by the flyback circuit (often called the FB circuit, but more properly known as the high-voltage power supply and horizontal-output circuit). The low-voltage supply consists essentially of a rectifier (typically a full-wave bridge), followed by a regulator (typically zener and transistor).

1.2.2 High-voltage power supply and horizontal output

Although these circuits have more than one function in all TV sets, the circuits do not have the same functions in all sets. The main functions are to provide a high voltage for the picture tube, and to provide a horizontal-deflection voltage (horizontal sweep) to the picture-tube deflection yoke.

In many sets, the circuits also supply a boost voltage for the picture-tube focus and accelerating grids. In other sets, the circuits supply a voltage for the video output transistor (which often requires a higher voltage than is available from the low-voltage supply). Often, the circuits also supply an automatic phase control and automatic frequency control (APC/AFC) signal to lock the horizontal oscillator frequency and an AGC signal for the tuner and IF modules.

The circuits generally have only one input, no matter what combination of outputs are provided. The circuits receive pulses from the horizontal oscillator and driver. These pulses are at 15,750 Hz and are synchronized with the TV broadcast.

1.2.3 Horizontal oscillator and driver

The horizontal oscillator and driver circuits provide drive signals to the HV-supply and horizontal output. These 15,750-Hz signals are synchronized with the picture transmission by sync pulses from the sync separator.

In addition to the one basic input and output, most horizontal oscillator and driver circuits have a feedback signal that is taken from the output and fed back to the input. The feedback signal provides an APC/AFC system from the horizontal sweep circuits. The APC/AFC system ensures that the horizontal sweep signals are synchronized (both for frequency and phase) with picture transmission, despite changes in line voltage and temperature or minor variations in circuit values.

The sync pulses are compared with the horizontal-sweep signals (for phase) in the phase-detector portion of the circuits. Deviations of the horizontal-sweep signals from the sync pulses cause the horizontal oscillator to shift in frequency or phase as necessary to offset the initial (undesired) deviation.

1.2.4 Vertical sweep

The vertical-sweep circuits provide a vertical-deflection voltage to the picture-tube deflection yoke. The vertical circuits also supply a blanking pulse to the picture tube. This pulse blanks the picture tube during retrace of the vertical sweep. The vertical signals are at a frequency of 60 Hz and are synchronized with the picture transmission by vertical-sync pulses from the sync separator.

1.2.5 Sync separator

The sync-separator circuits remove the vertical and horizontal sync pulses from the video circuits and apply the pulses to the vertical and horizontal sweep circuits, respectively. The sync-separator circuits also function as clippers and/or limiters to remove the video (picture) signal and any noise so that the sweep circuits receive only sync pulses.

1.2.6 RF tuner

The functions of an RF tuner, or tuner package, in a TV set are essentially the same as the RF sections in other AM/FM receivers. The basic TV tuner consists of an RF amplifier, mixer, and oscillator. The TV channels consist of two carriers, the FM-sound carrier and the AM-picture (or video) carrier. Both carriers are amplified by the RF amplifier, and are mixed with signals from the oscillator to produce two IF signals in the mixer. Thus, the basic RF tuner has one input and one output, even though there are two carriers.

The present trend in TV sets is to use some form of frequency synthesis (FS) in the tuner. This FS tuning is also known as digital tuning, quartz tuning, or synthesized tuning and is discussed in chapters 2 through 4. This section deals with basic tuner functions, whether the tuner is a replaceable package or discrete components in a shielded module.

The non-FS tuners (such as shown in Figure 1.3) receive AGC signals from the AGC circuits. These AGC signals are generally applied only to the RF amplifier (not the mixer and oscillator); they control gain of RF amplifier in the presence of carrier-signal variations. For example, if the carrier-signal strength increases, the AGC signals decrease the RF amplifier gain and thus offset the initial increase in carrier strength.

Some non-FS tuners also receive automatic fine-tune (AFT) signals. These AFT signals are applied to the oscillator and shift the oscillator frequency as necessary to keep the RF circuits tuned to the exact channel frequencies. For example, the AFT signals increase the tuner frequency and thus offset the initial frequency drift. In a typical non-FS tuner, the tuning is accomplished by slug-tuned coils, with one set of coils for each channel. Generally, there are three coils for each channel (one each for the RF amplifier, mixer, and oscillator). However, the coil arrangement varies with the type of RF tuner circuit.

In addition to (or instead of) the AFT circuits, many non-FS tuners have manual fine-tune circuits. The fine tune is a front-panel (user) adjustment control. Typically, the fine-tune control is a slug-tuned control in parallel with the oscillator coil. Some discrete-component tuners (as well as tuner packages) have other adjustments, such as traps and filters at the input and output. However, there is no standardization of such adjustments.

1.2.7 IF and video detector

The IF and video-detector circuits amplify both picture and sound signals from the RF tuner, demodulate both signals for application to the video and sound IF amplifiers (not shown), provide signals to the AGC circuits, provide AFT signals to the RF tuner, and trap (or reject) signals from adjacent channels.

The present trend is to combine most of the IF functions into a single IF module or IC. This is known as the VIF (video IF) module, or possibly the PIF (picture IF) module. (In some cases, the VIF or PIF functions are combined in a single module with the RF tuner.) In any event, the few parts external to the IC or module are adjustment controls (typically slug-tuned coils for transformers and traps, because it is not always practical to fabricate coils within an IC).

In many sets, both picture and sound outputs from the video detector are applied to the video-amplifier circuits where both signals are amplified by at least one stage of the video amplifier. At this point, the sound signals are applied to the input of the sound-IF amplifier (SIF), and the picture signals are applied to the remaining stages of the video amplifier (and to the sync separator). In other sets, the sound signals from the video detector are applied directly to the SIF amplifiers.

In addition to the signal input and output, most stages of the IF amplifier receive an AGC signal from the AGC circuits. These signals, which are in the form of a varying dc bias voltage, are the same as those applied to the RF tuner. The AGC circuits also receive video and sync signals from the IF amplifiers (taken from the last IF amplifier, before video detection).

In those sets with AFT, a portion of the video-detector output (in the form of an AFT voltage) is applied to the oscillator within the RF tuner. As discussed in chapters 2 through 4, these AFT signals lock the oscillator onto the frequency of the selected channel, and thus provide automatic fine tuning.

1.2.8 Video amplifier and picture tube

The video-amplifier circuits have several functions, and there are many circuit configurations in use. In a typical discrete-component set, the video-amplifier circuit has three inputs and three outputs. The primary input is the demodulated picture and sound signal from the video-detector output. Secondary inputs are blanking pulses from the vertical and horizontal sweep circuits. The primary output is the video signal (picture and sync) applied to the picture tube. The same video output is applied to the sync-separator and SIF circuits.

In the composite circuit of Figure 1.3, the three signals (picture information, and vertical and horizontal blanking pulses) are applied through the driver and output stages to the picture tube (usually at the cathode). These three signals control the picture-tube electron-beam intensity.

The pulses blank the picture tube (cut off the electron beam) during the retrace period of the horizontal and vertical sweeps. During the trace periods of both sweeps, the picture signal varies the intensity of the electron beam to paint a picture on the tube screen. As discussed, the picture-tube beam "follows" the broadcast camera tube beam, in both intensity and position, to reproduce the picture focused on the camera lens.

The sound signal is prevented from being applied to the picture tube by a trap between the video output stage and the picture-tube input. (In most sets, the input is at the cathode, but it might be a grid on a few picture tubes.) The trap (often called the sound, SIF, or 4.5-MHz trap) is usually a parallel fixed-capacitor and slug-tuned coil, tuned to 4.5 MHz, for discrete-component sets. In present-day sets using an IF module or IC, the sound trap is a crystal filter.

The sound signal is applied to the input of the SIF circuits after amplification by the first stage of the video amplifier (often called the driver or video amplifier). The picture and sync-pulse information does not pass to the audio section because the SIF circuits are tuned to 4.5 MHz (in a properly functioning set).

The picture signal and sync pulses are applied to the input of the sync-separator circuits after amplification by the video driver. The sound signal is also present at this point. However, because the sound information is FM and is at a frequency far removed from either the horizontal or vertical sweep frequencies (4.5 MHz compared to 60 and 15,750 Hz), the sound information does not affect the sync pulses. The picture signal is removed in the sync separator by clipping and limiting action. Only the sync pulses are applied to the sweep circuits.

In most black-and-white TV sets or monitors, the video amplifier and picture-tube circuits have at least two controls: contrast and brilliance (or brightness). There are many configurations for these two controls. The brilliance or brightness control determines the amount of fixed bias applied to the picture-tube cathode. This sets the intensity of the electron beam (with or without a signal present). The contrast control determines the amount of drive signal applied to the picture tube. An increase in drive signal produces an increase in contrast. In the circuit of Figure 1.3, the contrast control is, in effect, a volume or gain control between the video-driver and video-output stages.

1.2.9 AGC

There are many types of AGC circuits. Most discrete-component sets use keyed saturation-type AGC. In present-day sets, AGC is a function of the IF module. Either way, the RF-tuner and IF-stage transistors connect to the AGC line forward-biased at all times. On strong signals, the AGC circuits increase the forward bias, driving the transistors into saturation, thus reducing gain. Under no-signal conditions, the forward bias remains fixed.

Keyed AGC circuits have two inputs. One is from the IF circuit output and is at the IF center frequency. The other input is from the horizontal output circuit, at a frequency of 15,750 Hz. The AGC output is a varying dc bias voltage (controlled by bursts of IF signals).

A portion of the IF signal is taken from the IF amplifiers and is pulsed, or "keyed" at the horizontal-sweep frequency. The resultant keyed bursts of signal control the amount of dc voltage produced on the AGC line. The AGC output is applied to both the IF amplifiers and the RF tuner. A few AGC circuits are provided with an AGC control that sets the level of AGC action.

1.3 Basic Color TV Broadcast System

As shown in Figure 1.4, the basic color TV broadcast system consists of a transmitter, a color camera, a signal matrix, a brightness amplifier, a color amplifier, and a 3.58-MHz oscillator. The TV-program sound (or audio) is broadcast through an FM transmitter, as in the case of a black-and-white set. Likewise, the picture (or video) portion of the program is broadcast by means of amplitude modulation, or AM. The same channels and frequencies are used for color and for black-and-white set.

In addition to the horizontal and vertical sweeps, the color signal is made up of two components: brightness and color information. The brightness portion contains all the information pertaining to the picture details and is commonly called the Y or luminance (luma) signal. The color portion is called the chominance (chroma) signal and contains the information pertaining to the hue (or tint) and saturation of the picture.

The camera in Figure 1.4 contains three separate image pickup tubes, one tube for each of three colors (red, green, and blue). The relative intensity of the scene is also contained in this signal. The output signal from the camera contains the luma and chroma information.

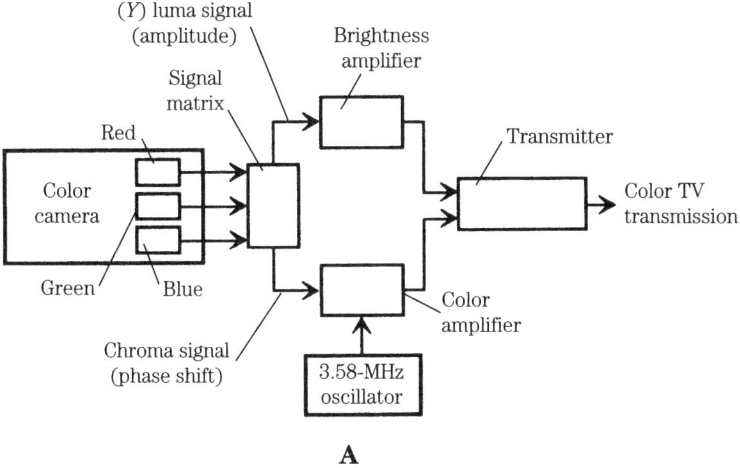

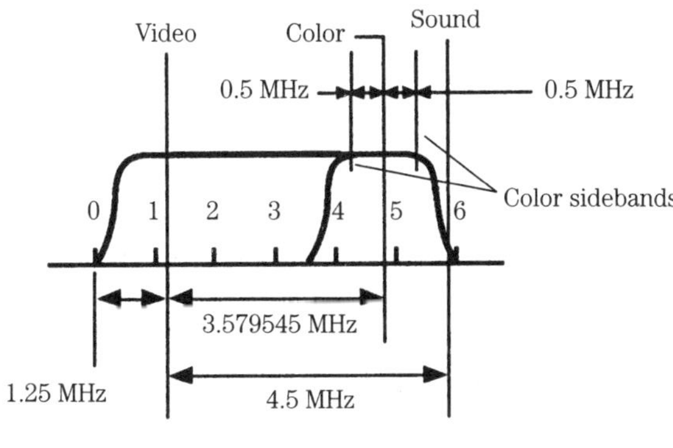

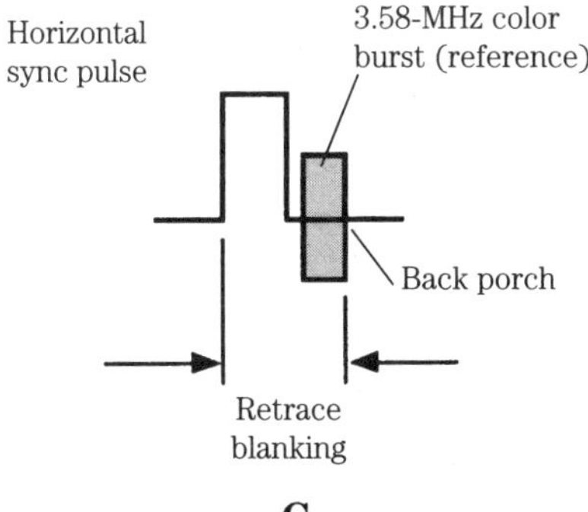

Figure 1.4 Basic color TV broadcast system.

The signal is separated in the signal matrix, and the individual components are amplified by the brightness and color amplifiers. The color information (chroma) is impressed on a 3.58-MHz subcarrier (in the form of phase shift) and is transmitted together with the luma signal (which is in the form of instantaneous amplitude shift or level). This method of color broadcast provides a compatible signal that can be reproduced on black-and-white sets as well as color sets. The black- and-white sets required only the luma signal (which is the equivalent of the picture information transmitted on the AM carrier of a black-and-white broadcast).

A compatible color signal must fit within the standard 6-MHz TV channel and have horizontal and vertical scanning rates of 15,750 and 60 Hz, plus a video bandwidth not in excess 4.25 MHz. The color information interleaves with the video and transmitted within the 6-MHz TV channel on the 3158-MHz subcarrier, as shown in Figure 1.4(b). The color subcarrier is actually 3.579545 MHz above the video carrier because this does not interfere with reception on black-and-white sets.

The color signals are synchronized by transmitting about eight cycles of the 3.58-MHz subcarrier signal (of the correct phase) along with the color signal. This signal is called the burst-sync signal (or color burst) and is added to the back porch of the horizontal-sync pulse, as shown in Figure 1.4(c). The location of the burst does not interfere with the horizontal sync on black-and-white sets because the burst occurs during the blanking or retrace portion of the sync pulse.

1.3.1 Black-and-white versus color signals

A black-and-white presentation requires only the transmission of the amplitude or luma signal. A color broadcast also requires the transmission of phase-shift information (representing colors). Before going into the color circuits, here's a review of the nature of the NTSC color video signal.

1.3.2 History of the NTSC color video signal

In 1953, the National Television Systems Committee (NTSC) of the Electronic Industries Association (EIA) established the color TV standards now in use by the TV broadcast industry in the United States and many other countries. The color system is compatible with the monochrome (black and white) system that previously existed.

The makeup of a composite video signal is dictated by NTSC specifications. These specifications include a 525-line interlaced scan operating at a horizontal-scan frequency of 15,734.26 Hz and a vertical-scan frequency of 59.94 Hz. Color information is contained in the 3.579545-MHz subcarrier. The phase angle of the subcarrier represents the hue or tint; the amplitude of the carrier represents the saturation.

1.3.3 Horizontal sync (NTSC)

A line of horizontal scan is normally considered as "beginning" at the leading edge of the horizontal blanking pedestal, as shown in Figure 1.5(a). In a TV set, the horizontal blanking pedestal starts as the electron beam of the picture tube reaches the extreme right-hand edge of the screen (plus a little extra in most cases).

The pedestal prevents illumination of the screen during retrace (until the electron-beam deflection circuits are reset to the left edge of the screen and ready to start another line of video display). The entire horizontal blanking pedestal is at the blanking level or the sync-pulse level. In a TV set, the blanking and sync levels are the "blacker-than-black" levels that ensure no illumination during retrace.

The pedestal consists of three parts: the front porch, the horizontal-sync pulse, and the back porch. The front porch is a 1.47-μs period of blanking level and is followed by a 4.89-μs horizontal-sync pulse at a –40 IEEE unit level. (An explanation of IEEE units is given in section 1.3.5.) When the horizontal-sync pulse is detected in a TV set, flyback is initiated, rapidly ending the horizontal scan and rapidly resetting the horizontal-deflection circuit for the next line of horizontal scan.

The horizontal-sync pulse is followed by a 4.40-μs back porch at the blanking level. When a color signal is being transmitted, 8 to 10 cycles of 3.579545-Hz color burst occur during the back porch. The color-burst signal is at a specific reference phase (depending on the color at that particular instant). In a color TV set, the color oscillator is phase-locked to the color burst reference phase before starting each horizontal line of video display. When a mono signal is being generated, there is no color burst during the back porch (because the burst is in no way required for black-and-white operation).

1.3.4 Vertical sync (NTSC)

A complete video image as seen on a TV screen is called a *frame*. A frame consists of two interlaced vertical fields of 262.5 lines each. The image is scanned twice at a rate of 59.94 Hz, and the lines of field 2 are offset to fall between the lines of field 1 (the fields are interlaced) to create a frame of 525 lines at a 30-Hz repetition rate.

At the beginning of each vertical field, a period equal to several horizontal lines is used, because the vertical-blanking interval prevents illumination of the picture tube during the vertical retrace. The vertical-sync pulse, which is within the vertical-blanking interval, initiates reset of the vertical-deflection circuit so that the electron beam returns to the top of the screen before the video scan resumes.

The vertical-blanking interval begins with the first equalizing pulse,

	H-sync	Burst	Gray 75% white	Yellow	Cyan	Green	Magenta	Red	Blue	Black
Luma (IEEE units)	−40	0	77	69	56	48	36	28	15	7.5
Chroma (IEEE units)	---	40	---	62	88	82	82	88	62	---
Vector (from B-Y)	---	180°	---	167°	284°	241°	61°	104°	347°	---

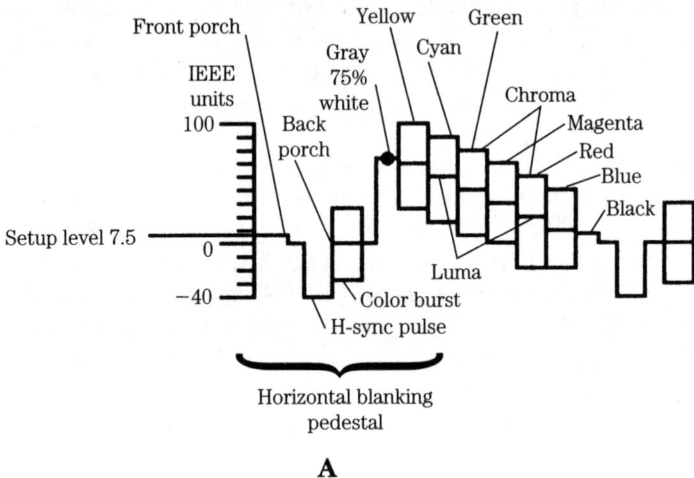

Figure 1.5 Composite video signal.

which consists of six pulses one-half the width of horizontal-sync pulses, but at twice the repetition rate. The vertical-sync pulse is an inverted equalizing pulse with the wide portion of the pulse at the –40 IEEE units level, and the narrow portion at the blanking level.

A second equalizing pulse occurs after the vertical-sync pulse. This pulse is then followed by 13 lines of blanking level (no video) and horizontal-sync pulses to ensure adequate vertical-retrace time before resuming video scan. The color-burst signal is present after the second equalizing pulse. Notice that in field 1, line 522 includes a full line of video, while in field 2, line 260 contains only a half line of video. This timing relationship produces the interlace of fields 1 and 2.

1.3.5 Amplitude (NTSC)

A standard NTSC composite video signal is 1 Vp-p, from the tip of a sync pulse to 100 percent white (Figure 1.5). This 1-Vp-p signal is divided into 140 equal parts, called IEEE units. The zero-reference level is the blanking level. The tips of the sync pulses are at –40 units, and a sync pulse is about 0.3 Vp-p. The portion of the signal that contains video information is raised to a setup level of +7.5 units above the blanking level.

A monochrome video signal at +7.5 units is at the black threshold. At +100 units, the signal represents 100 percent white. Levels between the +7.5 and +100 units produce various shadings of gray. Even when a composite video signal is not at the 1 Vp-p level, the ratio between the sync

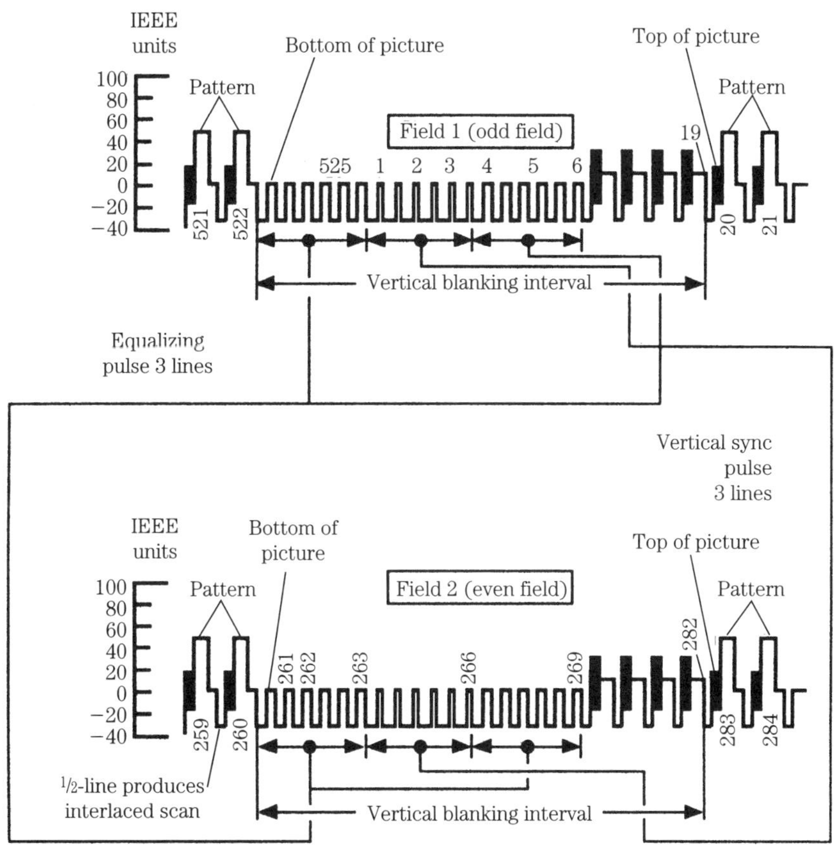

Figure 1.5 Continued.

pulse and video must be maintained: 0.3 of the total voltage for the sync pulse and 0.7 of the total voltage for 100 percent white.

There is also a specific relationship between the amplitude of the composite video signal and the percentage of modulation of an RF carrier. A TV signal uses negative modulation, where the sync pulses (–40 units) produce the maximum peak-to-peak amplitude of the modulation envelope (100 percent modulation), and white video (+100 units) produces the minimum amplitude of the modulation envelope (12.5 percent modulation).

This relationship of amplitude versus modulation is very advantageous because the weakest signal condition, where noise interference can most easily cause snow, is also the white portion of the video. There is an adequate amplitude guard band so that a peak white of +100 units does not reduce the modulation envelop to zero.

1.3.6 Color (NTSC)

The color information is an NTSC video signal consisting of three elements: luminance (luma), hue (or tint), and saturation, as shown in Figure 1.6.

Luma, or brightness perceived by the eye, is represented by the amplitude of the video signal. The luma component of a color signal is also used in black-and-white sets, where the luma is converted to a shade of gray. Yellow is a bright color and has a high level of luminance (it is nearer to white), whereas blue is a dark color and has a low level of luminance (it is nearer to black).

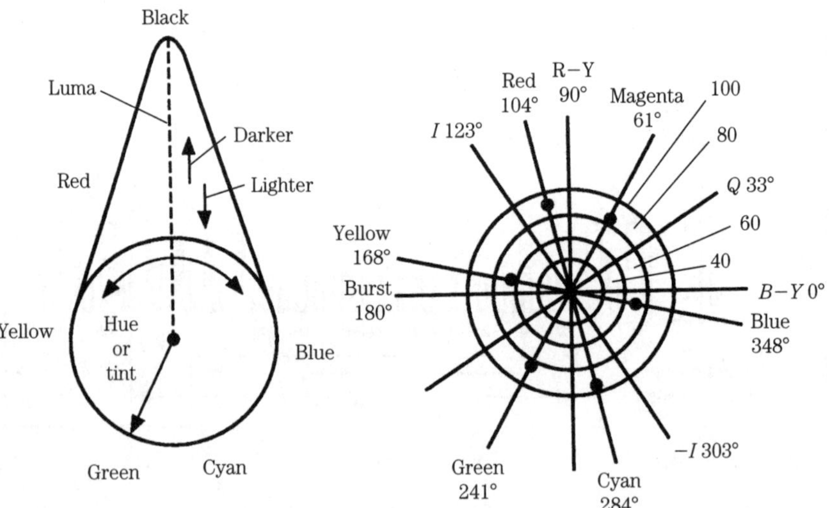

Figure 1.6 Elements of an NTSC color TV signal.

Hue, or tint, is the element that distinguishes between colors (red, blue, green, etc.). White, black, and gray are not hues. Hue information is contained in the phase relationship of the 3.58-MHz color carrier. The three primary colors (red, blue, and green) can be combined to create any hue. A phase shift through 360° produces every hue in the rainbow by changing the combination of red, blue, and green.

Saturation is the vividness of a hue and is determined by the amount that the color is diluted with white light. Saturation is often expressed in percentages: 100 percent saturation is a hue with no white dilation and produces a very vivid shade. Low saturation percentages are highly diluted by white light and produce light pastel shades of the same hue. Saturation information is contained in the amplitude of the 3.58-MHz color carrier. Because the response of the human eye is not constant from hue to hue, the amplitude required for 100 percent saturation is not the same for all colors.

The combination of hue and saturation is known as *chroma*. This information is usually represented by a vector diagram. Saturation is indicated by the length of the vector, and hue is indicated by the phase angle. The entire color-signal representation is three dimensional, consisting of the vector diagram for chroma and a perpendicular plane to represent the amplitude of luminance (Figure 1.6).

1.4 Basic Color Circuits

Figure 1.7 is the block diagram of a color TV set. Again, the diagram is a composite of several types of color TV sets, and is presented here as a point of reference. In most present-day sets, many of the individual blocks or circuits are contained in a single module or IC. For example, many TV sets have a single IC module containing all or most of the luma/chroma video-processing circuits. (The IC is often referred to as the Y/C Jungle chip, because of the great number of circuits involved.) However, no matter what configuration is used, the circuits of a color set are very similar to those of a black-and-white set, with two exceptions: the luma (Y) channel and the chroma (C) channel.

1.4.1 Luma channel operation

The luma channel consists of the detector, first video amplifier, delay line, and second video amplifier. (Note that a separate sound detector is used in the signal path prior to the video detector in many color sets. This provides good separation of the sound and video signals.)

The composite video signal is detected and amplified in the first stage of the luma channel. A portion of this signal is fed to the sync and color stages. The rest of the signal continues on to the delay line. This delay is required so that the luma information (amplitude) arrives at the picture tube at the same time as the color information (phase relationships).

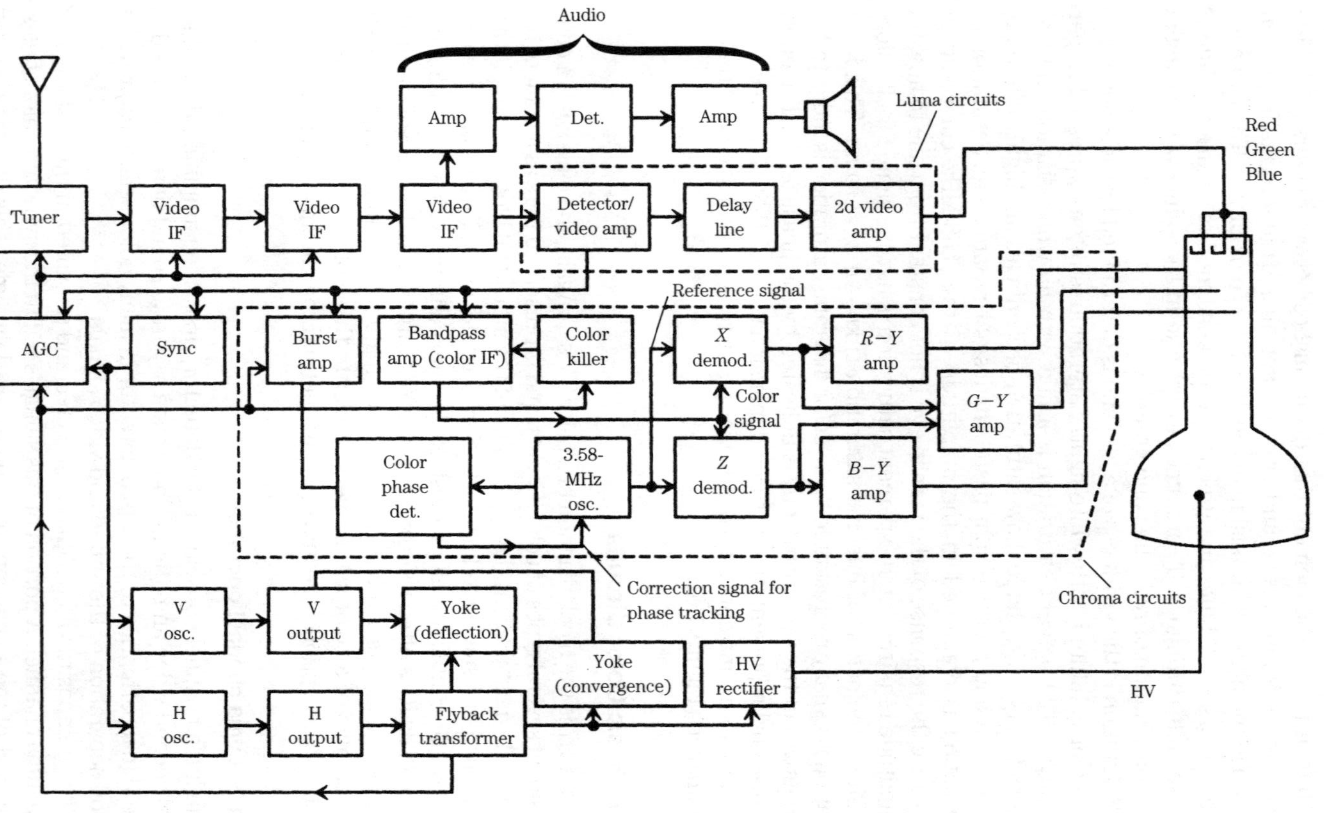

Figure 1.7 Basic color TV set circuits.

The narrow bandwidth of the color circuit causes the chroma signal to take longer to reach the picture tube. Additional amplification is given to the luma signal in the second video amplifier before luma is applied to the cathodes of the picture tube.

1.4.2 Chroma channel operation

The composite video signal is fed to the bandpass amplifier (sometimes called the color IF) from the first video amplifier. The chroma signal is separated from the composite color signal, amplified, and fed to the inputs of the demodulators. The burst amplifier removes the burst signal from the chroma signal and applies the burst to the color-phase detector.

A color-killer stage is used to prevent signals at frequencies near 3.58 MHz from passing through the color circuit during the reception of black-and-white broadcasts. In the absence of a color broadcast (no color-burst signals being transmitted), the color killer biases the bandpass amplifier to off, thus disabling all of the chroma channel. In most circuits, the color killer also receives pulses from the horizontal-sweep circuits (usually from the flyback transformer). The horizontal pulses operate the color killer to bias the bandpass amplifier to off during the horizontal-sweep retrace (so that no color information passes during the retrace period).

The 3.58-MHz oscillator provides a reference signal for demodulation of the color signal. The color-phase detector compares the 3.58-MHz oscillator signal with the burst signal being transmitted, and develops a correction voltage to keep the oscillator locked in phase with the burst signal.

In most circuits, the burst amplifier (or the burst-control circuit) receives pulses from the horizontal sweep. These pulses gate the burst amplifier to on only during the burst signal (immediately after the horizontal-sync pulses, during the retrace period, as shown in Figure 1.4). Thus, the 3.58-MHz oscillator is locked in phase with the burst at the beginning of each horizontal sweep.

As shown in Figure 1.7, the X and Z demodulators detect the amplitude and phase variations of the chroma signal to recover color information. At any given instant, the X and Z demodulators receive two signals: a 3.58-MHz signal from the reference oscillator (locked in phase to the reference burst) and a 3.58-MHz signal from the bandpass amplifier (at a phase and amplitude representing the color at that instant). The R-Y (red luma), B-Y (blue luma) and G-Y (green luma) amplifiers amplify the color signals and apply the signals to the picture-tube inputs (grids in this case).

1.4.3 Makeup of colors

Color picture tubes produce most colors by mixing three colors (red, blue, and green). (Operation of the color picture tube is described in section

1.4.4.) Any color can be created by the proper blending of these colors. That is, any color can be created if the three colors are present, each at a given level of intensity or brightness. This is because of a characteristic of the human eye. When the eye sees two different colors of the same brightness level, the colors appear to be of different brightness levels. For example, the eye is more sensitive to green and yellow than it is to red or blue.

The combination of any two primary colors produces a third color. In color TV, the process of mixing colors is additive and depends on the self-illuminating properties of the picture-tube screen (not on the surrounding light source, as found with the subtractive color-mixing process used with paints and printing).

If two primary colors, such as red and green, are combined, the primary colors produce a secondary color, yellow. The combination of blue and red produces magenta. Also, a secondary color added to the complementary primary color produces white.

For example, yellow + blue = white, cyan + red = white, and magenta + green = white. Thus, any color can be produced if the three colors (red, blue, and green) are mixed in the proper proportions (that is, if the amplitude of the three signals at the picture-tube inputs are at the proper levels).

For the purposes of this discussion, it is sufficient to understand that the three picture-tube inputs receive signals of the correct amplitude proportions (identical to the proportions existing at the color camera for any given instant). This occurs provided that (1) the set 3.58-MHz oscillator is locked in phase to the burst signals and (2) the demodulators receive a 3.58-MHz color signal that is of a phase and amplitude corresponding to the color proportions existing at the color camera.

1.4.4 Color picture tubes

Figure 1.8 shows a color picture tube and the related components. Remember that there are many types of color tubes; however, all color tubes have certain points in common (from a troubleshooting standpoint). Figure 1.8 shows the basic elements common to all color picture tubes.

The color picture tube is a cathode-ray tube (CRT) similar to that used for black-and-white sets and monitors. Like the black-and-white tube, the color tube has a magnetic deflection yoke that deflects the electron beam. The yoke contains two sets of coils, one for vertical deflection and one for horizontal. The vertical and horizontal sweep signals are applied to the corresponding set of coils so that the electron beam traces out a rectangular raster on the screen in the normal manner. Here, the similarity to a black-and-white tube ends.

The color picture tube has three complete electron guns. Thus, three electron beams are produced, one for each gun. The color picture tube also contains a shadow mask and a three-color phosphor-dot screen within the glass envelope.

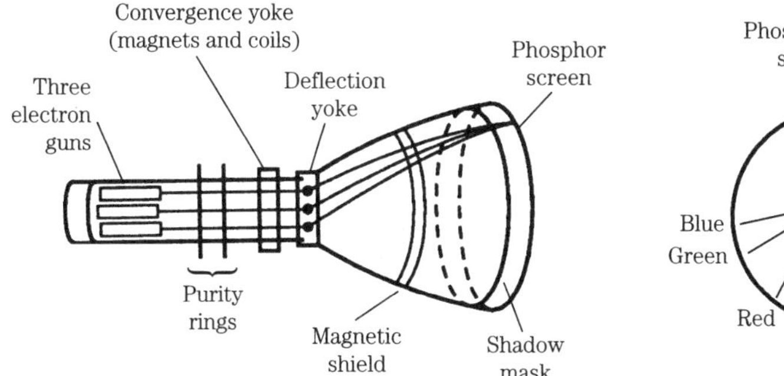

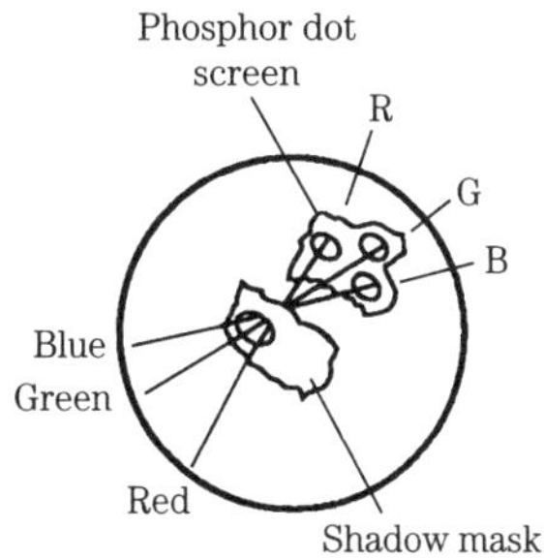

Figure 1.8 Color picture tube and related components.

The screen consists of several million phosphor dots of the primary television colors (red, green, and blue). These dots are arranged in trios and produce colored light when struck by electrons. Even though the dots are placed next to each other, they appear (to the human eye) to be superimposed so that the colors are (apparently) blended.

The shadow mask is a thin plate of metal with a number of small holes, each centered over a phosphor-dot trio. This mask is positioned in front of the screen and controls the landing of the electron beams on the phosphor dots.

The three electron guns (one for each color) are positioned approximately 120° apart and are precisely aimed so that the individual electron beams pass through the shadow mask at different angles to strike the three corresponding phosphor dots. The grids control the amount of light leaving the gun and thus regulate the amount of colored light emitted from the phosphor screen.

1.4.5 Color purity

For the color picture tube to reproduce color pictures correctly, the individual electron beams must strike phosphors of only one color. Each electron beam is then capable of producing a pure field of either red, blue, or green. Purity refers to the uniformity of the hue and brightness over the entire area of each color field and over the entire area of the combination of all three. For example, red is pure when the red display is uniform red, with no contamination from blue or green.

Color purity is determined by where the three electron beams strike the screen and can be adjusted by means of purity magnets and deflection yoke. Typical color purity adjustments are described in chapter 11.

1.4.6 Convergence

Convergence is the registry of the three beams at the same point in the shadow mask across the entire screen. The individual aiming of the beams must be corrected so that each passes through the same holes in the mask and strikes the same phosphor-dot trio.

Static convergence is controlled by convergence magnets that correct the travel of the individual electron beams. The static magnets converge the beams in the center of the screen but cannot compensate for the curvature of the screen to converge the beam at the edges (sometimes called the "pincushion effect").

Dynamic convergence, as this convergence compensation is known, is provided by electromagnetic coils mounted on the convergence yoke. Sawtooth voltages from the vertical and horizontal circuits are fed to the coils, which then alter the static dynamic field in unison with the scanning. Typical color convergence adjustments are described in chapter 11.

Chapter

2

Front-End Circuits

This chapter is devoted to TV-set circuits between the antenna or cable input and the IF stages. These include the electronic-tuner (ET), antenna-switching, and other front-end circuits. The IF circuits that follow the front end are discussed in chapter 3.

2.1 Front-End Basics

Unlike the discrete-component TV sets described in chapter 1, the tuning circuits of most present-day TV sets use some form of frequency synthesis (FS) tuning. This provides convenient push-button channel selection with automatic channel search and automatic fine tune (AFT) capability. The FS system is also known as *digital tuning* or *quartz tuning*. The key element in any FS system is the phase-locked loop, or PLL, which controls the variable-frequency oscillator (VFO) of the tuner. Generally, the VFO is some form of voltage-controlled oscillator (VCO). Quite often, the oscillator frequency is set by a varactor diode, which in turn is controlled by a variable voltage from the AFT line.

The tuner components (oscillator, RF amplifier, mixer) of the discrete-component set shown in Figure 1.3 are usually part of a tuner or front-end module in present-day sets. Figure 2.1 shows such a module, while Figure 2.2 shows the tuner channel-allocation frequencies for both the United States and Canada. Notice that the module of Figure 2.1 includes both the tuner and IF components. In other sets, there are separate tuner and IF modules. Before getting into some typical front-end circuits, here's a review of the basic FS and PLL system.

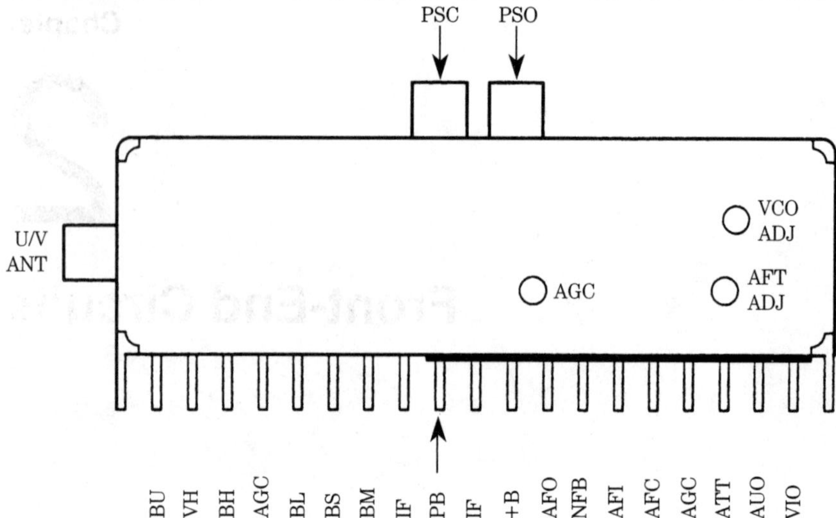

Figure 2.1 Typical tuner and front-end module.

2.2 FS and PLL Basics

Figure 2.3(a) shows the basic PLL circuit. *PLL* is a term used to designate a frequency-comparison circuit in which the output of a VFO is compared in frequency and phase to the output of a very stable (usually crystal-controlled) fixed-frequency reference oscillator. PLL is not unique to television; it is also found in many communications receivers.

In any PLL, should a deviation occur between the two compared frequencies or should there be any phase difference between the two oscillator signals, the PLL detects the degree of frequency or phase error. Then the PLL automatically compensates by tuning the oscillator up or down in frequency until both oscillators are locked to the same frequency and phase. (The loop is said to be locked at this time.) The accuracy and frequency stability of a PLL circuit depends on the accuracy and frequency stability of the reference oscillator (and on the crystal that controls the reference oscillator).

In the basic PLL of Figure 2.3(a), the VCO has a desired output frequency of 1 kHz. The actual output frequency depends on the tuning voltage produced by the phase comparator, which receives two input signals, both at 1 kHz. Any frequency or phase variation in the VCO output, when compared to the stable 1-kHz reference, causes the phase comparator to produce a correction voltage.

The magnitude of the correction voltage depends on the amount of frequency and phase deviation. The polarity of the correction voltage depends on the direction of phase and frequency deviation. The correction

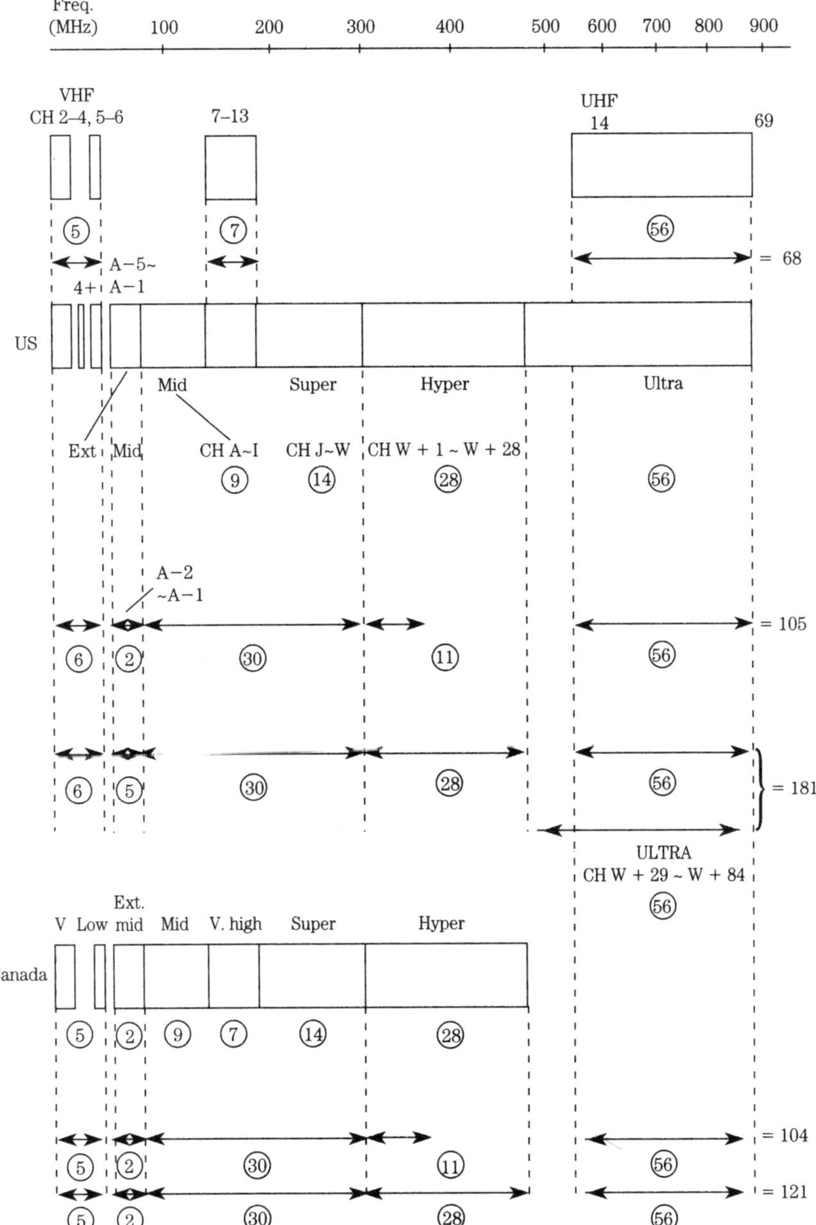

Figure 2.2 TV channel allocations for the United States and Canada.

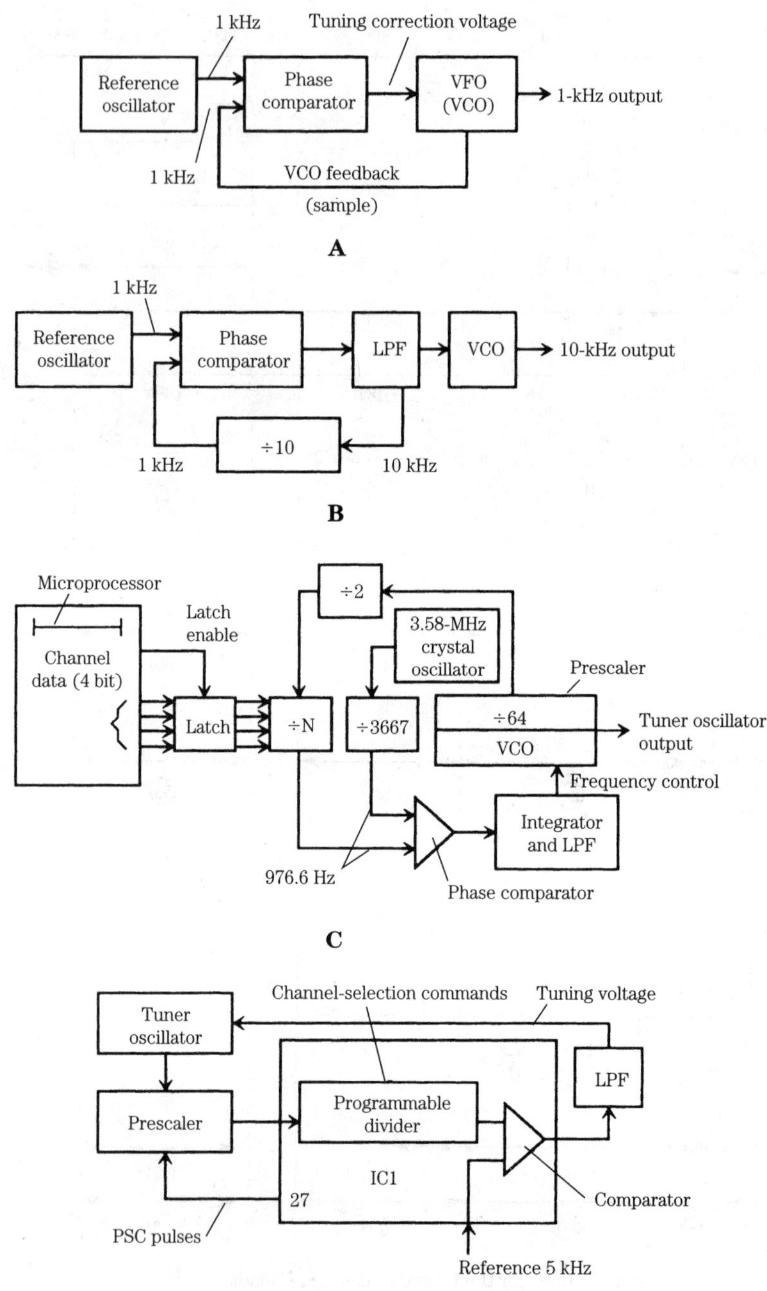

Figure 2.3 PLL basics.

voltage is applied to the VCO as an increase or decrease in tuning voltage. Changes in the tuning voltage alter the VCO frequency as necessary to make the VCO output of the same frequency and phase as the reference oscillator. When this occurs, the tuning voltage stabilizes and the PLL is said to be locked in.

Figure 2.3(b) shows a more sophisticated PLL circuit—one capable of comparing frequencies that are not identical. The circuit of Figure 2.3(b) includes a divide-by-10 element, which divides the VCO frequency by 10, and a low-pass filter, which acts as a buffer between the comparator and VCO. Notice that while the inputs to the comparator remain at 1 kHz when the loop is locked, the output frequency of the VCO is 10 kHz because of the divide-by-10 element.

Figure 2.3(c) shows a PLL circuit similar to that found in TV sets, but far less complex. The system is generally called an extended PLL and holds the tuner oscillator frequency to some harmonic (or subharmonic) of the reference oscillator (3.58-MHz in this case). The fixed-divide element of Figure 2.3(b) is replaced by a programmable variable divider (÷N) in Figure 2.3(c). A channel change is produced by varying the division ratio of the programmable divider with 4-bit data commands from the system-control microprocessor (which, in turn, is operated by front-panel push buttons and/or remote-control signals).

2.2.1 Pulse swallow control (PSC)

Most PLL tuning systems found in TV sets use some form of PSC, such as shown in Figure 2.3d. The PSC system uses a high-speed prescaler with variable-division ratio (instead of the fixed-division ratio prescaler in Figure 2.3c). The variable division ratio depends on the PSC signal from PLL IC1. As the number of PSC pulses increase, the division ratio increases. A specific number of PSC pulses are produced by IC1 in response to channel-selection commands.

The overall division ratio for a specific channel is the prescaler division ratio multiplied by the programmable-divider division ratio. The result of division at any channel is a precise 5-kHz output to the comparator when the tuner oscillator is locked to a given channel frequency. In effect, the programmable divider determines the basic channel frequency, and the prescaler performs the fine adjustment (AFT) to the channel frequency.

Notice that if you are troubleshooting any tuner with PSC and you cannot tune in a channel manually or with AFT, check the PSC line (pin 27 of IC1 in this case) for pulses and check the tuning voltage to the oscillator. If the pulses are missing, the tuner cannot lock onto any channel, even with a tuning voltage present.

2.2.2 Basic TV tuner operation

Figure 2.4 shows the circuits of a typical TV tuner (using PLL) in block form. Notice that the PLL IC1 receives channel commands from an IR remote-control transmitter after the commands have been decoded by a remote-control receiver circuit.

The multiband tuner is controlled by circuits within PLL IC1, (which, in turn, receives commands from the remote-control circuits). IC1 monitors signals from the IF demodulator circuits to determine when a station is being received. These signals are the AFT up and down (pins 35 and 36) and the station-detect signal (pin 34).

The detected video from the output of the IF demodulator is passed to a sync amplifier and detector Q7/Q9. The detected sync signal (station detect) is passed to pin 9 of IC3, amplified, and sent to the station-detect input at pin

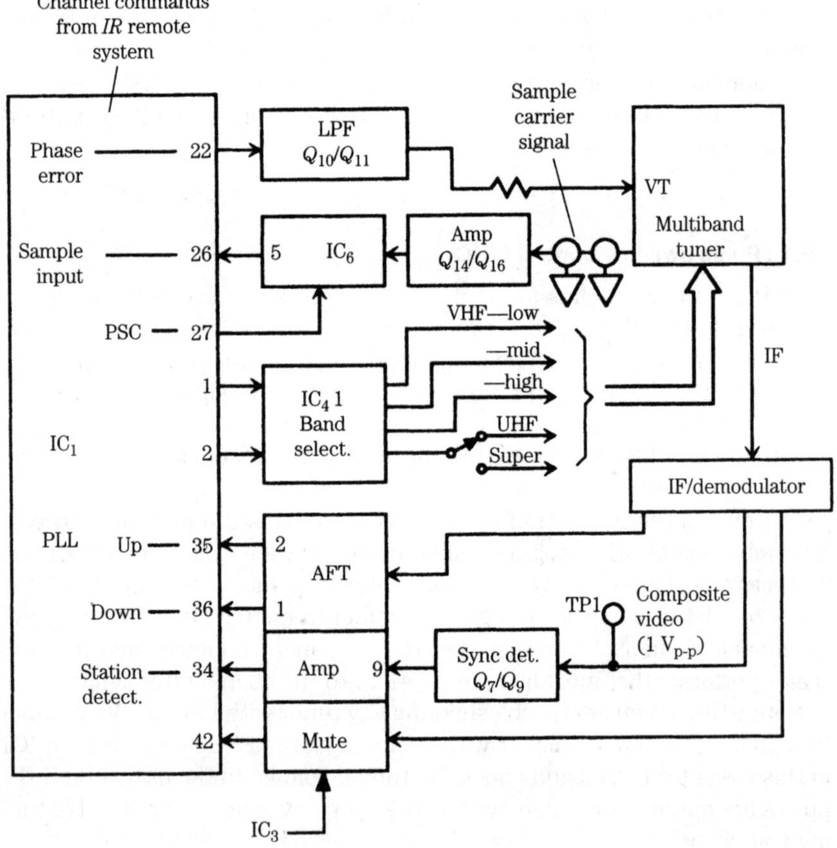

Figure 2.4 Typical TV tuner with PLL.

34 of IC1. When a station is tuned in properly, the sync is detected from the video signal and applied as a high to pin 9 of IC3. This applies a high to pin 34 of IC1, indicating to IC1 that video with sync is present.

To maintain proper tuning, IC1 monitors the AFT up and down signals (at pins 35 and 36) from IC3. The AFT circuit of IC3 is a window detector, monitoring the AFT voltage and outputting a high at pins 1 or 2, depending on the magnitude and direction of the AFT voltage swings (should the tuner-oscillator frequency drift).

When a channel is selected, band-switching information is supplied from IC1 (at pins 1 and 2) to IC4, which develops four band-switching outputs. (In this particular circuit, one of the four outputs is switched by a normal/cable switch to provide the five bands shown.)

The tuner oscillator passes a sample-carrier signal to oscillator amplifier Q14 and Q16. The amplified oscillator signal is then passed to prescaler IC6. The amount of frequency division is determined by PSC pulses from IC1. The frequency-divided output of IC6 (pin 5) is then passed to the sample input of IC1 (at pin 26). When a channel is selected, circuits within IC1 produce the appropriate number of PSC pulses at pin 27. The PSC pulses are applied to IC6 and produce the correct amount of frequency division.

The divided-down oscillator signal (sample input) at pin 26 of IC1 is divided down again within IC1 and compared to an internal 5-kHz reference signal. The phase error of these two signals appears at pin 22 of IC1 and is applied to low-pass filter (LPF) Q10 and Q11. The dc output from the LPF is applied to the tuner oscillator. This voltage sets the tuner oscillator as necessary to get the proper frequency for the channel selected.

2.2.3 Troubleshooting FS/PLL circuits

The most common symptoms for failure in a TV set with FS tuning are a combination of no stations received, failure to lock on channels, picture snowing, audio noisy, and color dropping in and out.

I have experienced all of these symptoms with a recently purchased TV set. The cause was traced to poorly soldered terminals on the tuner, IF module, and PLL IC. However, the following steps led to the defective solder junctions.

The first troubleshooting step is to isolate the problem to the tuner and IF modules or PLL system. Start by checking for power to all ICs and components. For example, in a typical FS circuit, the tuner requires both 12 V and 5 V, while the IF module and PLL IC require 5 V. Once you are satisfied that power is available to all FS components, start the isolation process.

Select a channel and confirm that the band-switching signal for that particular channel appears at the tuner input and band-select output. For example, if Channel 4 is available locally, select Channel 4 and check that the VHF-low band- switching signal is present (high). This signal appears at pin

1 of IC4 and at the tuner band-switch input. If the band-switching signal is not at pin 1 of IC4, suspect IC4 (or possibly IC1).

If the video signal does not appear at TP1, suspect a problem in the tuner and IF/demodulator circuits. In most present-day TV sets, this means replacing the complete assembly or possibly replacing individual tuner and RF/demodulator ICs.

If you can tune in channels using the substitute tuner-control voltage, the problem is most likely in the PLL components rather than in the tuner components.

If video appears at the test point with a substitute voltage, check for a high at the station-detect input of IC1 (pin 34). If the station-detect input is missing, suspect IC3, or the sync-detector circuits Q7 and Q9.

If the station-detect signal at IC1-34 is normal, monitor the AFT up and down inputs to IC1, pins 35 and 36, while changing the substitute tuning voltage. (Note that the substitute tuning voltage can come from any external source, but must match the normal voltage range at terminal VT of the tuner.)

If there are no logic changes at pin 35 and 36 as you tune through a station (as you vary the voltage at VT), suspect the window-detector circuit within IC3 (or possibly the AFT detector within the IF demodulator).

If the inputs at pins 35 and 36 appear to be normal, check for a sample oscillator signal at pin 26 of IC1. If the oscillator signal is missing, suspect IC6, the oscillator amplifier Q14/Q16, or possibly the shielded-cable from the tuner. (It is also possible that IC6 is not receiving proper PSC pulses from pin 27 and IC1, although you should still get sample signals from IC6 to IC1-26.)

If the input at pin 26 of IC1 appears to be normal, but the outputs at pins 1 and 2 are absent or abnormal, suspect IC1.

2.2.4 Typical FS/PLL Circuits

The following paragraphs describe some typical FS/PLL circuits.

2.3 Tuner with Separate PLL

Figure 2.5 shows the circuits of a TV tuner (a Sony 13 inch) where the PLL is separate from the tuner package. The purpose of this circuit is to interpret keyboard or remote inputs requesting channel changes. When such commands are received, microprocessor IC103 determines if a particular channel is to be tuned in or skipped (depending on the data in memory IC105). IC103 then selects the appropriate tuning band through operation of the band-switch circuits in IC102. This action tunes in the particular channel requested or, in the case of UP/DOWN tuning, tunes in the next channel from the one presently tuned. After the channel is tuned in, IC103 then monitors the AFT UP and AFT DOWN lines to maintain precise tuning.

2.3.1 Tuner

The BT-896 tuner TU101 is a multiband veractor-diode tuner capable of tuning channels 2–13, 14–83, and cable channels 1–125. TU101 operates from a single 9-V supply, and provides two prime outputs. The IF signal is output to the VIF circuits (chapter 3), and a sample of the tuner local oscillator is output the PPU-1A PLL unit TU102. The necessary inputs for proper operation are the band-select inputs L, M, H, and U from the band-select switch IC102, and the VC tuning voltage developed in TU102.

In addition to these basic inputs and outputs, there is an AGC correction voltage input from the VIF stage, and a control line (labeled CA) used to alter the tuning characteristics when tuning certain cable channels. Band switching for TU101 is performed by IC102. The state of the L,M,H, and J outputs from IC102 is determined by the binary logic applied to pins 3 and 4. In turn, the binary logic is controlled by IC103 outputs B-0 and B-1 (pins 11 and 12).

Table 2-1 shows the logic of the B-0 and B-1 outputs applied to IC102 from IC103. There are four tuner-voltage VC ranges as shown. For example, when the L output from IC102 to TU101 is high, the VC range is from 2 to 22 V. When the U output is high, the range increases (1.3 to 27 V). The level of the VC voltage determines what channel within a selected band will be tuned. The VC voltage level is controlled by PLL TU102.

2.3.2 PLL tuning unit

The PLL tuning synthesizer TU102 requires +5 V and +33 V. The +33 V is provided by zener D106 and is the source for the VC tuning voltage.

In operation, microprocessor controller IC103 transmits serial tuning data on the DAT (data) line from pin 5 to TU102.

This data stream is synchronized by negative-going CLK (clock) signals from IC103 at pin 7. At the end of each data transmission, a negative-going LAT (latch) pulse occurs at pin 9 of IC103, latching the transmitted data into PLL TU102. This sets the internal division ratio of TU102. In turn, TU102 monitors and divides the local-oscillator sample input from tune TU101, and develops the corresponding VC voltage output necessary to tune the particular channel requested.

2.3.3 Tuning memory

Tuning memory IC105 is a nonvolatile EAROM operating from standby 5 V and +36 V power. The IC105 chip is enabled when the chip-select line is made high by a CS signal from pin 22 of IC103. The initialize signal from pin 36 of IC103 is a positive-going pulse that occurs on power-up to reset the memory of IC105. The clock signal from pin 25 of IC103 is a negative-going pulse used to synchronize the transmission of parallel data between IC103 and IC105.

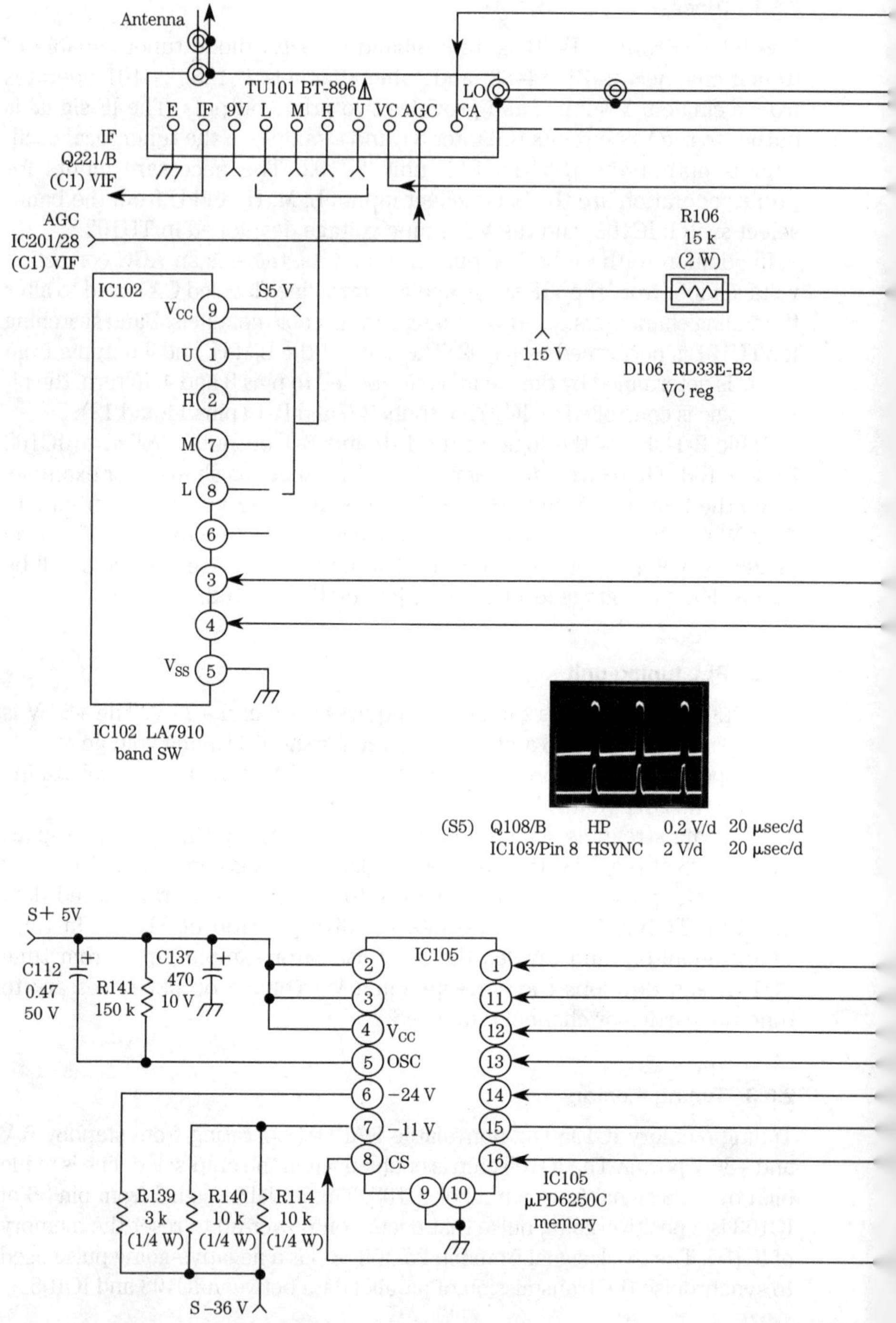

Antenna
TU101 BT-896
E IF 9V L M H U VC AGC CA
LO
IF
Q221/B
(C1) VIF
AGC
IC201/28
(C1) VIF
R106
15 k
(2 W)
IC102
S5 V
V_{CC} 9
U 1
H 2
M 7
L 8
6
3
4
V_{SS} 5
115 V
D106 RD33E-B2
VC reg
IC102 LA7910
band SW

(S5) Q108/B HP 0.2 V/d 20 μsec/d
IC103/Pin 8 HSYNC 2 V/d 20 μsec/d

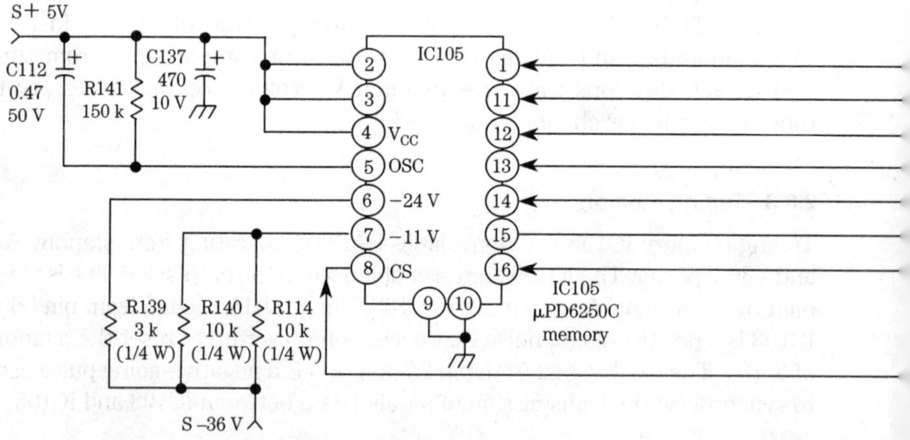

Figure 2.5 Tuner circuits with separate PLL (B3).

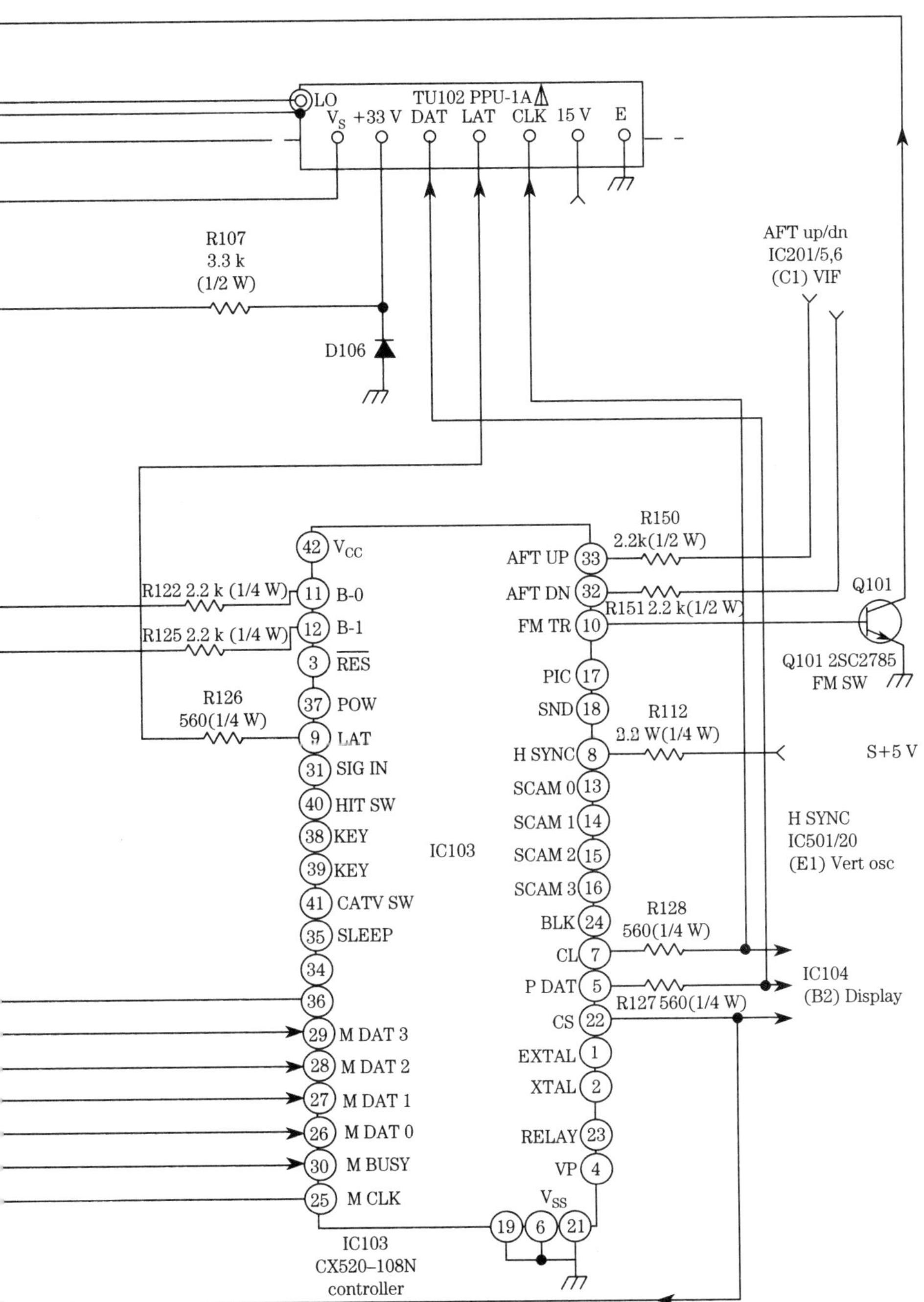

TU102 PPU-1A
LO
V_S +33 V DAT LAT CLK 15 V E
R107
3.3 k
(1/2 W)
D106
AFT up/dn
IC201/5,6
(C1) VIF
R150
2.2k(1/2 W)
42 V_{CC}
AFT UP 33
R122 2.2 k (1/4 W) 11 B-0
AFT DN 32
R151 2.2 k(1/2 W)
Q101
R125 2.2 k (1/4 W) 12 B-1
FM TR 10
3 RES
Q101 2SC2785
FM SW
R126
560(1/4 W)
37 POW
PIC 17
SND 18
R112
2.2 W(1/4 W)
9 LAT
H SYNC 8
S+5 V
31 SIG IN
SCAM 0 13
40 HIT SW
SCAM 1 14
H SYNC
IC501/20
(E1) Vert osc
38 KEY
IC103
SCAM 2 15
39 KEY
SCAM 3 16
41 CATV SW
BLK 24
R128
560(1/4 W)
35 SLEEP
CL 7
34
P DAT 5
IC104
(B2) Display
36
R127 560(1/4 W)
29 M DAT 3
CS 22
28 M DAT 2
EXTAL 1
27 M DAT 1
XTAL 2
26 M DAT 0
RELAY 23
30 M BUSY
VP 4
25 M CLK
V_{SS}
19 6 21
IC103
CX520–108N
controller

TABLE 2.1 Logic of B-0 and B-1 Outputs Applied to IC102

Pin	*3*	*4*	*8*	*7*	*2*	*1*	*Tuner*
Signal	*B0*	*B1*	*L*	*M*	*H*	*U*	*VC*
	1	1	1	0	0	0	2–22V
	0	1	1	1	0	0	2–23V
	1	0	0	0	1	0	1.5–25V
	0	0	0	0	0	1	1.3–27V

The M BUSY input from pin 15 of IC105 to pin 30 of IC103 is a negative-going pulse that occurs when memory operations within IC105 might prevent proper data transfer between IC103 and IC105. Such data transfers occur across the directional 4- bit data-bus M DAT 3. This 4-bit bus is used to transmit commands that set the operation mode of IC105, followed by data bits that are stored in IC105.

In the read mode, commands are sent over the bidirectional data bus to set IC105 to the read mode, and to select the appropriate address. Then the data bus is used to transfer data from IC105 into IC103. Memory IC105 is used during channel up/down operations to determine if the next higher or lower channel should be skipped or selected (tuned in). This type of operation is described as a skip-memory application.

2.3.4 AFT up/down

The AFT up and down input signals at pins 32 and 33 of IC103 are developed in the VIF stages (chapter 3) by IC201. The signals are positive-going pulses that inform IC103 when it is necessary to output correction data to PLL TU102 for automatic fine tuning operations.

2.3.5 FM trap

In most modes of operation, pin 10 of IC103 is high and Q101 is turned on. When cable TV is being received and the CATV selector switch is set to the CATV position, pin 41 of IC103 is low. This informs IC103 that cable is being used. With pin 41 low, cable channels 14, 15, 16, or 17 are tuned in, and pin 10 of IC103 also goes low, turning Q101 off. This allows the CA input of tuner TU101 to go high, enabling an FM trap within TU101. The frequencies of cable channels 14 through 17 are closed to the FM broadcast band. The FM trap prevents the FM broadcast signals from interfering with the cable signals.

2.3.6 H sync

The H-sync input at pin 8 of IC103 is developed in the sync-separator stage (chapter 5), and is present only when composite video is present, indicating that a channel is properly tuned. During the normal tuning sequence,

IC103 monitors the H-sync input to determine that the requested tuning operation has been accomplished. One receipt of the H-sync pulse, IC103 then monitors the AFT up and down signals at pins 32 and 33, and applies any fine-tuning corrections to the PLL TU102.

2.3.7 Typical tuning sequence

The following is a typical tuning sequence for the tuner with separate PLL, such as shown in Figure 2.5. When the AFT up and down request is received by IC103, either from the remote control (chapter 9) or the keyboard input (chapter 7), IC103 produces control signals to select the next higher or lower channel requested.

IC103 first outputs a chip-select signal at pin 22 to enable the skip memory of IC105. Then IC103 resets IC105 and places IC105 in the read mode. Finally, IC103 sends the address of the requested channel on the 4-bit data bus to select the next higher or lower channel.

When the address is selected, the data bits in that address of IC105 are applied to IC103. In turn, IC103 analyzes the data to determine if the next higher or lower channel is to be skipped or tuned. All of these operations on the data bus are synchronized by the clock signal applied to IC105 from pin 25 of IC103. If IC105 is busy and unable to accept or transmit data on the bus, the busy signal from pin 15 of IC105 is applied to IC103 causing IC103 to wait.

If the information at the selected address tells IC103 that the channel is to be skipped, IC103 continues on to the next higher or lower channel. This operation continues until a channel that is not to be skipped is reached. When this occurs, IC103 outputs B-0 and B-1 signals (at pins 11 and 12) to the band-switch circuits within IC102. In turn, IC102 applies the appropriate band-switching signals (L,M,H,U) to tuner TU101 as necessary for the selected channel. IC103 also outputs the necessary DAT (data), CLK (clock), and LAT (latch) signals to PLL TU102 to hold the tuning at the selected channel.

IC103 monitors the H-sync input at pin 8 to determine when the tuning operation is complete and composite video is present. When H-sync pulses are present (indicating that composite video is passing through the VIF stages), IC103 then begins to monitor the AFT up and down signals at pins 32 and 33 to determine if fine-tuning corrections is required. If necessary, IC103 outputs the appropriate DAT, LAT, and CLK signals to PLL TU102 until the AFT up and down inputs indicate that the channel is properly tuned. IC103 continues to monitor the AFT up and down inputs at pins 32 and 33, and produces fine-tuning corrections as necessary.

2.4 Tuner with Built-In PLL

Figure 2.6 shows the circuits of a TV tuner (a Sony 19 inch) where the PLL is part of the tuner package. The purpose of this circuit is to interpret key-

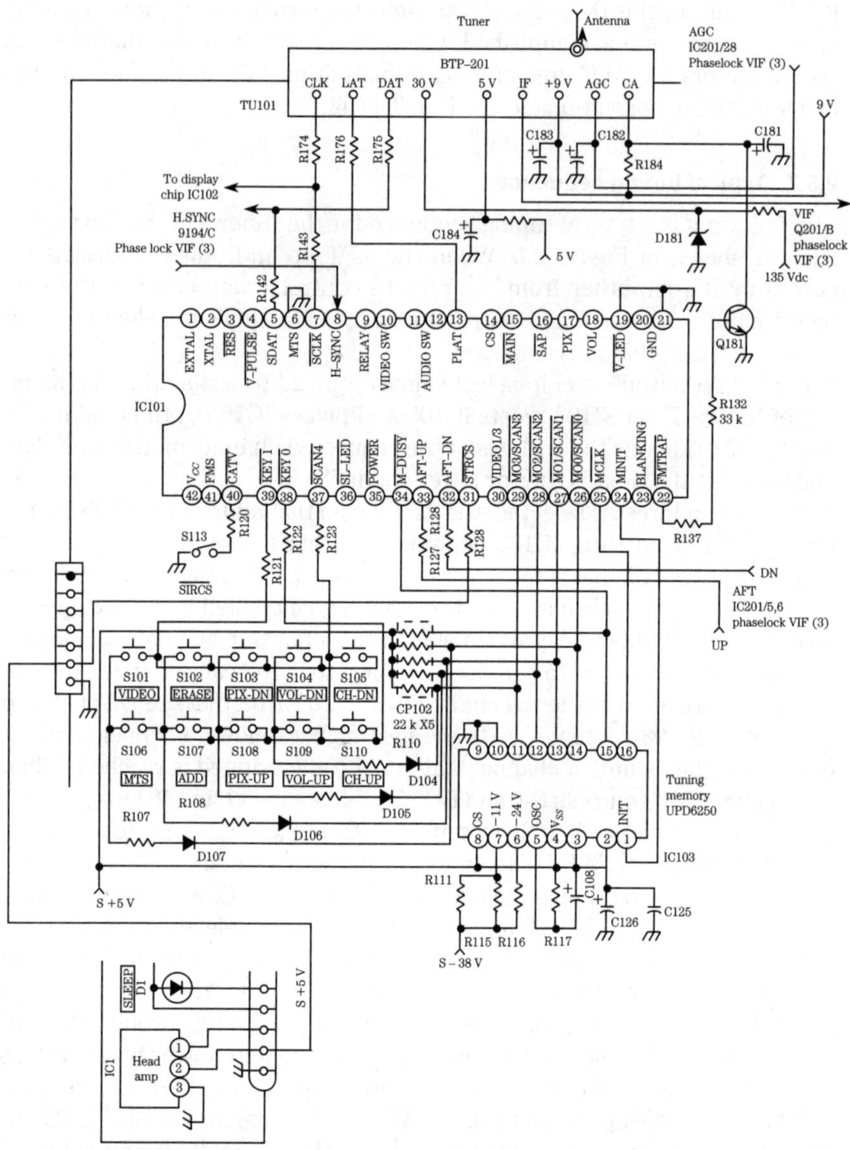

Figure 2.6 Tuner circuits with built-in PLL (2B).

board or remote inputs requesting channel changes. Tuner/PLL TU101 is controlled by microprocessor IC101, using SDAT (PLL data), SCLK (clock), and PLAT (PLL latch) signals at pins 5, 7, and 13 of IC101.

Tuner TU101 requires 5 V, 9 V, and 30 V. (The 30 V, provided by D181, is the source for the tuning voltage.) AGC correction is provided from pin

28 of IC201 in the VIF circuits (chapter 3). The tuner IF output is applied to the VIF stages through Q201. The CA terminal of TU101 is part of the FM-trap.

All of the TU101 tuning functions are under control of IC101. The following is a summary of such IC101 functions, on a pin-by-pin basis.

PIN	SIGNAL	FUNCTION
5	SDAT	Positive data to the TU101 PLL and to the IC102 display chip (chapter 9). This data stream is synchronized with the clock at pin 7, and PLL data bits are identified by the presence of the latch signal on pin 13.
7	$\overline{\text{SCLK}}$	Negative clock pulses to synchronize the serial data at pin 5. This line is also common to the display chip (chapter 9).
13	PLAT	This positive PLL latch signal identifies and transfers PLL data to the tuner.
40	$\overline{\text{CATV}}$	If S113 is closed, pin 40 is low, instruction IC101 that cable is in use.
22	$\overline{\text{FM TRAP}}$	This pin is normally high and Q181 is on. If pin 40 is low, indicating cable use, pin 22 goes low on channels 14, 15, and 16. These cable channels are very close in frequency to the FM band. With pin 22 low, Q181 is turned off to enable the FM trap (at pin CA of TU101) and prevent interference.
32, 33	AFT DN/UP	These positive-going AFT pulses occur on channel changes, or if the local oscillator within TU101 tends to drift. The AFT pulses are provided by the VIF circuits (chapter 3) and instruct IC101 to change PLL data (pin 5) as necessary to correct the tuning.
8	H-SYNC	These positive horizontal-sync pulses are derived from the VIF stages (chapter 3).
26–29	$\overline{\text{SCAN 0–3}}$	These lines are the negative-going keyboard scan lines, and are shared by the tuning memory IC103 by means of multiplexing.
37	SCAN 4	This line is always low, and serves as the fifth keyboard scan line.

PIN	SIGNAL	FUNCTION
38, 39	KEY 0, KEY 1	These keyboard inputs are interrogated by IC101 to determine keyboard closures (selection of functions at the TV-set front panel). When keyboard closures occur, keyboard scan pulses pull 38 and 39 low.
24	MINIT	A high occurs on this pin (during power-on) to reset tuning memory IC103.
34	$\overline{\text{MBUSY}}$	This line goes low during internal memory operations (within IC103) to indicate status of IC103 to IC101. (IC101 waits until IC103 operations are complete.)
26–29	MD0MD3	This is a 4-bit bus used by IC101 to obtain tuning status on any channel selected. The bus is multiplexed, and serves as an instruction bus, address bus, and data bus. The bus lines are also shared by the keyboard.
25	$\overline{\text{MCLK}}$	These pulses synchronize and multiplex the tuning-memory IC103 4-bit data bus. The low pulses can be observed during any tuning operation.
31	$\overline{\text{SIRCS}}$	SIRCS is the acronym for Sony Infrared Remote Control System. The infrared transmission from the hand-held remote is detected by head amplifier IC1. The output from IC1 is applied to pin 31, and causes IC101 to produce the necessary commands for tuning, picture, volume, channel addition/erasure, etc. (in place of the front-panel keyboard).

2.5 Antenna Circuits

Figure 2.7 shows the antenna-switching circuits found in the front-end stages of some TV sets (such as one of the Hitachi models). This circuit provides for selection of alternate inputs to the TV set (pay decoder, TV games, VCR, etc.) as shown, and improves reception of cable-TV signals.

The circuit is operated by a user-controlled switch. With the switch set to U/V IN, control terminal 3 is low, and Q01, Q01, and RL01 are off (RL01 is set to the A side). Under these conditions, the signal at U/V IN (A) is output to the tuner (D). When the switch is set to AUX, terminal 3 is high and Q01 and Q02 are on. This moves the contact of RL01 to the B side, connecting the AUX (B) signal to tuner (D). Simultaneously, D01 and D02 are turned on by the positive voltage through Q02, L04, and R02. This connects the U/V IN (A) input to converter (C).

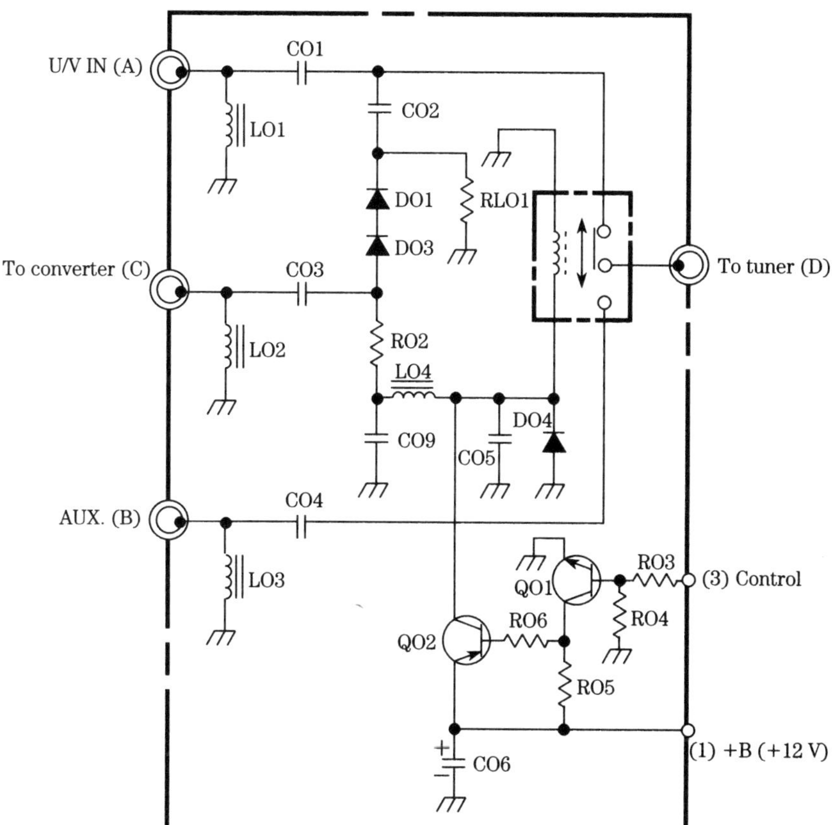

Figure 2.7 Typical antenna-switching circuits.

Chapter

3

VIF and SIF Circuits

This chapter is devoted to TV-set circuits between the front-end tuner and the audio and video processing circuits. The circuits are generally known as the VIF (video IF) and SIF (sound IF) stages, and usually include both video and sound detectors (although there are many different configurations in present-day IC sets). No matter what configuration is used, the basic function of the IF and video/sound-detector circuits is to amplify both picture and sound signals from the tuner, demodulate both signals for application to the video and sound processing circuits, and to trap (or reject) signals from adjacent channels.

3.1 VIF and SIF Basics

Figure 3.1 shows the signal path through the IF and video-detector stages of a typical discrete-component TV set. As discussed, the IF and video-detector functions are usually combined into one IC in present-day sets. In most cases the IF, video-detector, and sound-detector functions are combined into an IC such as shown in Figure 3.2, along with AFC, AGC, and APC functions.

In the circuit of Figure 3.1, all stages are forward biased and the first two stages Q1 and Q2 receive forward bias from the AGC circuit. This same AGC line is connected to the tuner. On strong signals, the forward bias is increased, driving Q1 and Q2 into saturation and reducing gain.

Each stage is tuned at the input and output by corresponding transformers T1 through T4. The stages are stagger tuned (each transformer tuned to a different peak frequency). This provides the overall IF circuit with a

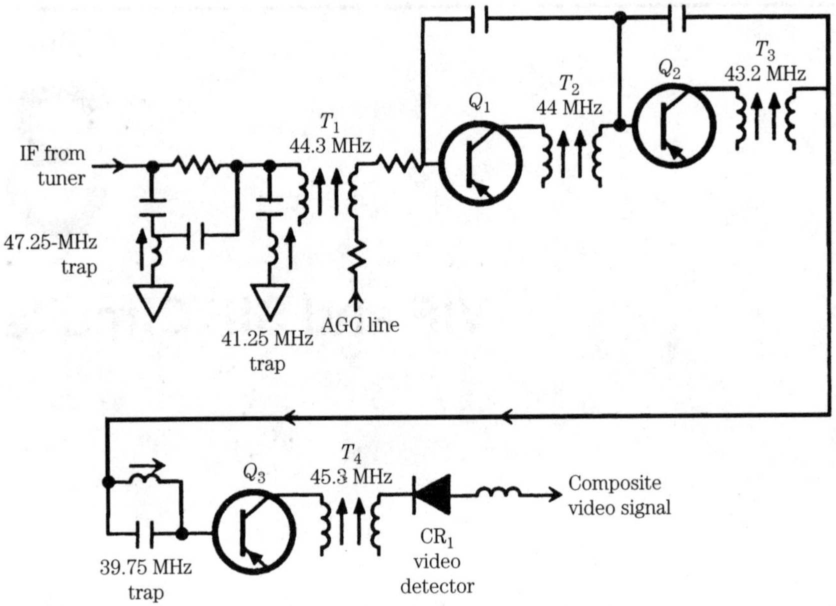

Figure 3.1 Signal path through IF and video detector in discrete-component set.

bandwidth of about 3.25 MHz. Three traps are used. The 41.25- and 47.25-MHz traps are series resonant, whereas the 39.75-MHz trap is parallel resonant. The traps are adjusted for a minimum output signal at the video detector when a signal of the corresponding frequency is injected at the IF input. Output of the IF stages can be measured at CR1. The video output is typically about 1 V, with the picture signals about 0.25 to 0.5 V.

In the circuit of Figure 3.2, IF output from the tuner is supplied to an IF amplifier within IC1 through SAW filter CP1 and T1. The IF output is applied to a video detector in IC1. The composite output from the detector is applied through filter CP4 to a video amplifier within IC1. (The detector output is also applied to a sound-IF amplifier in IC1, not shown.)

The output of the video amplifier is applied to the picture-tube and video/chroma circuits through filter CP5, L53, and Q3. The video-amplifier output is also applied to an AGC detector within IC1. This AGC circuit controls both the IF amplifier within IC1 and the tuner (through the RF amplifier in IC1). The AGC circuit is adjusted by R10.

The output of the IC1 video detector is applied to an AFC circuit in IC1. This circuit is adjusted by L6 and provides pulses to the horizontal AFC circuits (chapter 5). The IC video-detector output is also applied to an APC (automatic phase control) circuit that controls operation of the synchronous detector in IC1 (in conjunction with an IC1 lock detector which receives signals from the IC1 video amplifier). Notice that the synchronous detector is not always found in the IF IC; it might be located in a separate IC.

3.1.1 Recommended troubleshooting approach

The recommended approach for troubleshooting IF stages depends largely on the available test equipment and type of IF. Ideally, an analyzer or NTSC generator (chapter 11) is used because such generators duplicate the signals normally found at the tuner output (IF input) and video-detector output (as well as several other signals).

If the picture display is good with a signal injected at the video-detector output, but not good with a signal injected at the IF input (typically a coax cable from the tuner), the problem is in the IF. A possible exception is with defective AGC circuits. This problem can be eliminated by clamping the AGC line with the appropriate voltage. If the problem remains with the correct AGC bias applied, the problem is in the IF.

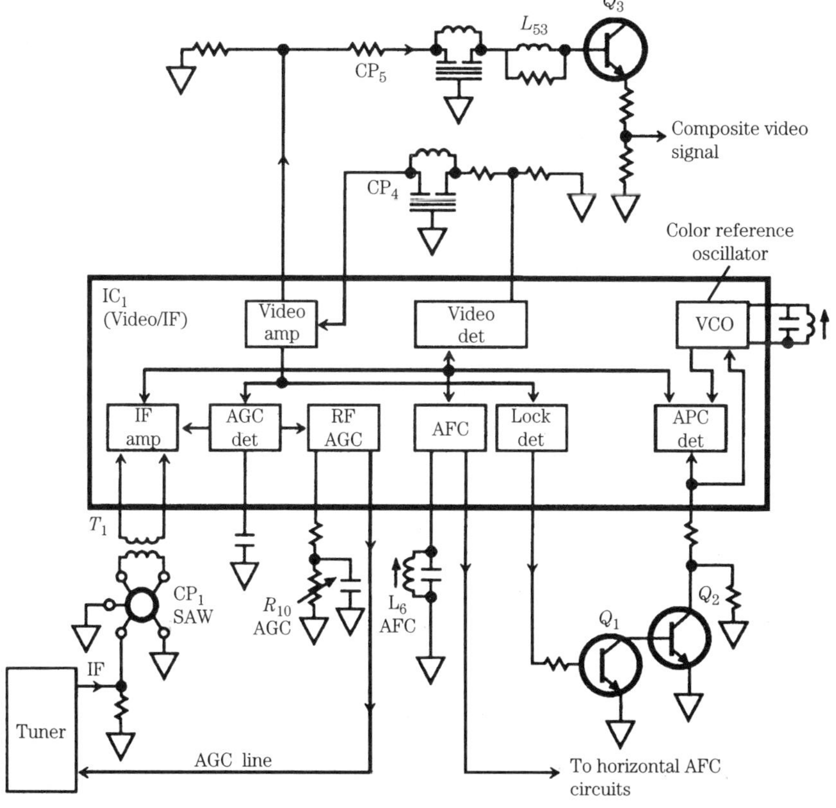

Figure 3.2 Signal path through IF and video circuits of IC set.

If an analyzer or NTSC generator is not available, the next recommended test setup is a sweep generator (with markers) and a scope. The sweep-generator signal is injected at the IF input (cable from tuner), and the signal is monitored at various points throughout the IF with a scope. Use a low-capacitance probe to monitor the video-detector output and a demodulator probe for individual IF stages.

The sweep generator is most effective when troubleshooting IF stages similar to those in Figure 3.1, because such stages require alignment and trap adjustment. When the IF stages are combined with other functions in an IC, such as shown in Figure 3.2, the analyzer or NTSC generator is far more convenient.

3.1.2 Typical troubles

The following sections discuss symptoms that could be caused by defects in the IF-amplifier and video-detector circuits.

No picture or sound. If a raster is present but there is no picture or sound, or the sound is very weak and noisy, inject an IF signal at the IF input. If operation is normal, the problem is in the tuner rather than in the IF. If the fault appears to be isolated to the IF stages (or IF IC), clamp the AGC line and repeat the test. If the problem is cleared, look for trouble in the AGC circuits. Check the adjustment or R10 (Figure 3.2). If the problem remains with the AGC properly adjusted, replace the IF IC (or check through each IF stage in the case of a discrete-component set).

Notice that the IF ICs in some sets provide AFT signals to the tuning system (such as the AFT up/down signals at pins 32 and 33 or IC103 in Figure 2.5). If these signals are absent or abnormal, the tuner will not fine tune a channel, even though the channel is locked in by the H-sync signal at pin 8 of IC103. This makes the tuner and/or tuning-system microprocessor appear to be bad, when the problem is in the IF IC. Obviously, if any circuit within the IF IC is defective, the entire IC must be replaced.

Poor picture or sound. The basic procedure for troubleshooting a poor picture and sound symptom (weak sound, poor contrast, etc.) is the same as for no sound or picture.

Hum bars or hum distortion. Hum is generally the fault of the power-supply filter. One possible exception is leakage of vertical-blanking pulses into the IF stages or IC (say through an open decoupling or bypass capacitor).

Picture smearing, pulling, or overloading. Generally, these symptoms are associated with the video processing circuits rather than the IF, particularly when sound is good. However, if the IF stages are improperly aligned or adjusted, or if there is a defective part making proper IF alignment impossible, the same symptoms can occur even though the sound might be good.

To eliminate doubt, inject a composite video signal from an analyzer or NTSC generator at the video-detector output. If the picture problems are eliminated, suspect the IC. Next, clamp the AGC (and/or adjust the AGC) to see if the problem is cleared. If the problem remains with good AGC, replace the IF IC (or try IF alignment in a discrete-component set).

Intermittent problems. When both the picture and sound are intermittent, suspect both the tuner and IF. However, the AGC circuit or the power supply could be at fault. Monitor both the tuner output and the IF output. Watch for changes when the intermittent condition occurs. As a minimum, monitor the video-detector output, AGC line, all dc voltages, and any tuning control signals (from tuner microprocessor, memory, bandswitch IC, etc.). As in the case of any intermittent condition, look for bad solder joints, breaks in PC wiring, intermittent capacitors (and even intermittent transistors). Also try tapping (not pounding) the IF and tuner modules. This can sometimes quickly pinpoint an intermittent module.

3.1.3 Typical VIF and SIF circuits

The following paragraphs describe some typical VIF and SIF circuits.

3.2 VIF and Audio

Figure 3.3 shows the sections involved in processing the IF signal (of a Sony 13 inch) from the tuner to develop the composite video output, as well as the sound output. The IF signal developed by the tuner is applied to the VIF stages within IC201. The video is processed, and composite video is separated from the carrier and from the sound signal. In this particular set, the composite video is applied to the video/chroma processing stages (chapter 4) through switches in the RGB interface stages (chapter 6).

The IF stages within IC201 return the RF AGC control voltage to the tuner, and thus control RF and IF gain. The VIF stages within IC201 provide AFT up and down correction signals to the tuning-control microprocessor IC103 (chapter 2). IC103 acts on these AFT correction signals to correct the PLL data and provide compensation for local-oscillator drift in the tuner.

The SIF signal is separated from the composite video output of IC201 and is returned to SIF circuits within IC201. These SIF circuits include an FM discriminator to remove FM audio from the 4.5-MHz sound carrier. In this particular set, the audio is then routed to the speaker and earphone jack through switches in the RGB interface stages (chapter 6), an audio preamplifier, and an audio power amplifier.

3.2.1 VIF

As shown in Figure 3.4, the VIF signal developed by the tuning system (chapter 2) is applied to VIF preamplifier Q221. The amplified IF signal is

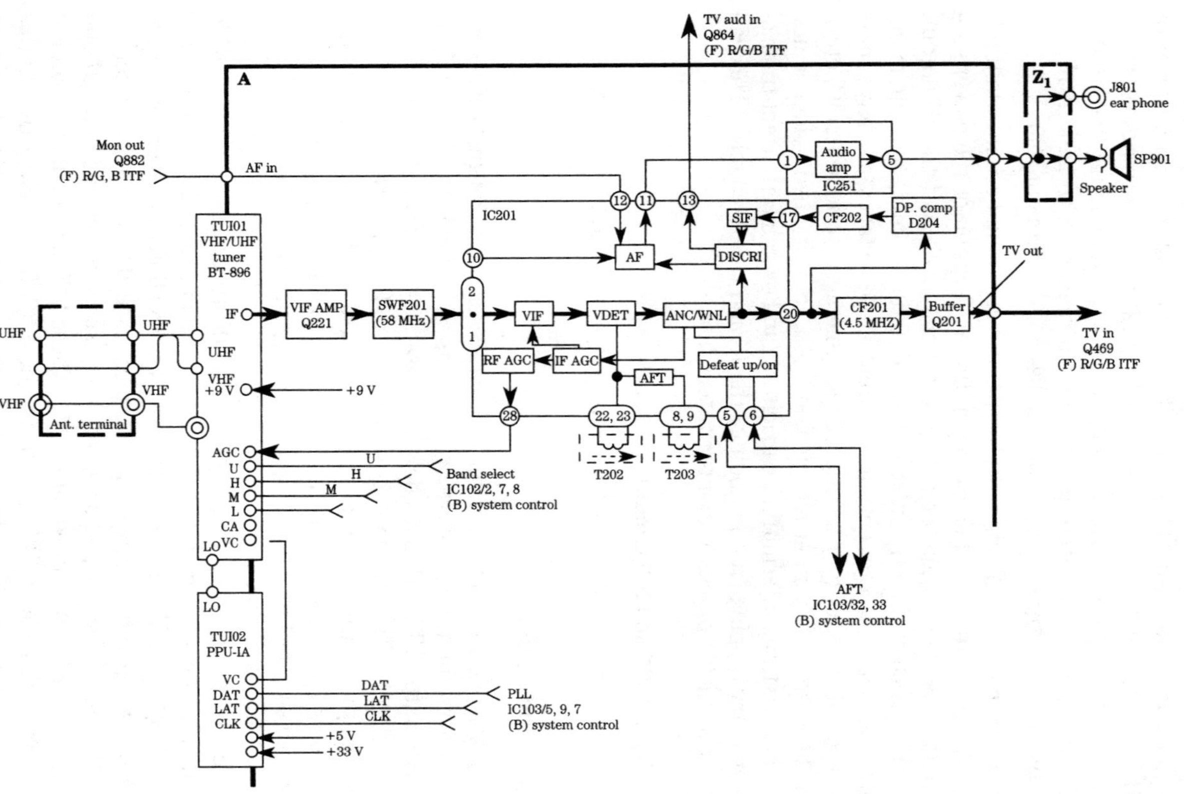

Figure 3.3 VIF and audio circuits (C).

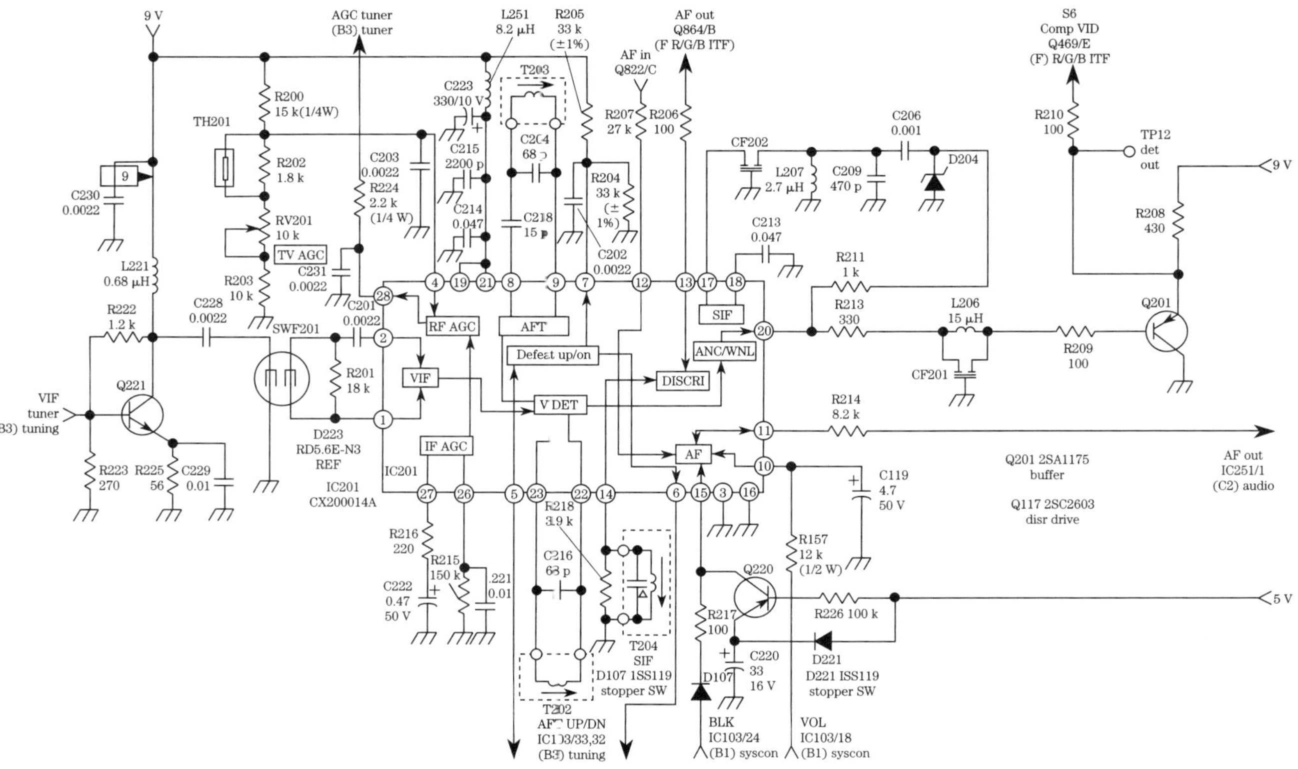

Figure 3.4 VIF circuits (C1).

then applied to SWF201 through C228. SWF201 is a 45.75-MHz SAW filter. (Most present-day sets use a SAW filter at this point because the superior bandpass characteristics eliminate the need for stagger tuning, such as shown in Figure 3.1.)

The VIF signal from SWF201 is applied to the VIF amplifier in IC201 through pins 1 and 2. The VIF signal is further amplified and sent to the synchronous-detector stage within IC201. The detector removes the 45.75-MHz carrier, leaving the composite video and the SIF signal. Both of these signals are processed by the ANC/WNL (automatic noise canceling white noise limiting) circuits, and are output from IC201 at pin 20. The VIF detector is tuned by T202.

The processed video/audio signal at pin 20 of IC201 is applied through CF201 to video buffer Q201. CF201 is a ceramic filter connected as a trap to remove the 4.5-MHz SIF signal from the composite video being buffered by Q201. The buffered output from Q201 is applied to the video/chroma stages (chapter 4) through switches in the RGB interface stages (chapter 6).

The output at pin 20 of IC201 is also applied through CF202 back to SIF circuits within IC201. CF202 is a ceramic filter connected in a bandpass configuration to pass only the 4.5-MHz SIF signal to pin 17 of IC201. Diode D204 serves to limit this signal.

The 4.5-MHz SIF signal is amplified and applied to an FM discriminator within IC201. The discriminator removes the 4.5-MHz carrier, leaving only the audio. This audio is applied to a preamplifier within IC201 through pin 13, switches in the RGB interface stages and pin 12. The preamplifier audio is then applied to audio power amplifier IC251 through pin 11 of IC201 and muting transistor Q251 (section 3.2.1). The discriminator is tuned by T204.

The audio preamplifier within IC201 has two control inputs. The input at pin 10 is the volume-control voltage developed by microprocessor IC103. As discussed in chapter 7, the control voltage at pin 10 varies from about 0 to 5 V, and sets the level of the audio at both the speaker and earphone jack.

The input at pin 15 is a muting signal. A high at pin 15 completely turns off the audio amplifier in IC201, removing the output signal at pin 11, to effectively mute the entire audio system. As discussed in chapter 7, a blanking signal is developed at pin 24 of microprocessor IC103 during channel changes. This positive signal is applied to pin 15 through D107, and mutes the audio.

Transistor Q220 provides audio muting when power is turned off.

When power is on, the +5-V line is high, Q220 is off, and C221 charges through D221. When power is turned off, the +5-V line goes low, Q220 is forward biased and turns on. This dumps the charge on C220 into pin 15, providing a momentary muting pulse to prevent speaker "pop" at power off.

The detected video is applied to the IF AGC stages within IC201. This is a peak AGC system that samples the level of the horizontal-sync pulses to determine the signal strength, and produces a corresponding dc correction voltage to the internal video IF amplifiers, as well as the RF AGC stage within IC201.

The RF AGC developed is applied to the tuner AGC terminal through pin 28 of IC201. The RF AGC controls gain of the RF amplifiers within the tuner, and varies between 2 and 8 V at pin 28. This is a reverse AGC system where an increase in IF signal increases the AGC voltage, driving the tuner RF amplifiers into saturation and reducing gain.

The filter networks at pins 26 and 27 of IC201 control response time of the IF AGC. In turn, this also controls the response of the RF AGC stages to prevent overreaction to sudden signal changes. The RF AGC level is set by RV201.

The center-frequency (45.75 MHz) VIF signal within IC201 is sampled and AFT correction pulses are produced at pins 5 and 6. If the IF signal varies in frequency above or below the 45.75-MHz center frequency, the AFT up and down correction pulses are directed to microprocessor IC103 at pins 32 and 33, as described in chapter 2. (IC103 performs AFT correction to the PLL as necessary to return the VIF signal to 45.75 MHz.) The AFT circuit is tuned by T203.

3.2.2 Audio

As shown in Figure 3.5, the audio recovered from the VIF signal in IC201 is applied to amplifier IC251 through C258. IC251 amplifies the audio to a level suitable for driving the speaker and earphone jack. The external RC networks at pins 1 through 4 and 8 of IC251 provide negative feedback to control gain and shape frequency response to the desired characteristics. Notice that the 14 V at pin 6 of IC251 is taken from a switching power supply (chapter 8), and is the only place where the 14 V is used. Thus, if there is a failure of audio only, check the 14-V source first. (When troubleshooting, always look for similar situations where there is a failure that can be isolated to a single source.)

Transistor Q251 and C259 provide for muting on power-up. When power is first applied, the relay switching voltage (chapter 8) is also applied through R111 to C259. This rising voltage is differentiated by C259, producing a momentary positive pulse to Q251. This turns Q251 on, pulling the audio signal down to ground through D251, and effectively muting the audio. This is a momentary effect that occurs for about one-half second on power-up.

3.3 Phase Lock VIF

Figure 3.6 shows the sections involved in processing the IF signal (of a Sony 19 inch) from the tuner to develop the composite video output, as well as the sound output. Although these functions are the same as described in section 3.2, the way in which the VIF operates is quite different.

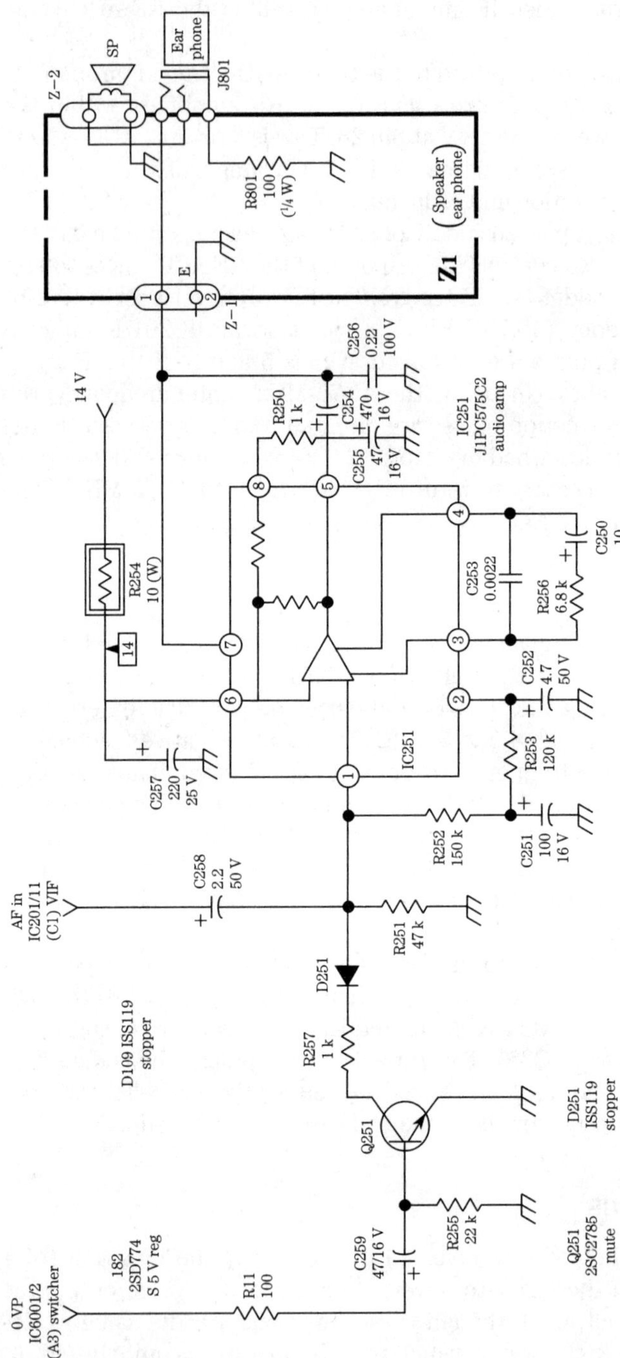

Figure 3.5 Audio circuits (C2).

The circuits described here involve the use of phase-locked synchronous video detectors with PLL. Figures 3.7 and 3.8 show details of the IF IC and PLL detector, respectively.

The IF signal from the tuner is input to IF IC201 through bandpass SAW filter SWF201 and preamplifier Q201. As shown in Figure 3.8, the VIF signal is amplified by the first, second, and third VIF amplifiers. The output of the third VIF stage is sent to the limiter amplifier and AM (video) detector. The limiter amplifier output, sharply tuned to 45.75 MHz (the video carrier), is applied to the AM detector, and used as the reference signal for the detector.

The detector video is applied to pin 20 of IC201 through a video amplifier, white-noise limiter, and noise-cancelling circuit. A portion of the video signal is also passed to the AGC system through a low-pass filter.

The peak AGC system detects the levels of the horizontal-sync pulses and develops a corresponding dc correction voltage. The output of the AGC circuit is filtered by external components at pin 27, and applied to the AGC drive stage. In turn, the drive stage provides correction voltages to the VIF amplifiers and RF AGC amplifier. The RF AGC voltage is applied to the tuner and controls gain in the normal manner. Typical range of the AGC voltage at pin 28 is 4 to 8 V. The RF AGC voltage can be adjusted by potentiometer RV201 at pin 4 of IC201.

The 45.74-MHz output from the limiter amplifier is also applied to the FM detector stage, and a 90° phase-shift stage. The detector monitors the frequency of the VIF 45.75-MHz signal and, if the frequency varies, produces a corresponding output via the AFT up/down circuit. Notice that the AFT output at pin 7 and the defeat function at pin 5 are not used in all models.

The AFT up and down stages generate positive pulses at pins 5 and 6, which are fed to the tuner microprocessor (IC101, pins 32 and 33, Figure 2-6). These pulses inform IC101 if the tuner oscillator is drifting high or low. IC101 responds by commanding the tuner PLL to retune the local oscillator, as described in chapter 2. The AFT circuit is tuned by T203.

3.3.1 PLL detector

As shown in Figure 3.6, the detected video output at pin 20 of IC201 is applied to pin 5 of the PLL detector IC251. Also, a sample of the VIF taken from the VIF amplifiers in IC201 is applied to pins 1 and 2 of IC251 through pins 24 and 25 of IC201. This IF signal is also applied to a synchronous detector and to a phase detector through a phase-shift network, as shown in Figure 3.7. The phase detector develops a voltage that is applied to the VCO in IC251 through a loop filter and amplifier. The VCO output is applied to the synchronous detector. The VCO is tuned by T251.

The synchronous-detector output is applied to pin 7 of IC251 through a switch and an ANC/WNL (automatic noise cancelling/white-noise limiting) circuit. The synchronous detector output is also applied to a beat detector,

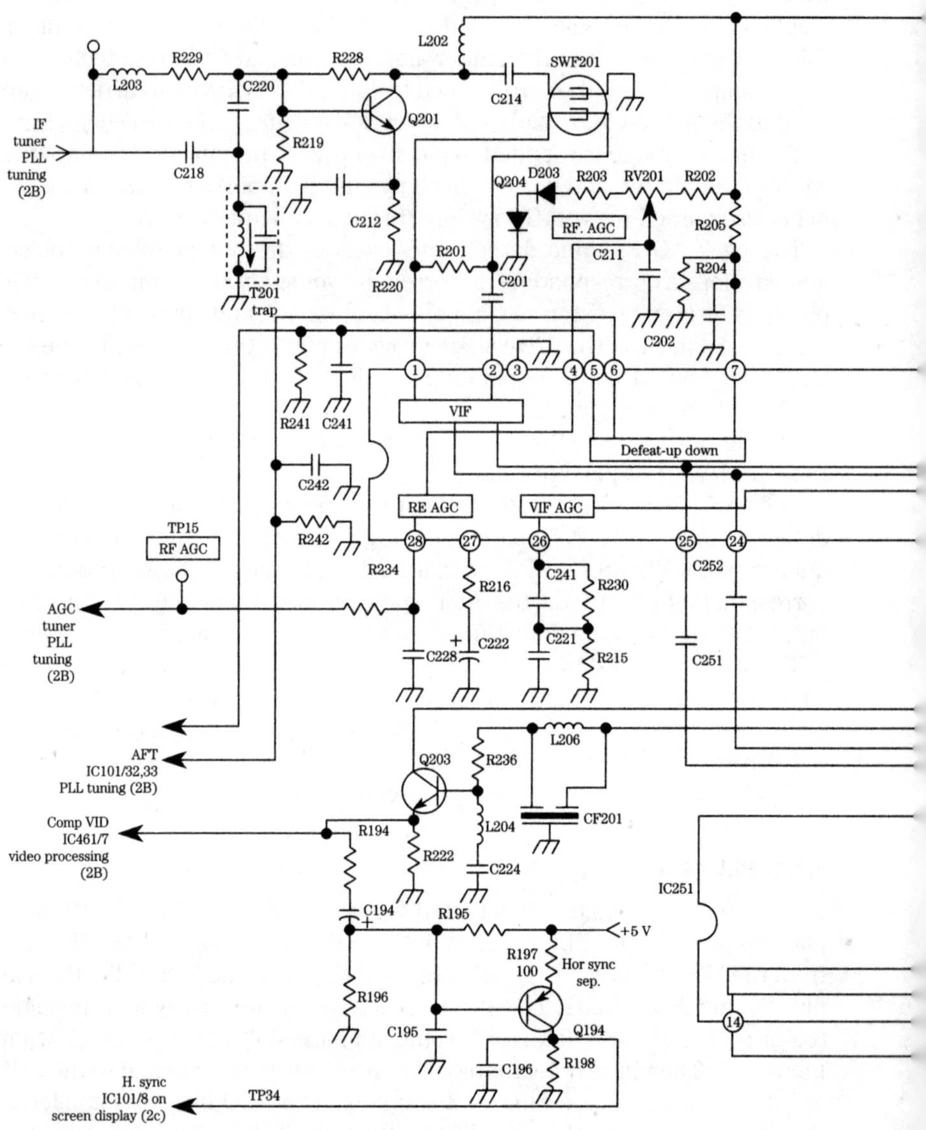

Figure 3.6A Phase lock VIF (3).

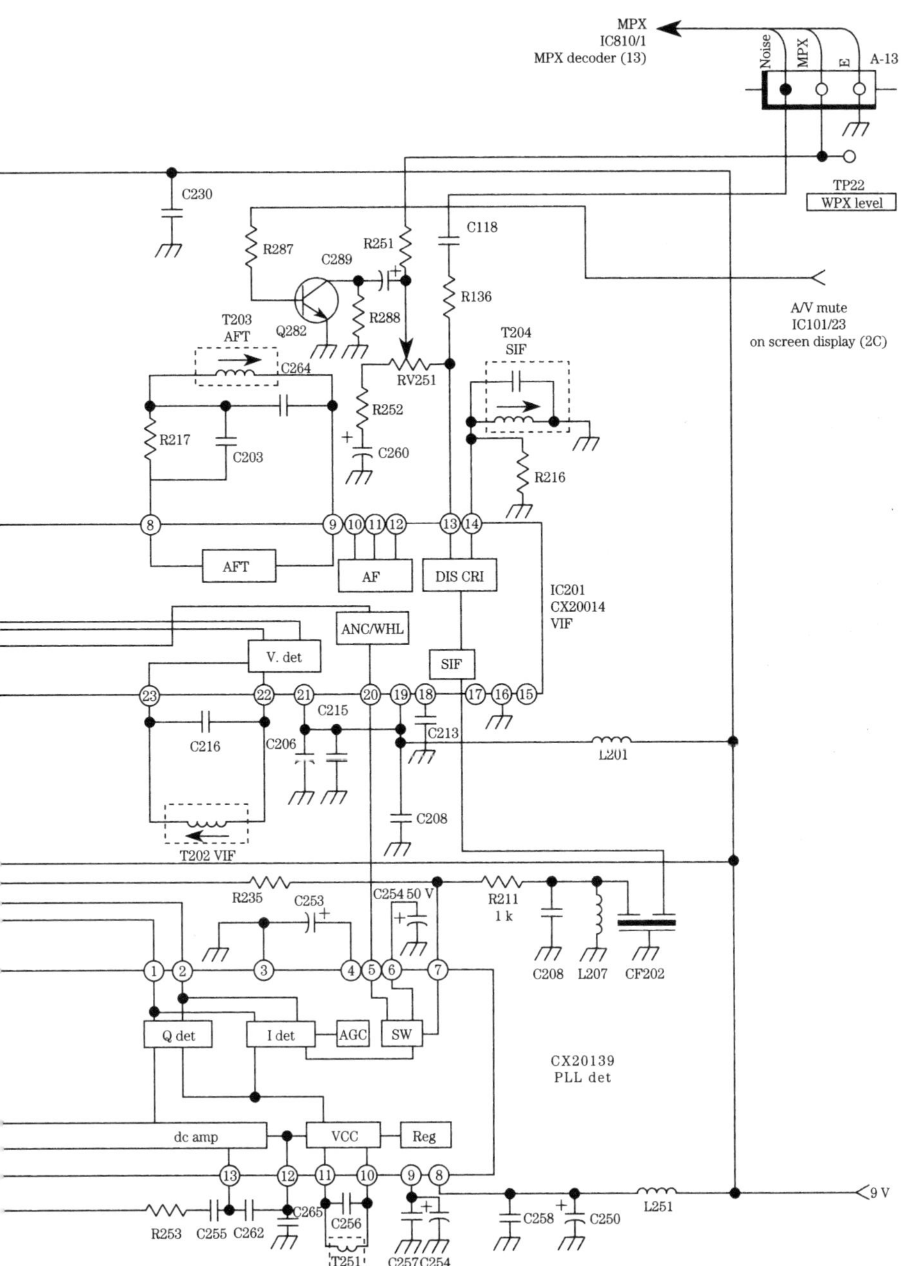
MPX
IC810/1
MPX decoder (13)
Noise
MPX
E
A-13
TP22
WPX level
C230
R287
C289
R251
C118
R136
A/V mute
IC101/23
on screen display (2C)
T203
AFT
Q282
R288
T204
SIF
C264
RV251
R252
R217
C203
C260
R216
AFT
AF
DIS CRI
IC201
CX20014
VIF
ANC/WHL
V. det
SIF
C215
C213
C216
C206
L201
C208
T202 VIF
R235
C253
C254 50 V
R211
1 k
C208
L207
CF202
Q det
I det
AGC
SW
CX20139
PLL det
dc amp
VCC
Reg
9 V
R253
C255
C262
C265
C256
T251
C257
C254
C258
C250
L251

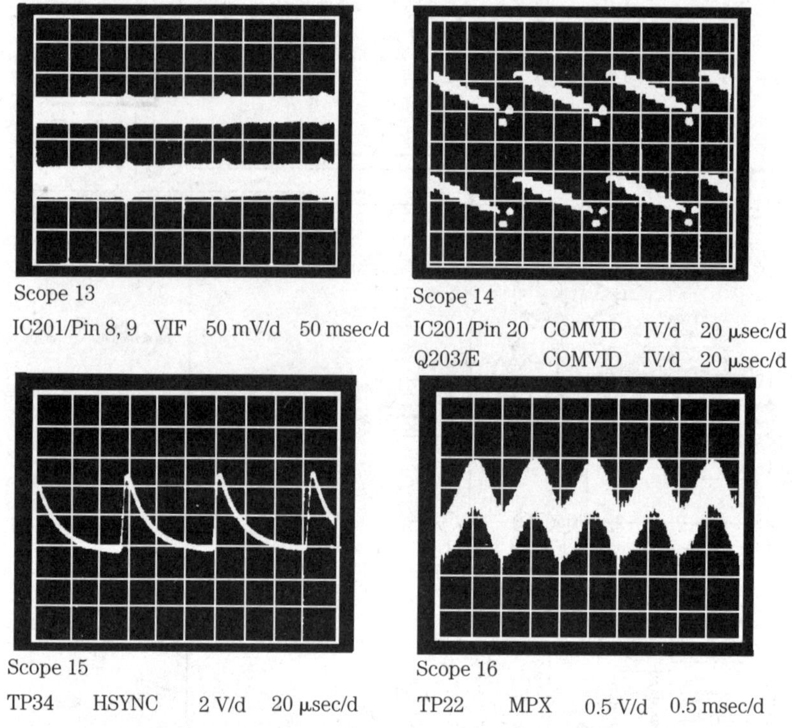

Figure 3.6B Phase lock VIF waveforms.

which develops an output when the PLL is locked. The beat-detector output operates the switch.

During normal operation, the switch passes the VIF signal from the synchronous detector in IC251 to pin 7. Should the PLL not lock up for any reason, the beat detector releases the switch, allowing the VIF signal from pin 20 of IC201 (and pin 5 of IC251) to reach pin 7. In effect, the set has two synchronous video detectors (the one in IC251 is phased locked), with the phase-locked detector used whenever possible.

3.3.2 Video and audio outputs

As shown in Figure 3.6, the output from pin 7 of IC251 is applied to the video-processing circuits (chapter 4) through a 4.5-MHz trap CF201 and buffer Q203. The composite video is also applied to horizontal-sync separator Q194, which removes the horizontal-sync pulses. These H-sync pulses are applied to the tuner microprocessor (IC101, pin 3, Figure 2.6) as described in chapter 2, and indicate that a channel has been tuned in.

The output from pin 7 of IC251 is also returned to the SIF circuits within IC201 through the usual 4.5-MHz bandpass filter CF202 at pin 17. The 4.5-MHz

SIF carrier is amplified in three stages before the audio is separated by an FM detector and applied to an MTS audio circuit (chapter 10) through pin 13. Notice that the volume-control circuit at pins 10 through 12 is not used in all models (such as when the set includes MTS stereo capability). Transistor Q282 is used to mute the audio signal during channel changes, and is operated by a muting signal from microprocessor IC101 in the on-screen display circuits (chapter 9). The FM detector or discriminator is tuned by T204.

3.4 PIF and SIF

Figure 3.9 shows the circuits involved in processing the IF signal (of some Hitachi models) from the tuner to develop the composite video output, as well as the sound output. (Notice that the video IF is called the picture IF, or PIF, in some Hitachi literature.)

Operation of this circuit is similar, but not identical to, that described for the VIF/SIF stages in this chapter. For example, 45.75-MHz IF from the tuner is applied through the usual SAW filter to video amplifiers within the IC. The amplified video is applied through a synchronous detector with PLL, and a 4.5-MHz trap to the video processing circuits. The 4.5-MHz sound carrier is returned to an FM detector through a limiter within the IC, and an external 4.5-MHz bandpass filter. The FM detector is crystal controlled, and extracts the audio in the usual manner.

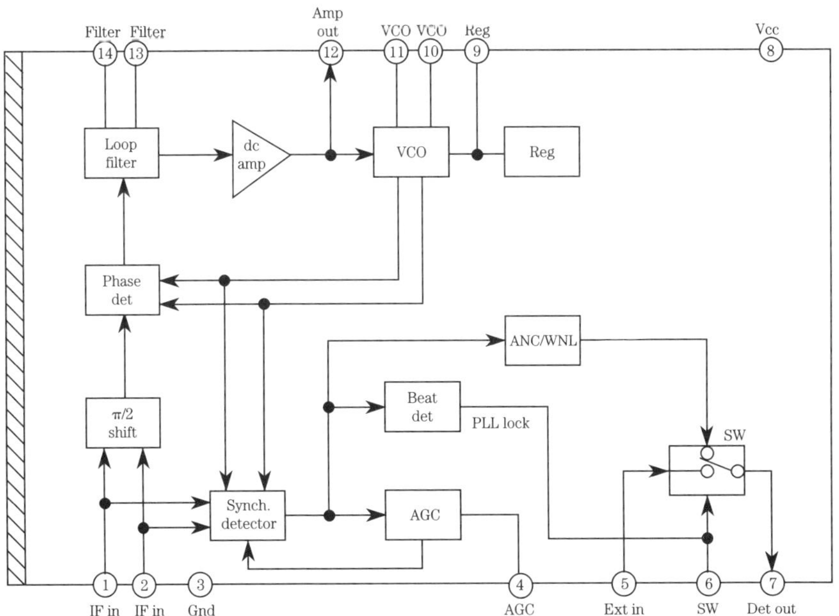

Figure 3.7 PLL detector.

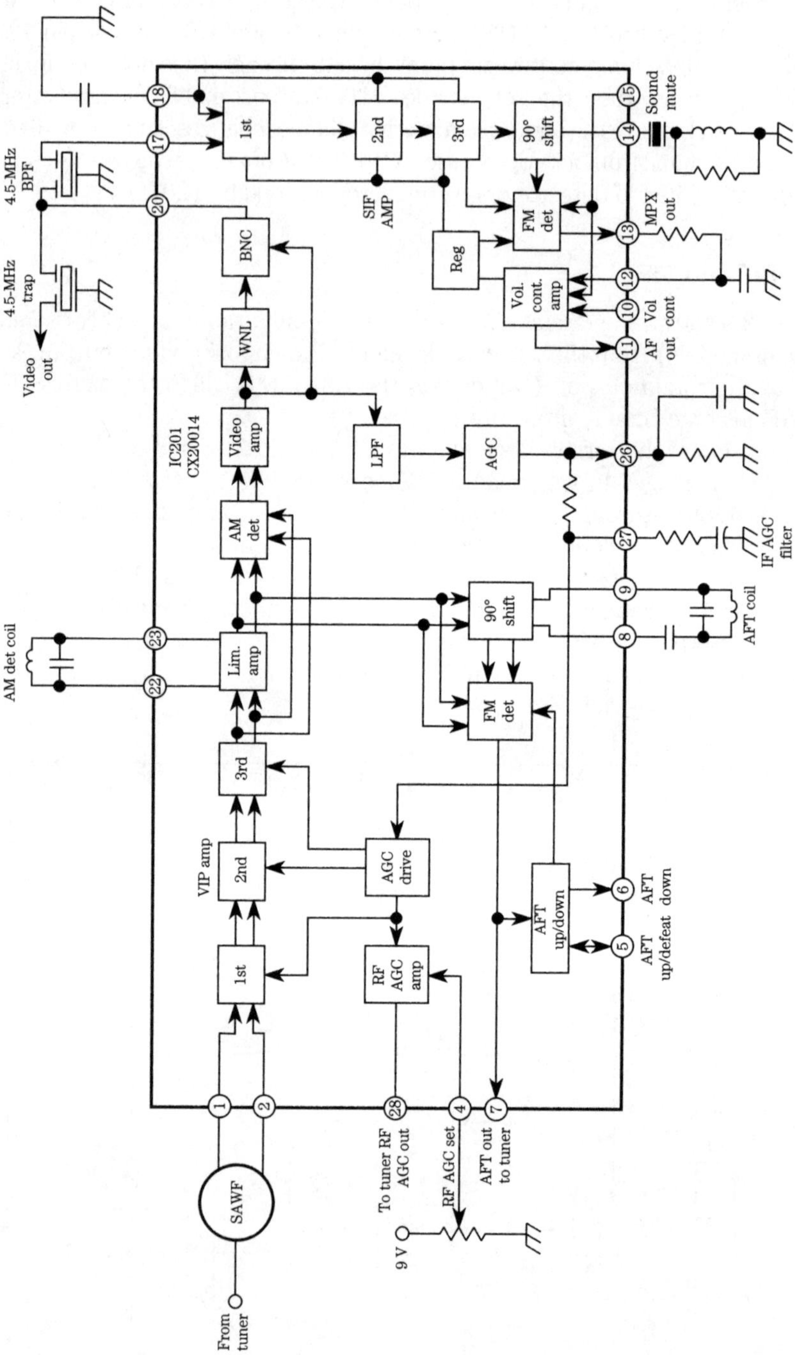

Figure 3.8 VIF IC.

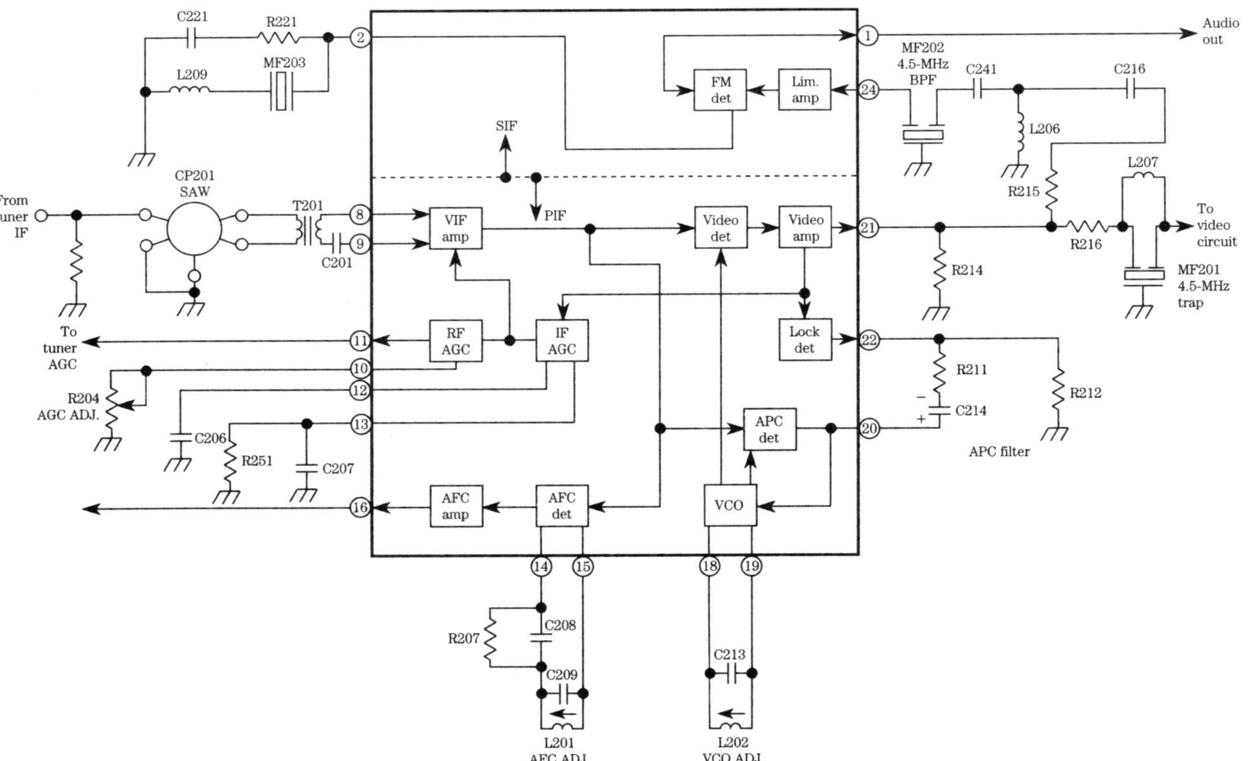

Figure 3.9 PIF/SIF circuit.

The detected video is also fed to IF AGC stages within the IC. This IF AGC controls the video amplifiers, as well as the RF AGC is returned to control the tuner in the usual manner, and is adjusted by R204.

The APC (automatic phase control) circuit consists of a VCO, a phase detector, and a low-pass filter. The 45.75-MHz VCO output is applied to the video detector along with the amplifier video from the tuner. The VCO output is also fed to the phase detector, after being shifted 90° in phase. The bandwidth of the low-pass filter (APC filter) is controlled by the lock-unlock detector, which compares the average level of the video detector with a referenced voltage. The APC VCO is adjusted by L202.

When the IF carrier generated by the VCO is locked by the PLL, the bandwidth of the APC filter becomes narrow. The bandwidth increases when the VCO carrier is unlocked. This makes it possible for the circuit to provide a pure IF carrier with comparatively short lock-up time (within 16 ms) and wide pull-in range (about ±2 MHz).

In this particular set, the AFC output at pin 16 is applied to the tuning microprocessor as a frequency-control signal. The AFC detector is tuned by L201.

Chapter

4

Video and Chroma Processing Circuits

This chapter is devoted to TV-set circuits between the VIF and SIF stages and the picture tube or CRT. Circuits that provide the horizontal and vertical sweep for the CRT are discussed in chapter 5. Chapter 6 describes additional color circuits (such as interface circuits used in video monitors).

Video and chroma processing circuits have several functions, and there are many circuit configurations. Thus it is very difficult to present a typical video/chroma processing circuit. However, most video/chroma circuits have certain inputs and outputs that can be monitored. If the inputs are normal but one or more of the outputs are abnormal, the problem can be localized to the video/chroma processing circuits.

4.1 Video and Chroma Processing Basics

Figure 4.1 shows the signal path of typical discrete-component video/chroma processing circuits between the VIF and SIF stages and the CRT. (Such circuits are often called the video-amplifier stages.) Typically, the signal input and output at Q1 are about 1 V. This signal, generally called the *composite video signal*, actually consists of sound, video, and sync pulses, as taken from the video detector that follows the VIF. The output of Q1 is applied to the input of Q2, to the input of the sync-separator (chapter 5), and to the input of the SIF (in this particular circuit). In the circuit of Figure 4.1, the output of Q1 is applied to Q2 through contrast control R3. In some circuits,

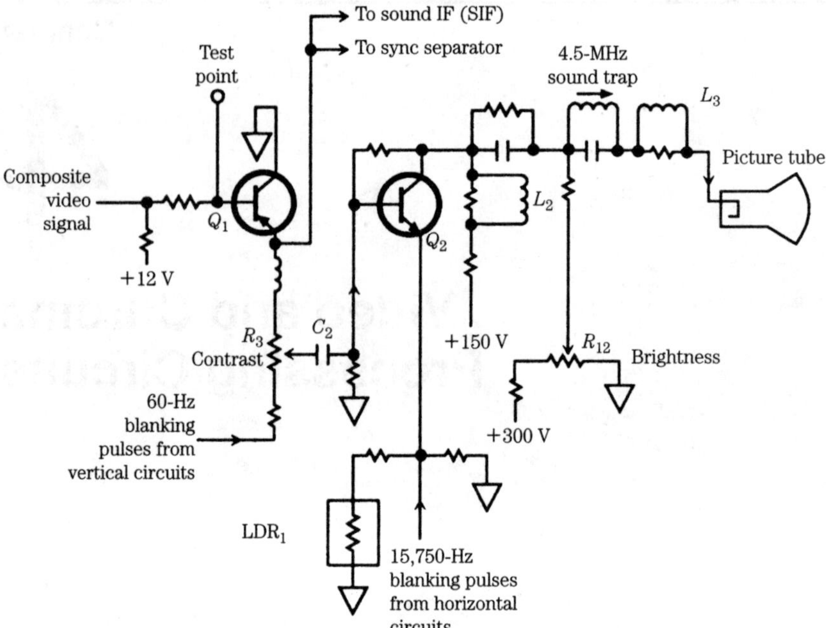

Figure 4.1 Signal path in discrete-component video-amplifier circuits.

the contrast control is part of the video-output circuit (typically in the emitter of the output transistor).

Transistor Q2 (typically a power transistor mounted on a heat sink) amplifies the 1-V output to about 25 to 50 V, depending on CRT screen size. Brightness control R12 sets the voltage level on the Q2 collector. The output of Q2 is applied through a 4.5-MHz sound trap to the cathode of the CRT.

The sound trap, when properly adjusted, prevents any sound from entering the video circuits. The Q2 output should contain only the video (picture information) and the horizontal and vertical retrace-blanking pulses. Because all of this information is applied to the CRT cathode, the video (picture) signal is negative, whereas the blanking pulses are positive. (A positive applied to the CRT cathode decreases brightness, whereas a negative increases brightness.)

In some circuits, the blanking pulses are applied to the first control grid of the CRT, rather than being mixed with video. In still other circuits, the mixed video and blanking pulses are applied to the control grid. In most present-day color sets, the blanking and video (both luma and chroma) are mixed in a single IC that also contains the color matrix, as discussed in section 4.3 and 4.4.

In the discrete circuit of Figure 4.1, the vertical retrace-blanking pulses from the vertical-sweep circuits (chapter 5) are applied to the emitter of

Q1. Horizontal-blanking pulses are applied to the emitter of Q2. Notice that both Q1 and Q2 are forward biased during normal operation.

The bias on Q2 is partially determined by light-dependent resistor LDR1, connected between the emitter and ground. As the ambient lighting around the set varies, the bias on Q2 (and thus the gain of the video amplifier) varies. This variable-gain feature provides the necessary changes in picture contrast to accommodate changing conditions of ambient light.

4.1.1 Recommended troubleshooting approach

The recommended approach for troubleshooting video processing stages depends largely on the available test equipment. Ideally, an analyzer or NTSC generator can be used to duplicate the signals normally found at the video-detector output (composite video, as well as horizontal/vertical blanking, as described in chapter 11).

As a quick check, the output of the video detector can be monitored with a scope (with a set tuned to an active TV channel). If there is a signal of about 1 V at the video-processing input (and the waveform resembles that shown in the service literature), picture problems are likely in the video-processing circuits.

Some technicians prefer to make a square-wave test of the video-processing circuits. A square wave of about 1 V is applied to the video-processing input, and the resultant output is monitored on a scope at the CRT. This checks the overall response of the video circuits, including the effect of contrast, brightness, and any other user controls. If square waves are passed without distortion, attenuation, ringing, and so on, the video circuits are operating properly.

4.1.2 Typical troubles

The following sections discuss symptoms that could be caused by defects in the video-processing circuits.

No raster. A no-raster symptom is usually associated with failure of the power-supply (chapter 8) or horizontal circuits (chapter 5). However, complete cutoff of the CRT can be produced by failure of the video-processing circuits, even though the high voltage and CRT voltages are normal.

For example, in the discrete-component circuit of Figure 4.1, if brightness control R12 is open or making poor contact on the ground side, a high positive voltage is applied to the CRT cathode. This cuts the CRT off (no raster), regardless of signal conditions or other voltages.

If such a condition is suspected, check all of the CRT voltages against the service literature. Pay particular attention to filament and cathode voltages. A low filament voltage can cause the filament to glow, but it does not produce enough emission for a picture. If any one of the voltages is not normal, trace the particular circuit and look for such defects as shorted or leaking capacitors, breaks in PC wiring, and worn controls. If all of the voltages appear normal, but no raster, suspect the CRT itself.

No picture and no sound. With a raster present, a no-picture/no-sound symptom is normally associated with failure of the tuner or VIF and SIF stages rather than in video. However, in the circuit of Figure 4.1, failure of Q1 could cut off both sound and picture, even though good signals are available at the video-detector output. This is true in present-day sets where there are often several discrete amplifier and buffer stages between the video-detector output and the video-processing IC input. Make the usual waveform and voltage checks if you suspect loss of signal in the stages between the detector and video processing IC.

No picture but normal sound. With a raster present and good sound, a no-picture symptom is definitely in the video-processing circuits (unless you are servicing a monitor or other set with video/TV switch). Monitor the signal at the Q2 collector and base, as well as at the CRT. Then measure the Q2 voltages. As in the case of any amplifier, total failure of the circuit is usually easy to locate. Look for such defects as an open coupling capacitor and coils, bad solder, open or worn controls, breaks in PC wiring, and the like.

In most discrete-component sets, Q2 is a power transistor and is subject to burn out if the heat sink is defective (bad thermal conduction between the sink and transistor). Also, other circuits can cause Q2 to burn out. For example, if C2 is shorted or badly leaking, the forward bias on Q2 can cause excessive current flow. Fortunately, power transistors are usually not required in the video-processing circuits of present-day IC sets.

Contrast problem. Too little contrast (or weak picture) that cannot be corrected by adjustment is the result of poor gain or lack of signal strength. In the circuit of Figure 4.1, contrast control R3 sets the signal level (or signal strength) applied to Q2. Poor contrast can also be the result of a weak picture tube. If you suspect the tube (because of age), try substitution (or rejuvenation, which is not always successful).

If the symptom is poor contrast, monitor the input to the video-processing stages (video-detector output). Always check the service literature for the correct amplitude of the video-detector signal. Typically, if the composite video signal is about 1 V, but contrast is poor, suspect the video circuits. If you suspect low gain in the video circuits, look for the usual problems (excessive collector/base leakage, leaking capacitors, low power-supply voltages, worn controls, etc.).

Too much contrast that cannot be corrected by adjustment is the result of too much gain or excessive signal strength. If the video-detector output is between 1 and 2 V with excessive contrast, suspect the video. If the detector output exceeds 2 V, suspect the tuner or VIF. It is also possible that the AGC circuits are failing to reduce excessive signals.

Excessive contrast is often associated with a defect in the contrast control R3. If all the voltages associated with R3 appear to be normal, monitor the signal voltages at the input and output of the video circuits. Pay particular attention to the signal voltage of the CRT input (cathode).

If heat sinks are used in any of the video circuits, there is an outside chance that a defective heat sink is causing excessive contrast. A defect in the heat sink (poor thermal conduction) can cause an increase in transistor gain (below the point where the transistor burns out), particularly if the set is used for short periods of time.

Sound in picture. When the picture appears to be modulated by the sound, suspect the tuner, VIF/SIF, and AGC circuits (including improper adjustment). If the 4.5-MHz sound trap is readily accessible, try to correct the problem with adjustment. As an alternative first step, monitor the video-detector output. If the picture display is fuzzy and/or varies with sound, the problem is ahead of the video amplifier.

If the problem is definitely in the video circuits, and cannot be corrected by adjustment, suspect the sound-trap components (open capacitor, shorted coil turns). Notice that in present-day IC sets, the sound is generally removed before the video stages, using some form of mechanical filter (such as CF201 in Figure 3.4). If an analyzer or NTSC generator with sound modulation is available, check the effectiveness of the filter by applying a modulated RF signal at the antenna, or a VIF/SIF signal at the VIF input (such as at Q221 in Figure 3.4). Monitor ahead and after the filter, while applying and removing sound modulation. For example, monitor at pin 20 of IC201 and at TP12 (Figure 3.4). The sound should appear at pin 20 of IC201, but not at TP12, when modulation is applied.

Poor picture quality. When the picture quality is poor (lack of detail, smearing, poor definition, etc.), the video circuits are usually suspected, particularly if the sound is good. However, these same symptoms can be caused by problems in the tuner and/or VIF.

If you have an analyzer or NTSC generator, inject a video signal at the video circuit input and check the CRT display. Suspect the video if the display is poor. As an alternate, check the video-circuit frequency response with a square wave as discussed. Look for such problems as leaking capacitors that change the resonant frequency of peaking coils, solder splashes that short out damping resistors across coils, or in rare cases, shorted turns in a coil (such as L206 and L207 in Figure 3.4).

Retraced lines in picture. In the circuit of Figure 4.1, both vertical and horizontal blanking pulses are fed to the video amplifier to blank the retrace lines. In most present-day IC circuits, the blanking is applied to the CRT through the same IC that contains the color matrix, bandpass amplifier, etc. (For example, in the circuit of Figure 4.6, the vertical and horizontal blanking are applied to pins 35 and 36 of IC301.)

No matter what system is used, the obvious first step is to monitor the blanking pulses at the point where the pulses enter the video circuits, or at the CRT. If the blanking pulses are absent, trace the particular circuit back to the pulse source.

If the blanking pulses are low in amplitude, the typical symptom is the presence of retrace lines only when the brightness control is turned up. If this is the case, check the pulse amplitude against the service literature. Then retrace the blanking pulses back to the source.

If you have an analyzer or NTSC generator with blanking pulses, substitute the generator pulses for the normal pulses. If this corrects the retrace problem, you have localized the trouble.

4.2 Color-Circuit Basics

The circuit of Figure 4.1 applies primarily to discrete-component black-and-white TV sets. As discussed in chapter 1, the processing circuits between the video detector and CRT are quite different for color sets. Figure 4.2 shows a matrix-demodulator circuit that is typical of discrete-component color sets. The circuit of Figure 4.2 represents the X and Z demodulators, as well as the R-Y, B-Y, and G-Y amplifiers shown in Figure 1.7. In

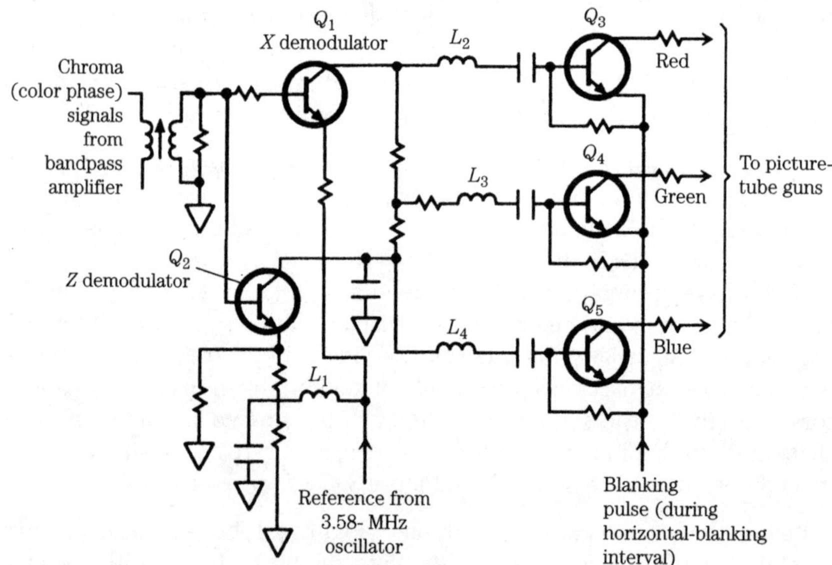

Figure 4.2 Typical matrix-demodulator circuit.

present-day sets, the circuits shown are found in a single IC (or possibly an IC with some external components).

In the circuit of Figure 4.2, the X and Z demodulators receive chroma signals (color-phase information) from the bandpass amplifiers and a reference signal from the 3.58-MHz oscillator. These signals are amplified and applied to the red, green, and blue CRT guns through the three matrix transistors. The relative phase of the signals at each gun depends on the phase relationship between the reference signal (3.58 MHz) and color-phase (chroma) signals.

Notice that the circuits are not adjustable, even for discrete-component sets. Thus, any failure must be corrected by replacement. Also notice that the output color signals from the matrix to the CRT are cut off by blanking pulses (during the horizontal blanking interval). Although the matrix and demodulator are not adjustable, the CRT drive circuits that follow the matrix are adjustable in most sets.

4.2.1 Localizing color troubles

Color troubles can be grouped into four classifications: no color, wrong color, weak color, or no color-sync. The following procedures describe the basic approach to color trouble localization, using a color-bar generator:

1. Connect the color-bar generator output to the antenna input.
2. Adjust the generator controls to produce an NTSC color-bar display (chapter 11).
3. Set the hue or tint control to midrange.
4. In turn, monitor each of the points shown in Figure 4.3 with a scope. Observe the following notes for each symptom. Notice that it is not always practical to monitor each of the points in Figure 4.3 in sets where many of the circuits are in a single IC. For example, as shown in Figure 4.6, the burst gate, color killer, bandpass amplifier, demodulator, and matrix are all in IC301. However, it is still possible to monitor the chroma (at pin 6), the reference oscillator (at pins 11 through 13), and the matrix output (at pins 32, 33, and 34).

No color. Start by checking the video-detector output. If there is no color output from the video detector, no color can be expected from the chroma circuits. Likewise, with a normal color signal at the detector, a no-color symptom is definitely localized to the chroma section that follows the video amplifier and/or detector.

Start at either demodulator stage (point A) and work back to the video detector. If the input to the demodulator is normal at A, check for the correct signal at B. The signal at B is generated by the reference oscillator, which is controlled by the phase detector. The signals at points A and B are

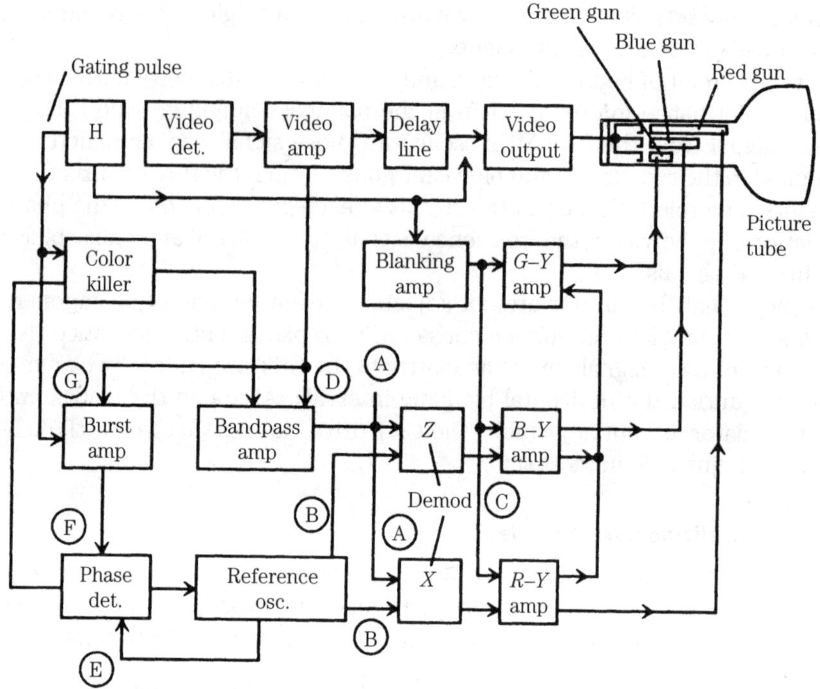

Figure 4.3 Localizing color troubles.

both at 3.58 MHz. (In the circuit of Figure 4.6, the reference oscillator is at 14.3 MHz, which is four times the 3.58-MHz burst signal.) The signal amplitude should be indicated in the service literature. If not, it is reasonable to assume that the amplitude is correct if both amplitude (A and B) are about the same.

The signal at A is passed through the bandpass amplifier which, in turn, is controlled by a bias from the color killer. If the bias is applied incorrectly (say because of some circuit failure), the color killer cuts the bandpass amplifier off, and there is no signal at A. Always check the bias and/or color-killer operation first when there is a no-color symptom.

As discussed, it is not practical to check all of the color circuits when they are part of an IC such as shown in Figure 4.6. However, you can check for all inputs to the IC. For example, if there is no-color symptom, look for chroma at pin 6, blanking at pins 35 and 36, reference-oscillator signals at pins 11 through 13. If any of these are absent or abnormal, trace back to the source. If all of the signals are good, suspect IC301. Also check the color-control signal at pin 21. If the user color control is set to one extreme, the color circuits will be cut off.

Wrong color. The demodulators are the most likely points to start localizing a wrong-color symptom. The procedure is essentially the same as that for a no-color symptom, except that the demodulator outputs must also be checked. These are shown in Figure 4.1 as point C. (In the circuit of Figure 4.6, you cannot check at the demodulator outputs, but you can check at the matrix output, pins 32, 33, and 34 of IC301.)

If the demodulator inputs are correct but there is no output from one demodulator, the trouble is localized immediately. If the demodulator outputs are present but not phased properly, the trouble is most likely improper alignment of the demodulator stages (if the demodulators are adjustable). If the demodulator outputs are correct, the trouble then can be localized to the B-Y, G-Y, or R-Y amplifiers (also known as the matrix amplifiers) or to the CRT. (Although the demodulators and matrix of the circuit in Figure 4.6 are not adjustable, the circuits between IC301 and the CRT are adjustable, as shown in Figure 4.7.)

Weak color. Again, the demodulator inputs are the most likely places to start. If the information is available, check the demodulator input amplitudes against those shown in the service literature. Note that the input from the reference oscillator is fixed (in amplitude), whereas the input from the bandpass amplifier can be varied by the user color control (at pin 21 of IC301 in Figure 4.6).

If it is necessary to advance the color control beyond the normal setting (about midrange) to produce the correct waveform amplitudes, or if the signals are still weak with the color control fully on, check the drive to the bandpass amplifier (at D in Figure 4.1, or pin 6 in Figure 4.6).

No color sync. Color sync is controlled primarily by the reference oscillator and the phase detector (the automatic phase control circuit (APC) of Figure 4.6). A phase difference between the oscillator and the generator (or TV station) bursts causes a loss of color sync. That is, color bars might not be in proper sequence, or the colors might be erratic (constantly changing).

The phase detector receives inputs from both the burst amplifier and reference oscillator (points E and F in Figure 4.1). If either input is incorrect in phase, the colors will be out of sync. (Although the phase of the reference oscillator in the IC of Figure 4.6 cannot be adjusted, the reference-oscillator frequency can be set to the crystal X301 frequency with capacitor CV301, using a frequency counter.)

4.2.2 Typical video/chroma processing circuits

The following paragraphs describe some typical video/chroma processing circuits.

4.3 Basic IC Video and Chroma Processing

Figure 4.4 shows the stages involved in processing both video (luma) and chroma from the VIF circuits to the CRT. In this particular IC set (Sony 13 inch), composite video from the VIF (chapter 3) is routed to the comb filter stages through the RGB interface circuits (chapter 6).

The comb filter, in conjunction with a 1H-delay line, separates the chroma and video information. The chroma information is passed through T352 (where low-frequency video information is filtered out), and applied to the chroma-processing circuits in IC301 at pin 6. The video information is applied through delay line DL301 to the video-processing circuits in IC301 at pin 3. Delay line DL301 compensates for the inherent chroma-processing delay.

The processed video and chroma are mixed in the RGB matrix of IC301, and are routed to the RGB driver board on the neck of the CRT (through the RGB interface, in this set). Several signals, such as the ABL (automatic brightness limiter), D sync (delayed sync), blanking and display, are input to the processing stages shown in Figure 4.4, as discussed in the following sections.

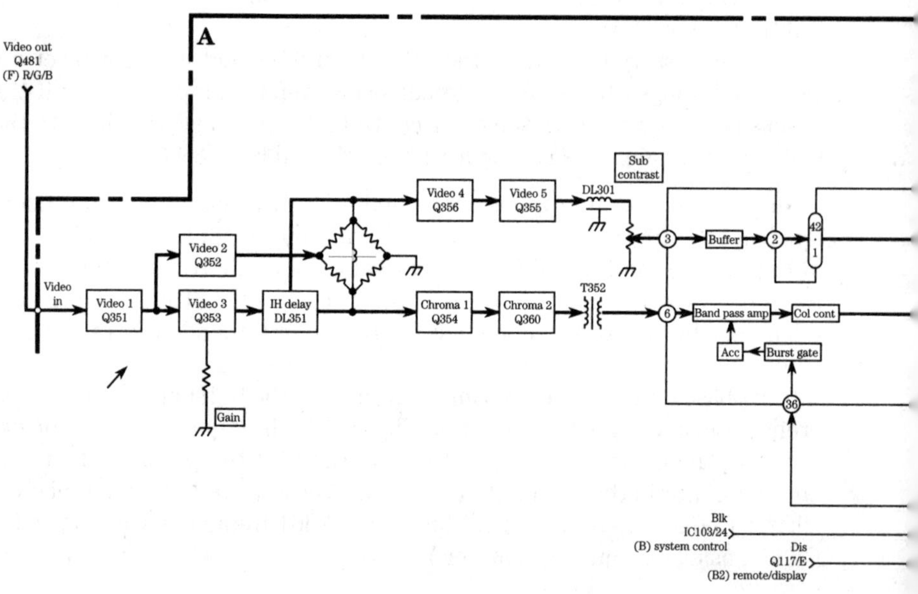

Figure 4.4 Video and chroma circuits (D).

4.3.1 Comb filter

Figure 4.5 shows the comb-filter circuits for a typical TV set (the Sony 13 inch). The purpose of the comb filter is to separate the luma information from the chroma information, and to direct each to the respective processing stages. In the NTSC color-transmission method, the phase of the chroma signal is inverted on every other line. The comb filter takes advantage of this characteristic, and operates on the assumption that luma and chroma content changes very little from line to line.

The comb filter incorporates a one-horizontal-line (1H) delay that allows the comb filter to compare a line of luma and chroma with the previous line of video and chroma information. (A 1H delay is about 63.5 microseconds.) Using the inversion characteristics of the NTSC signal, the chroma can be cancelled out, resulting only in luma output from one terminal. Likewise, by inverting line 1 and comparing it to an inverted line 2, the luma can be cancelled out, leaving only chroma at the other output.

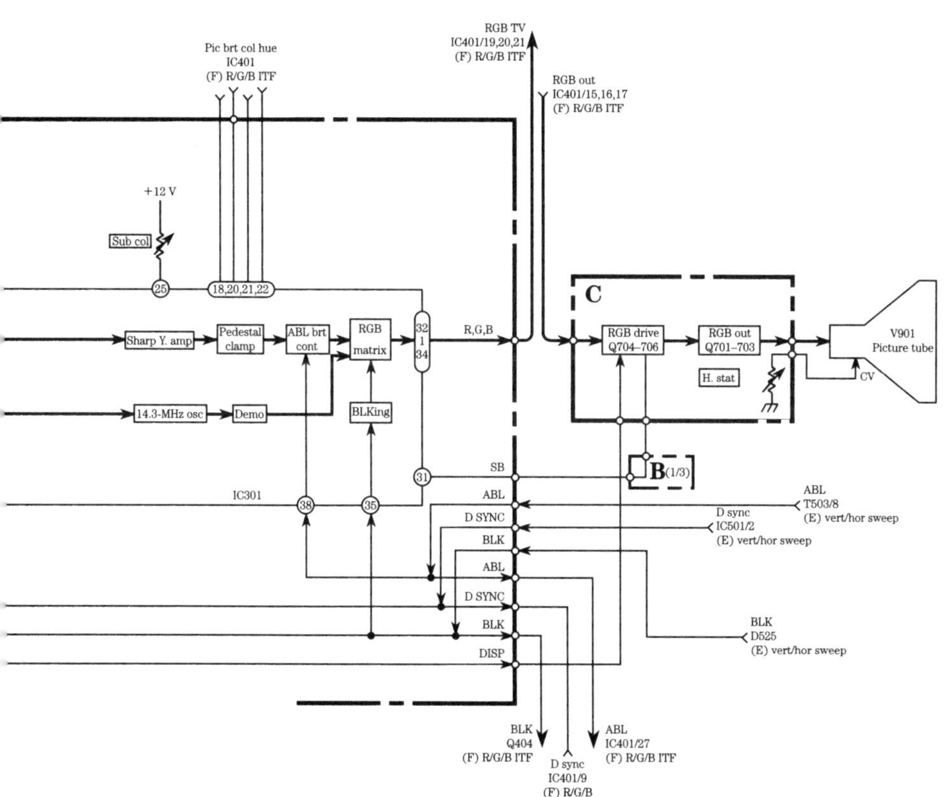

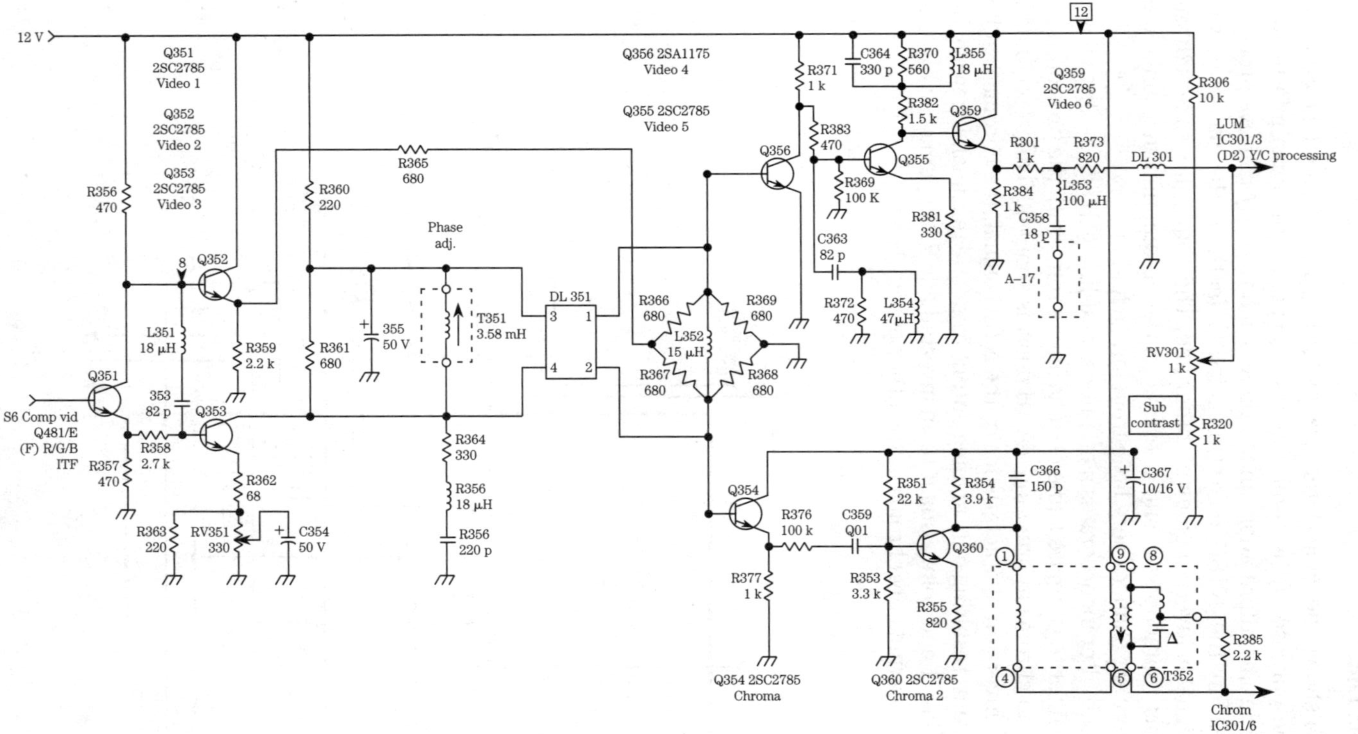

Figure 4.5 Comb filter circuits (D).

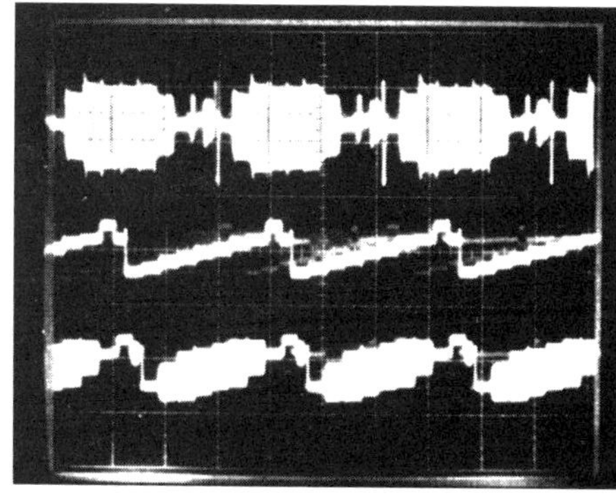

Figure 4.5 Continued.

(S17)	DL351/4	CHR+HFV	0.5 V/d	20 μsec/d
	DL351/1	LUM	0.5 V/d	20 μsec/d
	DL351/2	CHR+LFV	0.5 V/d	20 μsec/d
	Color bars			

The composite video from the VIF stages is applied to Q351 through the RGB interface circuits (chapter 6). Transistor Q351 is a phase splitter that provides buffered composite video to Q353, and inverted composite video to Q352. Transistor Q352 buffers the inverted composite video and applies it to one leg of the resistive bridge network at the output of delay line DL351. Transistor Q353 inverts the composite video before application to the DL351 input. RV351 sets the level through Q353.

The tank circuit at pin 4 of DL351 is series-resonant at about 2.5 MHz. This effectively removes most of the lower-frequency video information from the composite video at pin 4, leaving primarily high-frequency video and the chroma at the DL351 input. The bandpass of DL351 is at 3.58 MHz.

The inverted chroma and high-frequency video passing through DL351 appears at pin 2, while the noninverted chroma and video appears at pin 1. Because the inverter (pin 2) chroma/video is delayed by 1H, the pin 2 information arrives at the bridge simultaneously with the next line of information through Q352. T351 adjusts the phase so that the inverted and noninverted signals are exactly 180° out of phase.

The upper leg of the bridge (across R366) provides cancellation of the out-of-phase chroma signal, leaving only the video at Q356. The lower leg of the bridge (across R367) provides cancellation of the out-of-phase video information, leaving only the chroma at Q354.

The video or luma information is amplified by Q356, Q355, and Q359. The resonant circuits associated with these transistors condition the video signal to restore the normal high-frequency/low-frequency content. The conditioned video is then passed through another delay line DL301, which provides an approximate 0.8-ms delay. The purpose of DL301 is to compensate for inherent delay encountered by the chroma signal so that the chroma and luma signals arrive at the IC301 matrix (Figure 4.6) simultaneously. RV301 provides for internal contrast adjustment by setting the dc level of the luma signal.

The combed out chroma information is buffered by Q354, amplified by Q355, and applied to the 3.58-MHz bandpass transformer T352. This effectively removes any remaining low-frequency video from the chroma signal, presenting relatively pure chroma information to the Y/C (luma/chroma) processing stages of IC301.

4.3.2 Y/C processing

Figure 4.6 shows the circuits involved for processing the combed-out luma and chroma signals before application to the CRT output or drive circuits. Compare this to the basic circuits of Figures. 4.1 through 4.3.

The combed luma information is applied to pin 3 of IC301, buffered, and then applied to a sharpness amplifier through an external video-peaking network. The sharpness amplifier enhances the high-frequency characteristics of the video signal. The processed and amplified video signal is then passed through a pedestal clamp for dc restoration, and through an ABL brightness-control circuit to the RGB matrix.

The combed chroma information is applied to the color demodulator through pin 6 of IC301, a bandpass amplifier, and a color-control circuit. The chroma information is demodulated using the 14.3-MHz oscillator signal as described in section 4.2. The demodulated R-Y, B-Y, and G-Y signals are applied to the matrix. The resulting RGB signals from the matrix are applied to the CRT drive circuits through the RGB interface (chapter 6). The reference oscillator can be adjusted to the frequency of crystal X301 by CV301.

Several external signals are input to the Y/C processing stages. The PIX (picture) signal is a dc voltage developed in the system-control through Q102. This signal controls the gain of the video and color amplifiers in IC301. There is also a user picture control (in the RGB section) connected to pin 18.

The brightness, color, and hue inputs from user controls on the RGB interface (chapter 6) are applied to IC301 at pins 20, 21, and 22. These dc inputs set the brightness level for the luma signal, the color saturation of the color signal, and the phasing or hue (also called *tint* on some sets) of the color signals. RV303 provides for internal adjustment of the color level.

The blanking signal at pin 35 of IC301 is developed in the vertical and horizontal output stages (chapter 5). These are positive-going vertical and horizontal blanking pulses used to turn off the RGB outputs during the vertical and horizontal retrace periods. The blanking signals are also applied to the dc-restoration pedestal clamp within IC301. The D sync (delayed sync) pulse at pin 36 is used to trigger the burst gate. This allows the burst amplifier to operate only when the burst signal is present (when color information is being transmitted).

The ABL or automatic brightness limiter input at pin 38 of IC301 is derived from the horizontal-output or flyback (FBT) transformer T503 in the horizontal stages (chapter 5). Beam current through the CRT is reflected in the ABL winding of T503. If the beam current becomes excessive, the dc level at pin 38 decreases lowering the brightness level. This limits the maximum beam current to prevent overdriving the CRT.

4.3.3 RGB output

Figure 4.7 shows the circuits involved between the Y/C processing and the CRT. In this particular set (Sony 13 inch) all of the circuits are on a board (the C board) located at the neck of the CRT.

Drive/output transistors Q701 through Q706 amplify the RGB signals from pins 32 through 34 of IC301 (Figure 4.6) for application to the CRT cathodes. Notice that the blue and green circuits have both a drive and a background adjustment (RV702 through RV705), while the red circuit has only a background adjustment (RV701), in this particular set.

The RGB outputs from IC301 are applied to the drive/output circuits of Figure 4.7 through switches on the RGB interface board (chapter 6). The switches permit either the normal TV RGB signals from IC301, or external RGB signals (from computers, VCRs, etc.) to be reproduced on the CRT, thus making the set a monitor TV.

The display signal from the system-control microprocessor (chapter 7) is applied through connector C-5 to the green driver/output circuits (Q702/Q705). This signal generates a channel indication and bar-graph display on the CRT.

Flyback transformer T503 in the horizontal-output stages provides the focus, screen heater, and high voltage for the CRT. The screen and focus voltages are derived from the +800 V at pin 1 of connector C-4, while the heater voltages H1/H2 are applied through connector C-2. T503 also supplies the +190 V required by the CRT drive transistors Q701 through Q706.

The CRT focus voltage at pin 1 (grid 4) is set by RV707, while the screen (CRT brightness) voltage at pin 9 (grid 2) is set by RV706. There is also a signal brightness control RV709 that sets the dc voltage at pin 31 of IC301 (Figure 4.6).

The convergence voltage for the internal convergence plates of the CRT is derived within the CRT from a high-voltage second-anode lead across a resistance network built into the CRT. The return path is from CRT pin 2 through RV708 to ground. RV708 provides for adjustment of the static con-

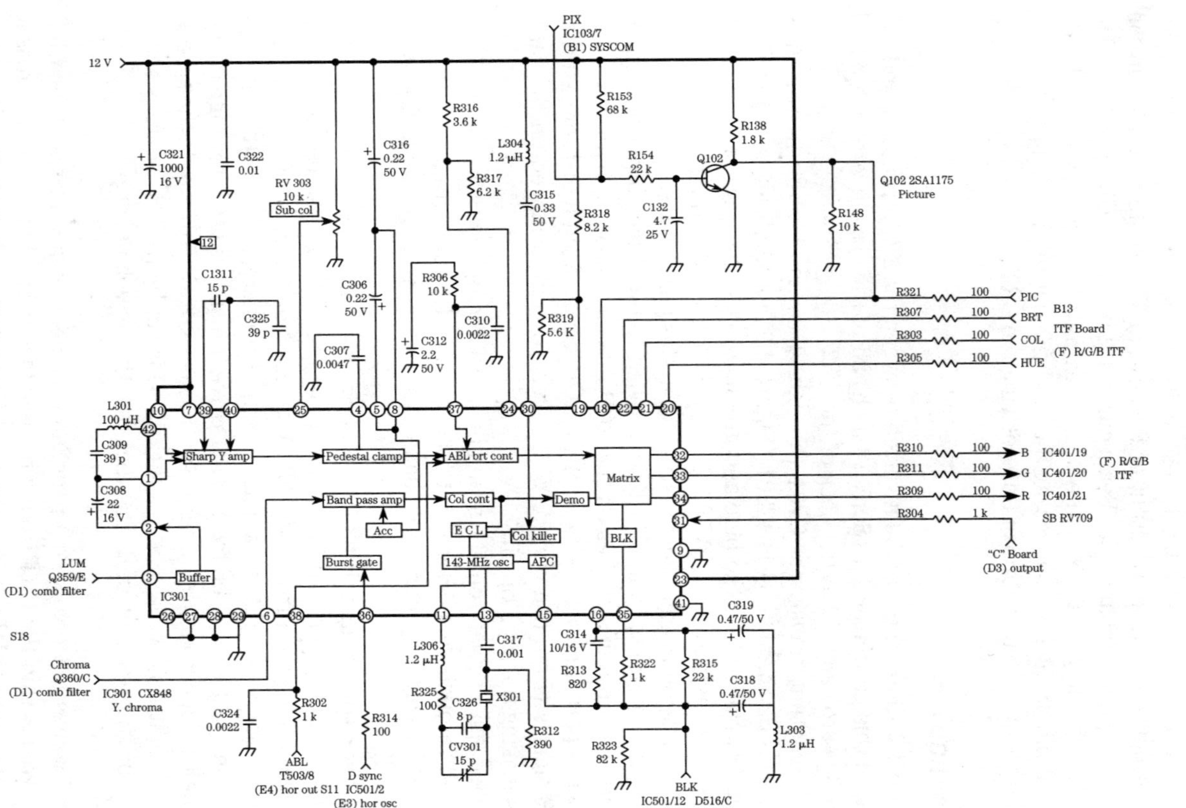

Figure 4.6 Y/C processing circuits (D2).

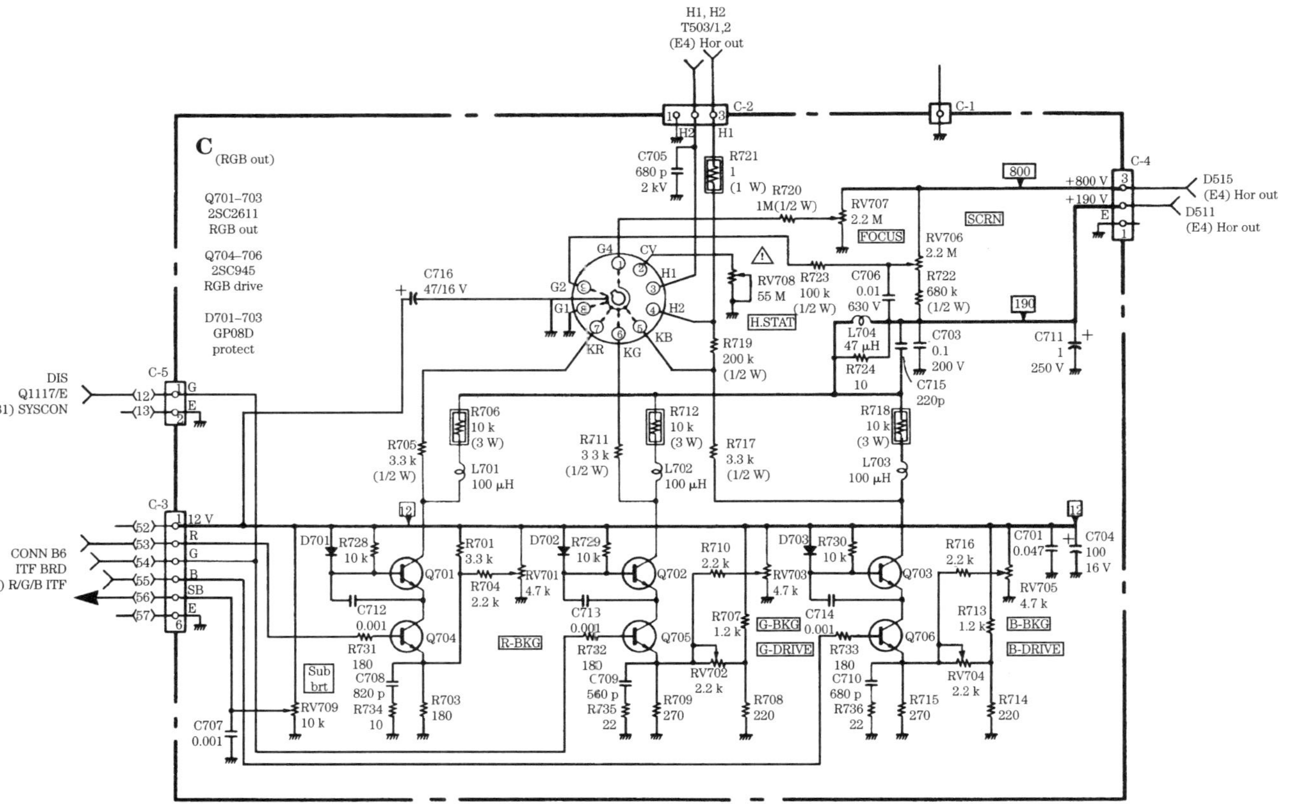

Figure 4.7 RGB output circuits (D3).

vergence. (Notice that this internal static-convergence voltage is unique to Sony Trinatron.)

4.4 Video and Chroma Processing with a Jungle Chip

This section describes video and chroma processing circuits that use a so called Y/C jungle chip, such as shown in Figure 4.8. Such chips are multipin and incorporate many of the functions found in a discrete form on other sets.

As an example, the chip of Figure 4.8 includes such luma (Y) functions as brightness control, picture control, sharpness control, video muting, pedestal clamp, video delay line, and dynamic picture, as well as the chroma (C) functions of dynamic color, automatic color control (ACC), 3.5-MHz reference VCO, automatic phase control (APC), color killer, color demodulator, matrix, and chroma/video miser.

The chip of Figure 4.8 also includes some of the primary CRT sweep functions normally found in the horizontal/vertical sections (chapter 5). The sweep functions include sync separator, horizontal oscillator, vertical oscillator, horizontal blanking, vertical blanking, and high-voltage hold-down or X-ray protection.

Table 4.1 lists the function, pin number, and description for the chip of Figure 4.8 (which is designated as IC301 in a Sony 19 inch). The following sections describe operation of circuits associated with IC301.

4.4.1 Video processing input

Figure 4.9 shows the input circuits to the jungle-chip processing. IC461 contains switches that select the input to be processed, as determined by signals from the system-control microprocessor (chapter 7). The input can be composite video from the PLL detector (section 3.3.1), or external video (chapter 10). Composite video is selected when pin 3 of IC461 is high, as shown in Figure 4.8.

The signals at pin 4 of IC301 are applied to a vertical separator Q352 through CRT clamp Q358/Q359 and video buffer Q351. The composite video from Q352 is applied to 1-H delay line DL351 through sync expander Q353/Q354. The composite video is also applied to jungle chip IC301 at pin 42 through R360. (This is the sync input to IC301.)

DL351 and the associated circuit parts form a comb filter. The chroma output is applied through Q357 to the chroma input of IC301 at pin 39. The luma output is applied through equalizer Q355/Q356 to the luma input of IC301 at pin 41. A 3.2-V reference voltage is applied to the base of Q356 from pin 40 of IC301.

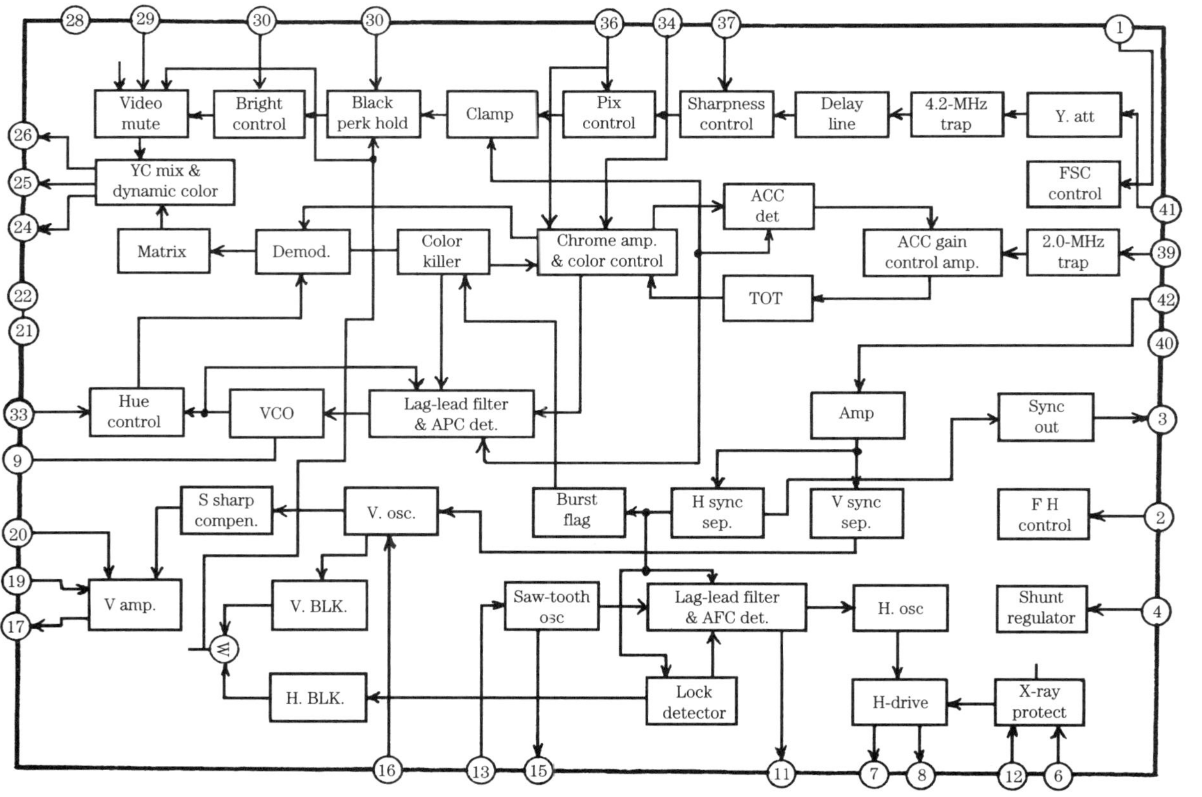

Figure 4.8 Y/C jungle chip.

TABLE 4.1 Y/C Jungle Pinout

Function	Pin no.	Name	Description
X-ray protection	6	V ref.	Applies the X-ray protection voltage.
	12	Hpl	X-ray protection input terminal. Switches the levels of pin (7) Hd output and RGB output of pins (26), (25) and (24) to low when the level exceeds and the reference voltage of pin (6).
Horizontal output	7	Hd	Horizontal output. A pulse of approx. 9 Vp-p is output, and the level becomes nearly GND level during HOLD DOWN.
	13	Hp2	Horizontal flyback input. Is compared with the video sync input at pin (42) and forms the AFC loop.
Vertical output	17	Vd	Vertical output
	20	V_{FB}	Vertical NF output
RGB output	24 B	B output	Drops to approximately 0.5 V during HOLD DOWN.
	25 G	G output	Drops to approximately 0.5 V during HOLD DOWN.
	26 R	R output	Drops to approximately 0.5 V during HOLD DOWN.
User control	32	BRT	Brightness control pin.
	33	HUE	Hue control pin.
	34	COL	Color control pin.
	36	PIX	Picture control pin.
	37	SHR	Sharpness control pin.
Video input	39	Cin	Chroma input pin. Comb filter output is input.
	41	Yin	Y input pin. Comb filter output is input.
	42	Jin	Sync input pin. Comb filter input circuit.

4.4.2 Video processing (jungle chip)

Figure 4.10 shows the circuits involved for processing the combed-out luma signals before application to the CRT drive circuits (the RGB output circuits). Compare this to the circuits of Figure 4.6. The following is a description of the principle video-processing inputs and outputs to and from IC301. As discussed, the luma is applied at pin 41, and the sync at pin 42. The chroma-processing circuits of jungle chip IC301 are discussed in section 4.4.3. The horizontal and vertical sweep inputs and outputs using IC301 are covered in chapter 5.

The sharpness-control circuit in IC301 affects the high-frequency detail of the picture. This circuit is adjusted by RV304. As a point of reference, the dc range at pin 37 is from zero to 5.5 V.

The PIX (picture) control circuit is adjusted by the voltage at pin 36. This voltage is developed at pin 17 of system control IC101 as a PWM (pulse width modulation) signal, and is filtered to produce a dc level at Q301. The level of this signal is adjusted by RV306 to provide an internal picture control.

The output of the matrix is also rectified by CP301 through CP303 and applied through Q302 to the picture-control circuit as a form of feedback to maintain a constant picture level (after internal adjustment by RV306 and user selection through system control). The voltage range at pin 36 is 1.6 to 1.96 V. Notice that the picture-control signal at pin 36 also goes to the IC301 chroma amplifier, and thus affects both video and chroma portions of the picture.

The brightness circuit is adjusted by the voltage at pin 32. This voltage is controlled by the user brightness control RV303, and an internal brightness control RV706 in the RGB output circuit (Figure 4.12). The voltage range at pin 32 is from 2.61 to 2.72 V. The brightness-control input at pin 32 also receives a signal from the horizontal oscillator should the X-ray protection circuit be activated, as discussed in chapter 5.

The video-mute block is controlled by a signal at pin 29. This is a blanking signal generated by the system control IC101 during channel changes to blank the CRT as new channels are tuned. The signal at pin 29 is a positive-going pulse of about 5 Vp-p with a duration of about 1 second.

The on-screen display data is applied through D350 to the green input of the RGB output circuits. Such data originates from the on-screen display circuits (chapter 9).

The picture ABL circuit Q304 and Q305 samples the current through the ABL winding of the flyback transformer T503 in the horizontal output (chapter 5). The resulting voltage controls the conduction level of Q304 and Q305, which apply a corresponding signal to the picture-control input at pin 36 (along with the input from Q301).

The AFC compensation circuit Q303 monitors the output levels of the RGB signals and produce a corresponding correction signal to pin 11 of IC301. The RGB signals from the matrix at pins 24 through 26 of IC301 is applied to the CRT through the RGB output circuits.

4.4.3 Chroma processing (jungle chip)

Figure 4.11 shows the circuits involved for processing the combed-out chroma signals before application to the CRT drive circuits (RGB output). Compare this to the circuit of Figure 4.6. The following is a description of the principle chroma-processing inputs and outputs to and from IC301. As discussed, the chroma is applied to pin 39, and the sync at pin 42.

The chroma signal is amplified by an ACC (automatic color control) amplifier within IC301 and applied to other color circuits, along with a burst signal taken from the horizontal-sync separator. The burst signal (or flag) controls the APC (automatic phase control) detector. In turn, the APC detector controls the color killer and chroma amplifier.

The amplified chroma is demodulated using the VCO signal controlled by crystal X301. (Notice that the service manual cautions any monitoring at

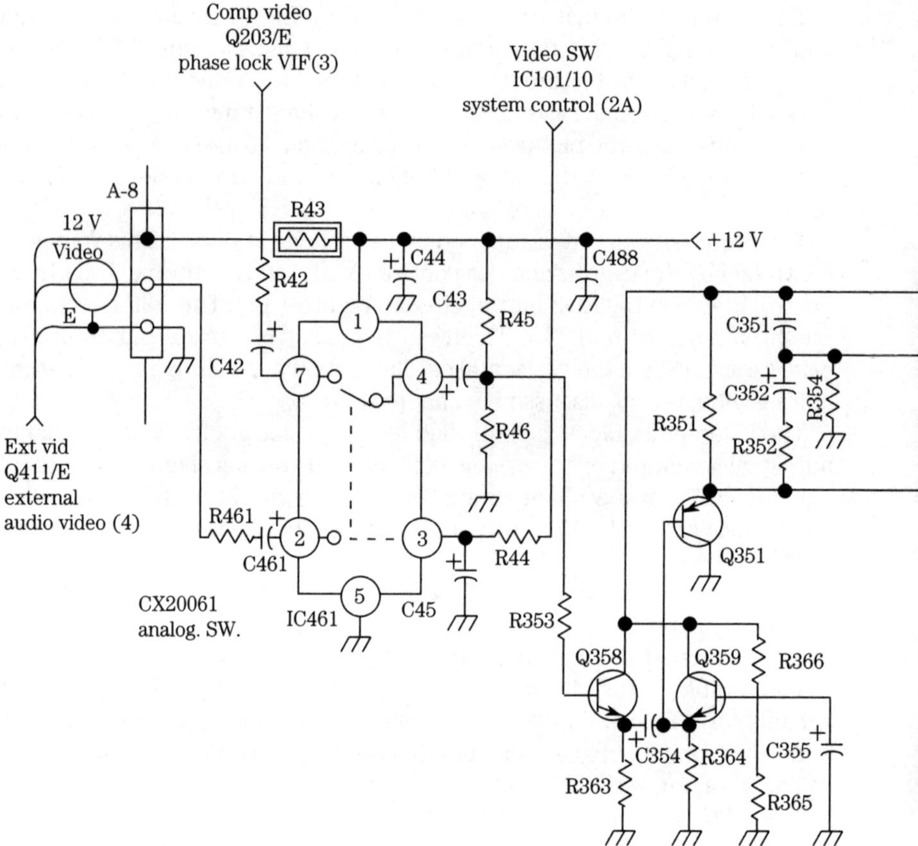

Figure 4.9 Input circuits to jungle-chip processing (5).

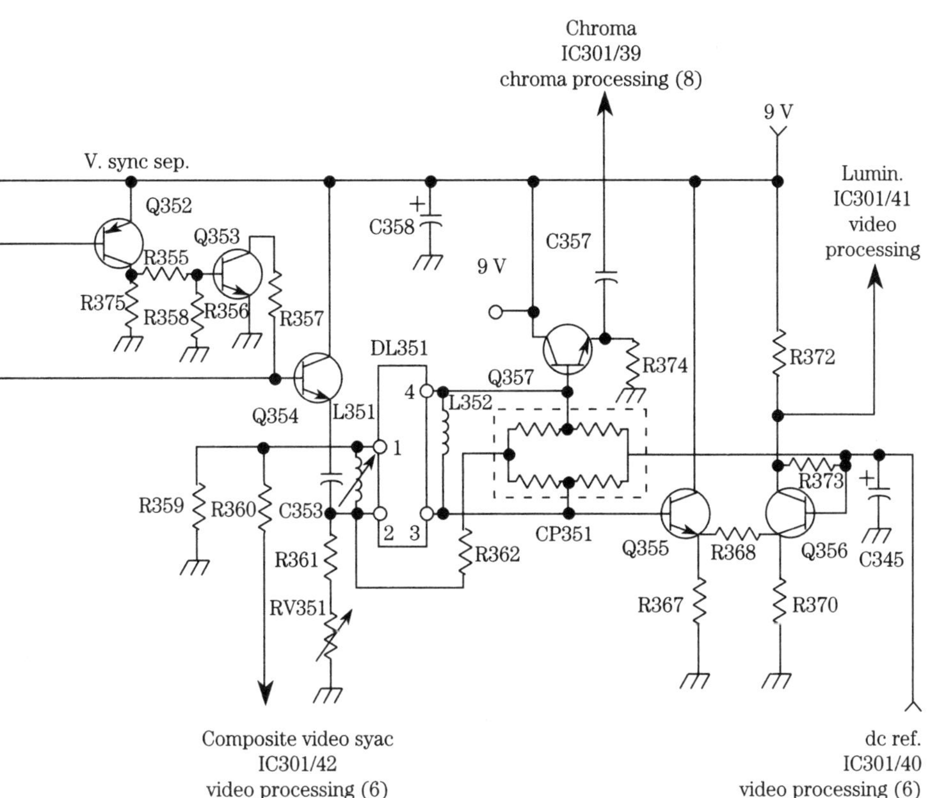
Chroma
IC301/39
chroma processing (8)
9 V
V. sync sep.
Lumin.
IC301/41
video
processing
Q352
Q353
C358
C357
9 V
R355
R375
R358
R356
R357
R374
R372
DL351
Q357
Q354
L351
L352
R359
R360
C353
R373
CP351
Q355
R368
Q356
C345
R361
R362
R367
R370
RV351
Composite video syac
IC301/42
video processing (6)
dc ref.
IC301/40
video processing (6)

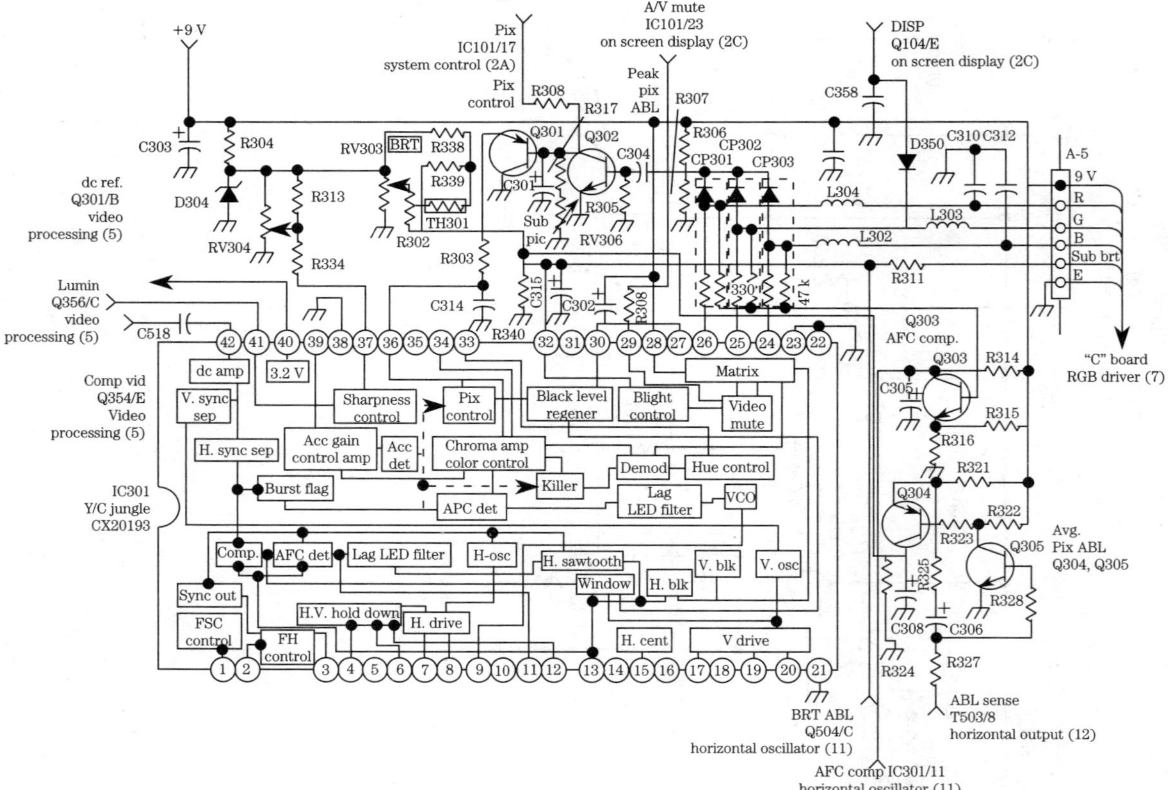

Figure 4.10 Video processing of combed-out luma signals.

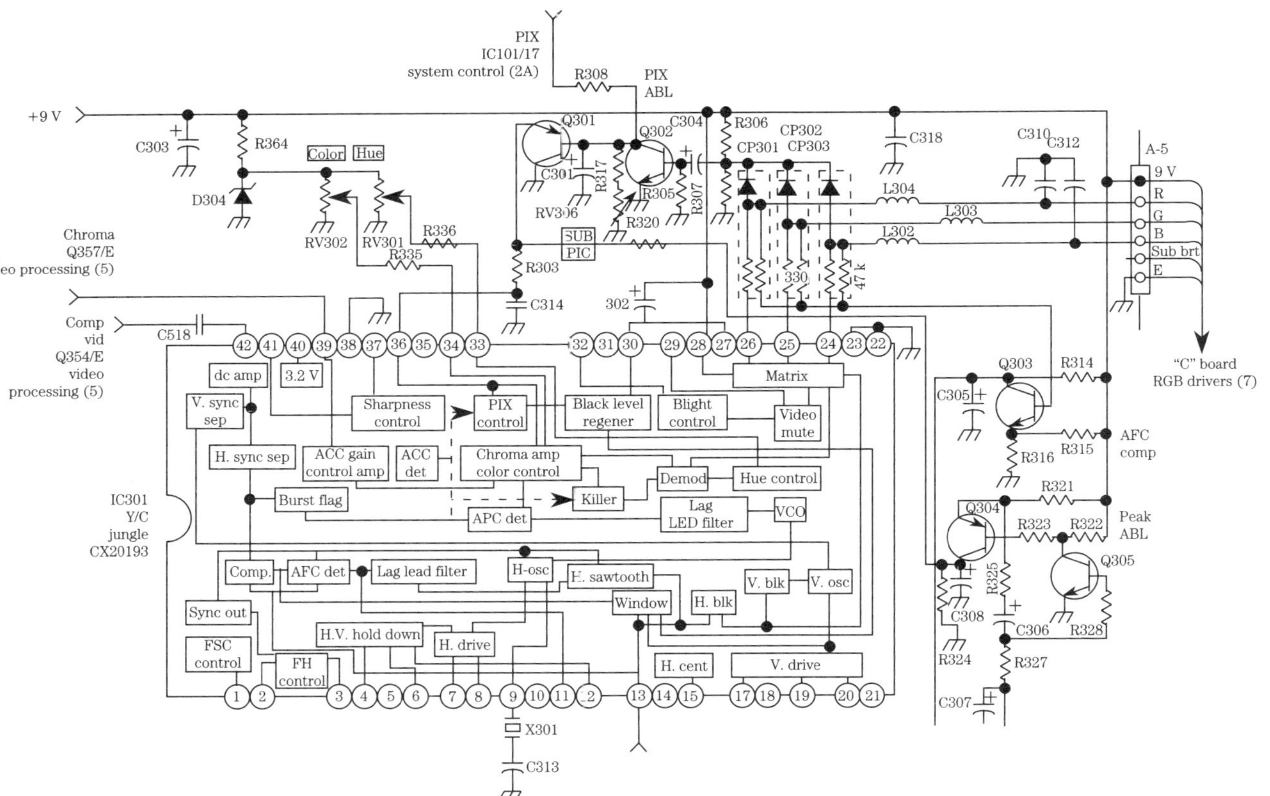

Figure 4.11 Video processing of combed-out chroma signals.

pin 9 because the scope might disable the VCO within IC301.) Typically, the VCO signal is about 60 mV, at a frequency of 3.58 MHz. The demodulated chroma is applied to the matrix within IC301 and mixed with video information (Figure 4.10) to form the RGB outputs.

RV301 and RV302 provide for user control of the hue and color, respectively. Notice that the picture signal at pin 36 also affects the color-signal level.

4.4.4 RGB drivers

Figure 4.12 shows the circuits involved between the jungle-chip video/chroma processing and the CRT. In this set (Sony 19 inch) all of the circuits are on a board (the C board) located at the CRT neck.

Drive/output transistors Q701 through Q703 amplify the RGB signals from pins 24–26 of IC301 (Figure 4.10) for application to the CRT cathodes. Notice that the blue and green circuits have both a drive and a background adjustment, while the red circuit has only a background adjustment.

Flyback transformer T503 in the horizontal-output stages provides the focus, screen, heater, and high voltage for the CRT. The screen and focus voltage are derived from the +1000 V at connector C-2, as are the heater voltages H1 and H2. T503 also supplies the +200 V required by the CRT drive transistors Q701-Q703.

The CRT focus voltage at pin 1 (grid 4) is set by RV707, while the screen (CRT brightness) voltage at pin 3 (grid 2) is set by RV708. There is also a signal brightness control RV706 that sets the dc voltage at pin 32 of IC301 (Figure 4.10).

The convergence voltage for the internal convergence plates of the CRT is derived within the CRT from a high-voltage second-anode lead across a resistance network built into the CRT. The return path is from CRT pin 13 (CV) through RV709 to ground. RV909 provides for adjustment of the static convergence. (Notice that this internal static-convergence voltage is unique to Sony Trinatron.)

4.5 Sony Trinitone Switching

Figure 4.13 shows the circuits involved in a Sony Trinitone set (a 25 inch in this case). The Trinitone feature allows the user to alter the characteristics of dynamic-color circuits to suit individual preference. The dynamic-color circuits detect relatively large areas of white in the picture and, during such times, boost the blue level to present "whiter" whites on the screen. The Trinitone feature also allows the user to shift the whites to blue, or red tone, or to accept the normal dynamic-color circuits have little or no effect on the chroma information (only on white). Notice that the Trinitone feature is not found on all Trinatron sets.

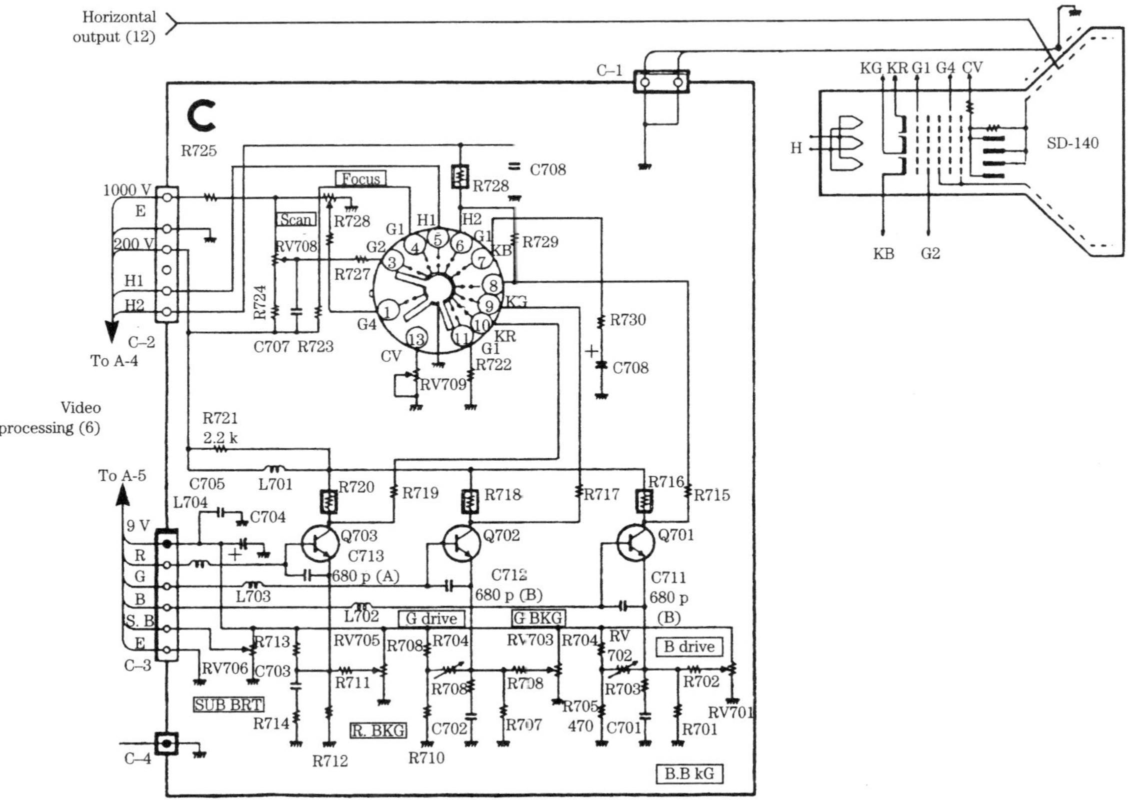

Figure 4.12 RGB drivers (7).

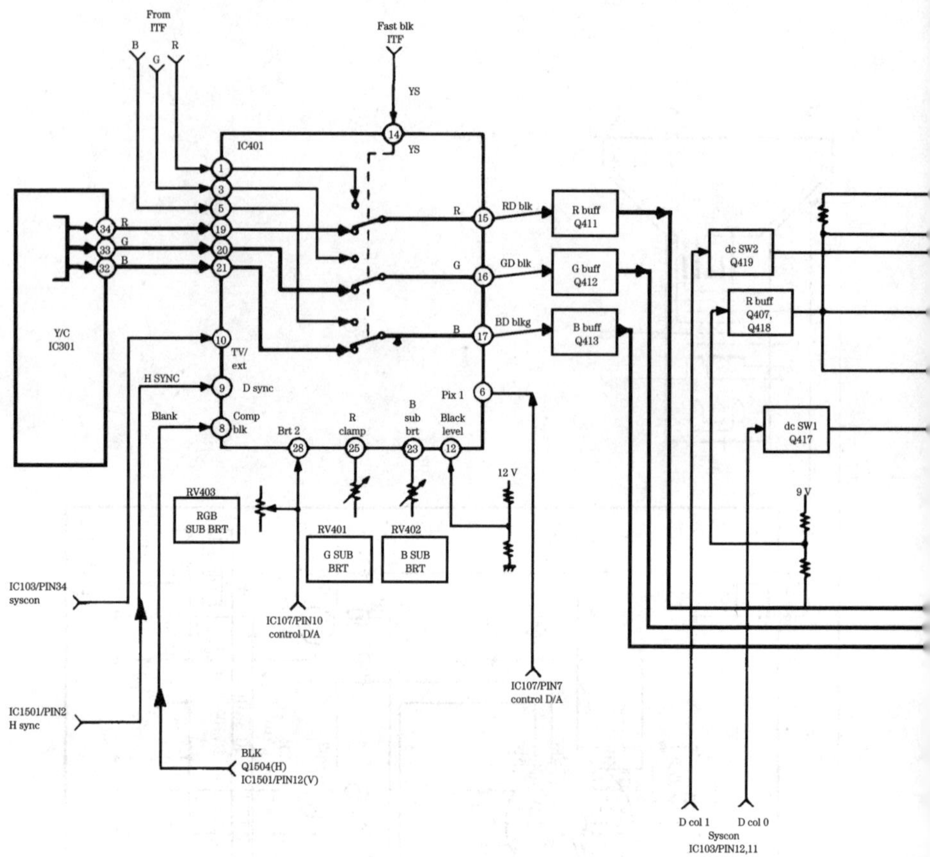

Figure 4.13 Trinitone switching circuits (F).

4.5.1 Selector IC401

IC401 is the switch for selecting either TV RGB or external RGB. If pin 10 is set high, the external mode is used. In this mode, the input of such signals as sync and blanking, and the dc levels set by brightness, picture, black level, and sub-brightness controls affect the external RGB display. Also, the YS or fast-blanking signal applied at pin 14 is usable. This YS input is from an external microcomputer and is used to rapidly switch the IC401 internal switching between external RGB and TV RGB, without affecting the selec-

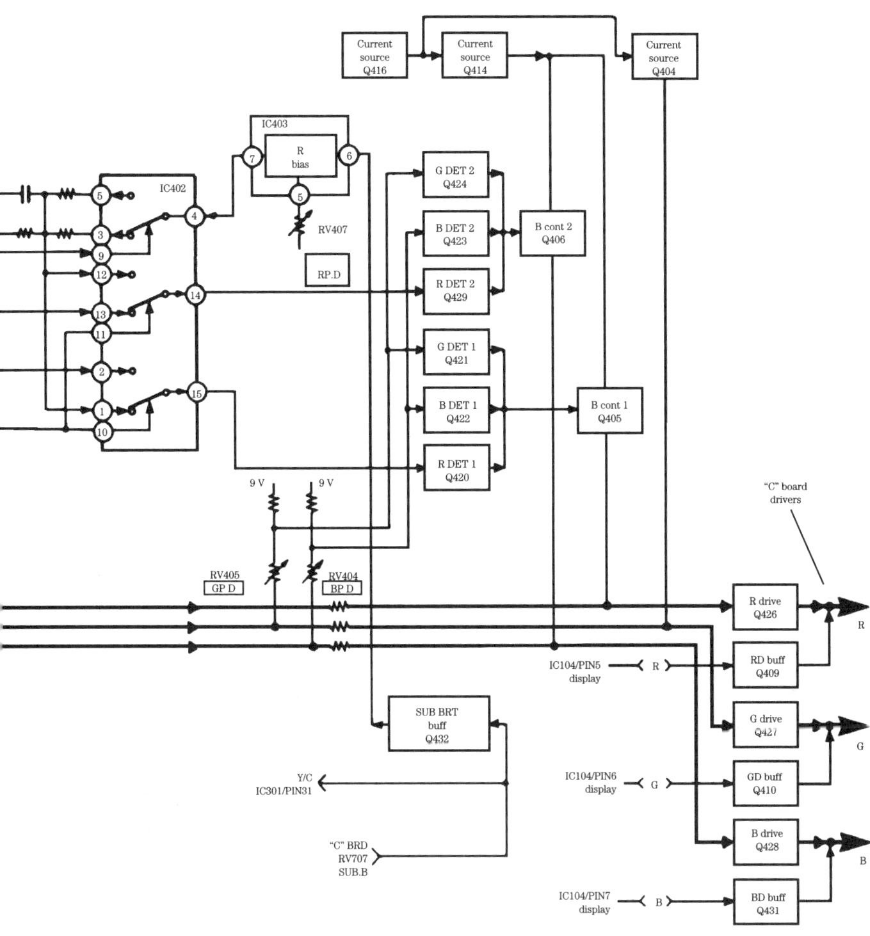

tor-control input. Such rapid switching creates the effect of superimposing computer graphics over the TV or external-video images.

4.5.2 Trinitone switching

The operating mode of the dynamic-color circuit is set by the logic stages of the two inputs D COL 0 and D COL 1. These control signals are from the system controller IC103 at pins 11 and 12, respectively. In turn, the switch-

TABLE 4.2 Trinitone Switching Logic

	IC103		IC402		
Mode	Pin 11	Pin 12	Pin 9	Pin 10	Pin 11
White	1	1	0	0	0
Blue	1	0	1	0	0
Red	0	1	1	1	1

ing state of IC402 is set by the logic levels at pins 9, 10, and 11 of IC402. Table 4.2 shows the logic states for both IC103 and IC402

4.5.3 Color control

Transistors Q416, Q414, and Q404 are constant-current sources for RGB drivers Q426, Q427, and Q428. The green driver Q427 has a constant bias supplied by Q404, and is unaffected by the dynamic-color operation. The bias for the red and blue drivers is set by the conduction level of Q405 and Q406. In turn, Q405/Q406 are controlled by the outputs of the RGB detectors.

The green and blue detectors vary from the preset in accordance with the signal present on the green and blue signal lines.

The red detector is controlled by preset RV407 and can be changed remotely through the switching action of IC402. The red detector output also varies with the signal content of the red-signal line and the red-bias network selected.

Although the dynamic-color circuits are somewhat complex, troubleshooting can be simplified by starting with adjustment procedures found in the service literature. This should pinpoint any circuit defects to a small section of the dynamic-color stages.

Chapter

5

Vertical and Horizontal Sweep Circuits

This chapter is devoted to TV-set circuits that provide the vertical and horizontal sweep for the CRT. Because such circuits are closely associated with the high-voltage and flyback-transformer (FBT) stages, as well as the separator, these circuits are also included in this chapter. The low-voltage power-supply circuits are covered in chapter 8.

Vertical/horizontal circuits have several functions, and there are many circuit configurations. Thus it is very difficult to present a typical configuration. However, most vertical/horizontal circuits have certain inputs and outputs that can be monitored. If the inputs are normal, but one or more of the outputs are abnormal, the problem can be localized to the vertical/horizontal circuits.

5.1 Sync-Separator Basics

Figure 5.1 shows the signal path through the sync-separator circuits of a typical discrete-component TV set. In most present-day IC sets, the sync separators are within an IC, although they might also include external circuits (such as shown in Figure 5.6).

The function of the sync separator is to remove the vertical and horizontal sync pulses from the video circuits and apply the pulses to the vertical and horizontal sweep circuits. The sync-separator circuits also function as clippers and/or limiters to remove the video signal (picture) and any noise. Thus, the sweep circuits receive sync pulses only and are free of noise and/or video signal (in a properly functioning set).

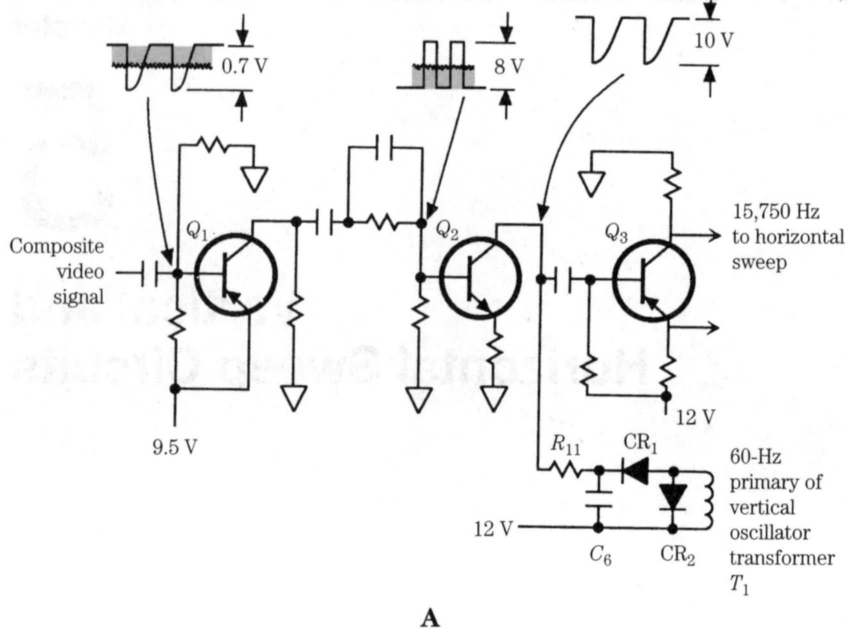

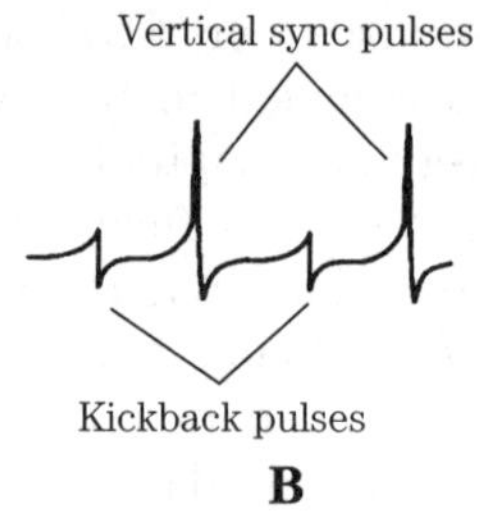

Figure 5.1 Signal path through sync-separator circuits.

In the circuit of Figure 5.1, the video input is negative going and the input transistor Q1 is pnp. Thus, both the sync pulses and the signal or noise turn Q1 on. However, with Q1 biased near zero, the large sync pulses drive Q1 into saturation at a level never reached by the signal and/or noise. The second sync separator Q2 is reverse-biased so that the first portion (about 0.3 V typically) of the Q1 output is clipped. This further removes any signal and/or noise.

The output of Q2 is applied to the low-pass filter (called the vertical integrator) of C6 and R11. The 60-Hz vertical signals are applied to the vertical-sweep circuit input through CR1 and CR2. These diodes pass the sync pulses to the vertical-sweep input, but prevent the vertical oscillator pulses from passing into the sync circuits. The Q2 output is also applied to Q3, which acts as a phase splitter or phase inverter. The output from Q3 is applied to the horizontal circuits.

5.1.1 Recommended troubleshooting approach

Failure of the sync-separator circuits can produce symptoms similar to those produced by failure in other circuits. For example, if the low-pass filter (vertical integrator) fails, there is a loss of vertical sync or poor vertical sync. This same symptom can be produced by failure of the vertical-sweep circuit as well. Also, operation of the sync-separator circuits depends on signals from other circuits. For example, if the video output is slow, the sync-separator output is low (or possibly absent).

The first logic step in troubleshooting the sync circuits is to try to adjust the horizontal and/or vertical controls (such as the horizontal frequency and vertical hold controls of Figure 5.6). If this fails, proceed with the usual voltage and waveform measurements. Remember, that if it is necessary to set a control to either extreme, or if the control must be reset repeatedly, look for marginal component failure.

5.1.2 Typical troubles

The following sections discuss symptoms that could bc caused by defects in the sync-separator circuits.

No sync. If both horizontal and vertical sync are absent, the first step is to check for proper sync pulses at the input of the sync separator (from the video detector and possibly through a video amplifier, such as at pins 1 and 22 of IC501 in Figure 5.5). If these pulses are abnormal (particularly if low in amplitude), both the horizontal and vertical sync outputs from the separator will be abnormal.

If the sync pulses to the separator are good, check the sync pulses at the end of the last point common to both horizontal and vertical sync. In the circuit of Figure 5.1, such a point is the collector of Q2. If the pulses at the Q2 collector are not good, with a good input from the video detector, the problem is in Q1 or Q2. Further, waveform measurements and/or voltage measurements can be used to isolate the problem. In the circuit of Figure 5.6, check at pins 19 and 20 of IC501. There should be both vertical and horizontal sync at pin 20, but only vertical sync at pin 19.

No vertical sync. If vertical sync is absent or critical (requires frequent adjustment) but there is good horizontal sync, the problem is in the sync separator or vertical sweep. Measure the waveform at the vertical-integrator output (the primary of T1 in Figure 5.1). If the pulses are good at this point, the problem is in the vertical sweep. If the pulses are abnormal at the input to the sweep, suspect the integrator.

In the circuit of Figure 5.6, if there is vertical sync at pin 20 of IC501, but not at pin 19, suspect Q506. On the other hand, if there is vertical sync at pin 19, but the vertical drive at pin 14 is not synchronized, suspect IC501. Of course, before you pull IC501, try correcting the problem by adjustment.

When measuring waveforms of vertical and horizontal sync pulses, it is convenient to set the scope sweep to 30 Hz (for vertical) and 7827 Hz (for horizontal). This displays two cycles of the corresponding pulses.

When measuring vertical pulses (30 Hz) after the vertical integrator or separator, there should be no horizontal pulses (because the vertical integrator acts as a low-pass filter). However, when checking vertical pulses ahead of the integrator (say at Q1 or Q2 in Figure 5.1), the horizontal pulses might appear on the scope display. When the scope is set to measure horizontal pulses (7875 Hz), the vertical sync pulses should not appear because vertical pulses are slow in relation to the horizontal pulses. In the circuit of Figure 5.6, there should be both vertical and horizontal sync at pin 20 of IC501, but only vertical sync at pin 19.

When measuring ahead of the vertical integrator or sync separator (the last point common to both horizontal and vertical sync), look for any noise or video signal that might have leaked through. For example, in the circuit of Figure 5.6, there should be little noise or video at pin 22 of IC501 and no noise or video at pin 20. If there is excessive noise at pin 22, suspect the low-pass filter formed by R506/C508 (this is not the vertical integrator). If there is no reduction in noise/video at pin 20 from that for pin 22, suspect IC501.

When measuring the output of the vertical integrator or sync separator, look for kickback voltages or spikes from the vertical-sweep circuit. Such kickback produces a display similar to that shown in Figure 5.1. Diodes CR1/CR2 in Figure 5.1 are included on some discrete circuits to prevent the kickback from reaching the separator. In the circuit of Figure 5.6, where the vertical oscillator is part of IC501, kickback is usually not a problem. However, if you are experiencing mysterious sync problems, look for vertical-oscillator signals that might have leaked from pin 19 back through Q506 to pin 20.

No horizontal sync. If horizontal sync is absent or critical but there is good vertical sync, the problem is in the horizontal portion of the sync separator or in the horizontal sweep. In the circuit of Figure 5.1, measure the waveform at the collector of Q3. If the pulses are good at this point, the problem is in the horizontal sweep. If the pulses are abnormal at the input to the horizontal sweep, suspect the horizontal portion of the separator stages. In the circuit of Figure 5.8, check for horizontal sync at pins 1, 2, 4, 5, and 10. If there is good vertical sync, but the horizontal-sync pulses at pin 1 are absent or abnormal, suspect C510, R521, and C511. (Again, R521 and C511 form a low-pass filter to remove noise and video from the sync line.) If there are good horizontal sync pulses at pin 1, but not at pins 2/4/5, or no drive output at pin 10, suspect IC501. (Try adjustment before you pull IC501.) Also look for problems in external components, such as R520 between pins 4 and 5.

Picture pulling. There are several forms of picture pulling. Often, the nature of the picture-pulling symptoms can pinpoint the trouble. For example, when picture pulling appears to be steady and there is a bend in the image, this usually indicates that the horizontal AFC circuits are receiving distorted pulses. A likely cause is vertical kickback entering the sync separator and distortion of the horizontal output. Check the related pulses against those shown in the service literature.

If the picture pulling appears unsteady, and particularly if the pulling tends to follow the camera signal, this indicates poor sync separation. That is, the video output is not being clipped and limited sufficiently to remove all camera signals from the circuits. This problem is more frequently in discrete-component sets than in sets where the corresponding separation circuits are IC. For example, the problem is more likely in a circuit where the sync separator is discrete, such as shown in Figure 5.6, than in a circuit where the separator is in IC301 (Figure 5.12).

Do not confuse picture pulling with distortion of the raster. If the raster is bent, the problem is not in the sync separator but in thc horizontal or vertical sweep circuit (or possibly the yoke). If the raster edges are sharp but the picture is pulling or bent, the problem can be in either the sweep or separator circuits. Check the raster alone by switching to an unused channel (or switch to the video function if the set is a TV monitor).

5.2 Vertical-Sweep Basics

Figure 5.2 shows the signal path through the vertical sweep circuits of a typical discrete-component TV set. In most present-day IC sets, the vertical sweep circuits (vertical oscillator and vertical output or drive) are within one or two ICs, although they might also include external circuits (such as shown in Figures 5.6 and 5.7).

The vertical-sweep circuits provide a vertical deflection voltage (vertical sweep) to the CRT deflection yoke. The vertical circuits also supply a blanking pulse to the CRT. This blanks the CRT during retrace of the sweep. The vertical-sweep signals are at a frequency of 60 Hz and are synchronized with the picture transmission by means of the sync signals taken from the sync separator. The vertical circuits of most IC sets also provide for distortion correction (such as pincushion correction).

In the circuit of Figure 5.2, the vertical oscillator is of the blocking type,

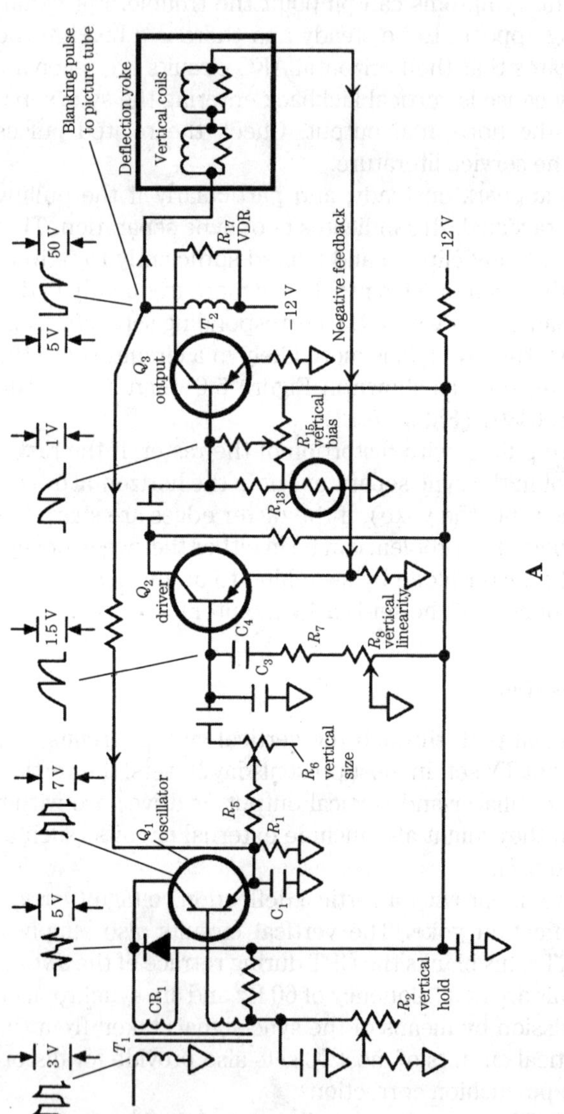

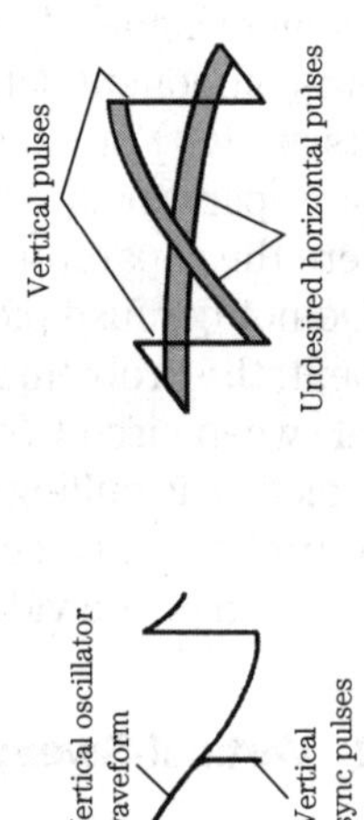

Figure 5.2 Signal path through vertical-sweep circuits.

operating at a frequency of 60 Hz, and producing a pulse output of about 1 to 2 V. The oscillator pulse output is modified into a sawtooth sweep and is applied through the driver to the output stage. The final output is a peaked sawtooth waveform. The sawtooth portion (about 4 to 5 V) is applied to the CRT deflection yoke, while the peaked portion (about 50 V) is applied as a blanking pulse to the CRT.

The oscillator is locked in frequency by the sync pulses. Vertical hold R2 is set so that the vertical sweep is locked with picture transmission. The Q1 pulse output is converted to a sawtooth sweep by the C1R1 emitter network. The sweep is made linear through feedback, and is adjusted with linearity control R8. The amplitude of the sawtooth waveform to driver Q2 (and thus the vertical size of the picture raster) is set by vertical-size control R6. The emitter of Q2 receives negative feedback from the yoke to provide a more linear output.

The deflection yoke requires heavy current. For this reason, output transistor Q3 is of the power type mounted on a heat sink (in discrete sets). Because there is always the danger of thermal runaway in power transistors, thermistor R13 is included in the base circuit. Note that if it is necessary to replace Q3, it might be necessary to adjust vertical-bias control R15. Most discrete sets have a vertical-bias control, because the characteristics of replacement transistors can be different from the original. In present-day sets, the output transistor is replaced by an IC (such as IC502 in Figure 5.7).

5.2.1 Recommended troubleshooting approach

Complete failure of the vertical sweep is an easy symptom to recognize and it is usually easy to locate. With complete failure, there is no vertical sweep and the CRT display is a horizontal line. (With any TV set or monitor, do not operate the set when there is complete failure in vertical sweep. The bright horizontal line can burn out the CRT screen.)

In the case of complete vertical-sweep failure (all other functions normal), the logical first step is to check the waveforms at all points. If the oscillator waveform is bad, the problem is quickly localized to the oscillator stage. (However, in some vertical circuits, the oscillator does not go into oscillation unless there are sync pulses present, so always check for sync pulses first.) If all waveforms appear normal, but there is no vertical sweep, the deflection yoke is a logical suspect.

In IC sets, it is usually necessary to trace vertical-sweep waveforms through at least two ICs. For example, as shown in Figure 5.5, the vertical oscillator and drive can be checked at pin 14 of IC501 (or at pin 4 of IC502). The vertical-sweep output to the yoke can be measured at pin 2 of IC502. (In this particular set, there is a vertical-output test point TP82, as shown in Figure 5.7.)

In the case of marginal failure (loss of sync, critical sync, distortion, vertical nonlinearity, line pairing or splitting, poor interlace, or lack of vertical height) the first step is to try correcting the problems with adjustment. If the problem is not eliminated or if it is necessary to set a control to an extreme, suspect marginal component failure. Look for leaking capacitors, transistors, or diodes, worn potentiometers, or open capacitors.

5.2.2 Typical troubles

The following section discusses symptoms that could be caused by defects in the vertical-sweep circuits.

Insufficient height. The first step for this symptom is to adjust the vertical height control (sometimes called the drive or size control). If the height control must be fully on or nearly fully on to get proper height, suspect the vertical drive. Notice that in some sets, there are two vertical-size adjustments. For example, as shown in Figures 5.5 and 5.6, there is a vertical-size adjustment on the B board and a subvertical-size adjustment RV504 on the main D board. If it is impossible to get the proper vertical height by adjustment, look for leaking capacitors, worn vertical-height control, and leaking transistors. Again, check the vertical waveforms from oscillator to yoke.

Vertical-sync problems. The first step for this symptom is to adjust the vertical hold (sometimes called the lock or sync control), such as the vertical-hold control shown connected to pin 19 of IC501 in Figure 5.5. If the hold control must be fully on or fully off to get sync, or if it is necessary to readjust the control frequently, there is a problem in the vertical-sync circuits. Look for leaking capacitors, a worn vertical-hold control, and leaking transistors.

In the case of a very critical sync (difficult to adjust or does not hold after adjustment), observe the amplitude of the vertical-sync pulse riding on the sawtooth portion of the vertical-oscillator or drive waveform (Figure 5.2B). If the sync amplitude does not remain constant, look for problems in the sync separator (Figure 5.6) rather than in the vertical-sweep circuit.

Vertical distortion (nonlinearity). There are several forms of vertical distortion. Some are easy to recognize, such as keystoning, where one side of the picture is much larger than the opposite side. Keystoning is generally caused by a defect in the deflection yoke or related parts (Figure 5.7), but it could be caused by a defect in the vertical output IC502.

Other forms of vertical distortion are not so easy to recognize. In the extreme, there is compression of the picture at the top with picture spreading at the bottom, or vice versa. Often, the compression or spreading is slight. Use a crosshatch pattern from a TV generator (chapter 11) to check linearity. For a quick check of linearity, adjust the vertical hold to produce slow rolling. Watch the blanking bar as the bar moves up or down on the screen. The blanking bar should remain constant (in vertical height) at the bottom, middle, and top of the screen.

Vertical distortion can be caused by improper adjustment, as well as defective circuits. One problem here is the interaction of controls and components. For example, if the sawtooth sweep is low in amplitude (because of a defect not associated with nonlinearity), the height control can be adjusted to get proper height. These extremes in circuit resistance might make it impossible to get a linear picture no matter how the linearity control is adjusted.

If the problem cannot be cured by adjustment, the next step is to localize the fault with voltage and waveform measurements. Remember that the vertical sweep cannot be linear if the sawtooth waveform is not linear. Of course, if the sweep output from the final stage (such as at pin 2 of IC502 in Figure 5.7) is good, but the vertical display is not linear, suspect the yoke and related parts.

Line splitting. The problem of line splitting (line pairing or poor interlace) is often associated with vertical sweep. Although the trouble appears in the vertical sweep, the actual cause is often in related parts. For example, the trouble can be caused by an open capacitor in the sync separator or by horizontal pulses leaking back into the vertical sweep.

If you suspect horizontal pulses in the vertical sweep, check the sweep waveform with the scope retrace-blanking disabled. Any horizontal pulses mixed with the sweep will appear as a pulse train on some portion of the vertical sweep output waveform (Figure 5.2B). If the horizontal pulses are present, the most likely causes are in the horizontal circuits, such as corona discharge from the high-voltage rectifier, leakage between leads (not too common in PC wiring, except where there are solder splashes or arc burns), and breakdown in the high-voltage section (resulting in a spark discharge being picked up by the vertical circuits).

If there are no horizontal pulses in the vertical circuits, but there is definite line splitting, suspect the sync circuits, particularly the sync separator. Generally, when line splitting is constant, the problem is in the sync separator. When splitting is intermittent, suspect the high-voltage section (spark discharge, radiation from high-voltage lead, etc.).

5.3 Horizontal-Sweep Basics

Figure 5.3 shows the signal path through the horizontal sweep circuits of a typical discrete-component TV set. In most present-day IC sets, the horizontal sweep circuits (horizontal oscillator, AFC, and horizontal drive) are within one or two ICs, while the horizontal output is discrete (possibly a power transistor with built-in damper diode), as shown in Figures. 5.5, 5.8, and 5.9.

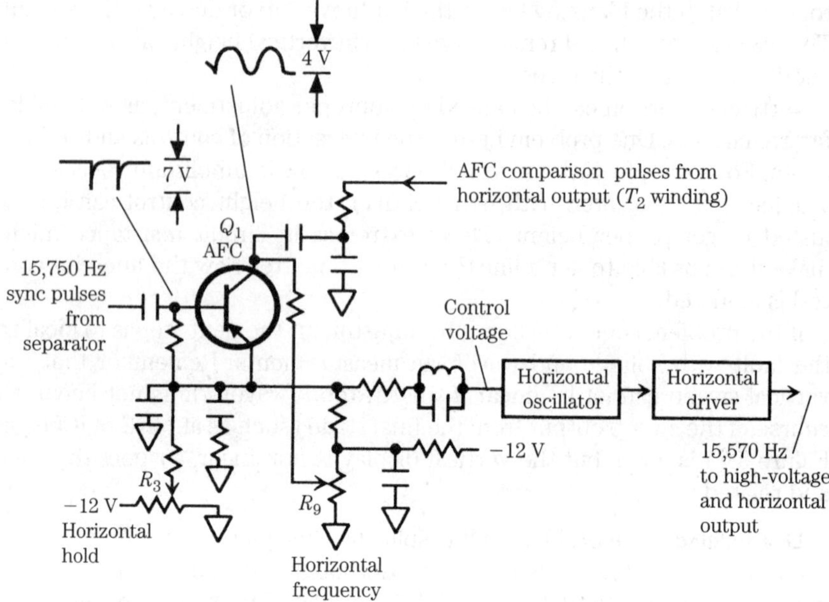

Figure 5.3 Signal path through horizontal-sweep circuits.

The horizontal oscillator and driver circuits provide drive signals to the horizontal output and high-voltage circuits (see Figure 5.4). The signals are at a frequency of 15,750 Hz, and are synchronized with the picture transmission by sync signals from the sync-separator circuits (see Figure 5.1).

Both discrete and IC horizontal oscillator circuits include some form of AFC system to ensure that the horizontal sweep signals are synchronized with picture transmission, despite changes in line voltage and temperature (or minor variations in circuit values). The AFC action is produced by comparison of the sync signals with horizontal-sweep signals.

Deviations of the horizontal-sweep signals from the sync signals cause the horizontal oscillator to shift in frequency or phase as necessary to offset the initial (undesired) deviation. For example, if the horizontal-sweep increases in frequency from the sync signals, the horizontal oscillator is shifted in frequency by a corresponding amount but in the opposite direction (a decrease in frequency to offset the undesired change).

In the discrete circuit of Figure 5.3, the Q1 base is driven by the sync pulses, whereas the collector receives comparison pulses from the horizontal output. The emitter waveform is a combination of both pulses. When there is a change in phase or frequency between the two sets of pulses, the emitter current changes. In turn, the control voltage applied to the horizontal oscillator changes in a direction that corrects the frequency and phase of oscillation.

In the IC circuit of Figure 5.8, the AFC circuit in IC501 receives both sync pulses from pin 1 and the AFC comparison pulses from the horizontal output. The dc output from the AFC circuit is applied to the horizontal oscillator through pins 4 and 5 (and R520). This dc output keeps the horizontal oscillator locked to the sync pulses. The horizontal-oscillator frequency can be adjusted (within limits) by RV506.

5.3.1 Recommended troubleshooting approach

Failure of the horizontal oscillator and related circuits (AFC and horizontal drive) can produce symptoms similar to those produced by failure in other circuits. For example, if the horizontal oscillator stops oscillating, there is no drive to the horizontal output. Thus, there is no high-voltage or boost-voltage output, and the CRT screen is dark. The same symptom can also be produced by failure of the horizontal output circuit, the low-voltage supply (chapter 8), and the CRT. Also, in IC sets such as shown in Figure 5.8, there are circuits (high-voltage holddown) that shut the horizontal oscillator down when the CRT high voltages reach dangerous levels (and could produce X-ray radiation).

Also, operation of the horizontal oscillator depends on signals from other circuits. For example, the horizontal AFC circuit must have sync signals from the sync separator and comparison signals from the horizontal output. Because of these conditions, the only practical troubleshooting approach is to check waveforms at all outputs and inputs, followed by voltage measurements.

If the driver output is absent or abnormal with good sync and comparison pulse inputs, the problem is in the horizontal oscillator or driver sections. A waveform measurement at the input of the horizontal driver or horizontal oscillator output localizes the problem further. Notice that in the circuit of Figure 5.8, IC501 contains both the horizontal oscillator and a predriver, while a horizontal driver and output are in discrete form (Q501/Q502 of Figure 5.9). Generally, if the driver output is present but the symptoms point to a horizontal circuit failure (horizontal pulling, jitter, distortion, etc.), the problem is most likely in the AFC.

5.3.2 Typical troubles

The following section discusses symptoms that could be caused by defects in the horizontal-oscillator/driver circuits.

Dark screen. If the CRT screen is dark (no raster) but sound is normal (indicating that the low-voltage supply is probably good), the first step is to measure the waveform at the horizontal-drive output (and at the predrive output, pin 10 of IC501, Figure 5.8). If the waveform is normal, the problem is in the horizontal output and high-voltage circuit or the CRT. If the waveform is not normal (weak, distorted, etc.) the problem is likely in the horizontal oscillator (or the high-voltage hold-down circuit has turned the oscillator off).

Before checking individual parts, there are some checks that help isolate the problem: waveforms at sync input (from sync separator), comparison pulse input, horizontal oscillator (or predriver) output, and voltages at all transistor elements and corresponding IC pins.

If the sync pulses are absent or abnormal, the problem is in the sync separator rather than in the horizontal section. If the comparison pulses are not normal, with a good output from the horizontal drive, the problem is in the horizontal output and high-voltage section.

If the sync pulses and comparison pulses are both normal, but the horizontal output is absent or abnormal, the problem is likely in the AFC section or the horizontal oscillator. Obviously, if any section in IC501 of Figure 5.8 is defective, IC501 must be replaced (at great expense to the customer).

Narrow picture. The most logical cause of a narrow picture that cannot be corrected by adjustment is an insufficient horizontal drive. In circuits such as those shown in Figures 5.8 and 5.9, the first step is to isolate the problem to the IC (monitor at pin 10 of IC501), or to the discrete components (monitor at Q501, T501, and Q502).

Horizontal pulling or improper phasing; loss of sync. Horizontal pulling is present when the picture pulls and appears in diagonal form. If the picture pulls completely into diagonal lines, this indicates a complete loss of sync. The direction of the slant can provide a clue to the problem. If the lines slant to the right, the oscillator frequency is high, and vice versa. If the picture shifts to the right or left so as to be decentered, suspect incorrect phasing. That is, the horizontal oscillator is on frequency but not in exact phase with the sync signals.

It is possible that any of these problems can be the result of improper adjustment, so the first step is to adjust all horizontal controls. If this does not clear the problem, check all of the waveforms, transistor voltages, and related IC pins. Pay particular attention to the sync pulses and comparison pulses applied to the AFC. If either of these is absent or abnormal, the AFC circuits cannot operate properly (even though the IC is good). For example, if the comparison pulses at pin 3 of IC501 in Figure 5.8 are absent, horizontal sync might not be completely lost. However, the horizontal frequency adjustment (RV506) will become very critical and phase shift occurs (the picture is decentered).

Horizontal distortion. There are many forms of horizontal distortion that can be caused by defects in the horizontal oscillator and driver circuits. The so-called "pie-crust" distortion is a typical example. With pie crust, the picture image appears to be made up of wavy lines, even though there is no pulling, loss of sync, or jitter. Such distortion is almost always the result of marginal performance in a component rather than complete failure. The most likely defects are in capacitors, particularly the capacitors that filter the control voltage from the AFC to the horizontal oscillator (such as C516 in Figure 5.8).

5.4 High-Voltage and Horizontal-Output Basics

Figure 5.4 shows the signal path through the high-voltage and horizontal output circuits of a typical discrete-component TV set. Most present-day IC circuits include a few additional features, as discussed in sections 5.5 and 5.7. However, the basic elements are covered in Figure 5.4. Figure 5.5 shows the relationship of the circuits to other stages in a typical IC set (a Sony 13 inch).

The circuit of Figure 5.4 provides a high voltage for the CRT and a sweep voltage to the horizontal yoke. In most cases, the circuit also supplies a boost voltage for the CRT focus and screen or accelerating grids, and possibly a voltage for other components that require voltages higher than that which is available from the low-voltage supply. The horizontal output circuits also supply an AFC signal to control the frequency of the horizontal oscillator (section 5.3). In most present-day sets, an AGC winding is not required, because the AGC signal is developed by the PIF/SIF circuits, rather than by a keyed-AGC circuit found in discrete sets.

Although there is little standardization, most horizontal output circuits have certain characteristics in common that must be considered during the troubleshooting process. The circuit receives pulses from the horizontal driver (at 15,740 Hz) synchronized with picture transmission. Transistor Q1 is normally biased at or near zero so that one edge of the pulse drives the transistor into heavy conduction (near saturation); the opposite swing of the pulse cuts off Q1. In effect, Q1 is operated in the switching mode.

Notice that Q1 in Figure 5.4 corresponds to Q502 in Figure 5.9. Also, T1 corresponds to T501, T2 corresponds to T503, and damper diode CR2 is built into Q502.

The Q1 collector current is applied through a winding on the flyback transformer T2, resulting in pulse waveform at the other windings. The high voltage is rectified and applied to the CRT anode; the boost output is rectified and applied to the CRT focus and accelerator grid. On some sets, the horizontal-output transistor current is applied through the deflection yoke. On other sets, the transformer has a separate winding for the yoke. Either way, the horizontal-output pulses produce the horizontal sweep.

Because of the great variety in horizontal-output circuits, it is difficult to arrive at a typical theory of operation. However, the important point to consider in practical troubleshooting is that the damping diode CR2 (built into Q502) starts to conduct when Q1 is cut off. Diode CR2 continues to conduct until Q1 conducts.

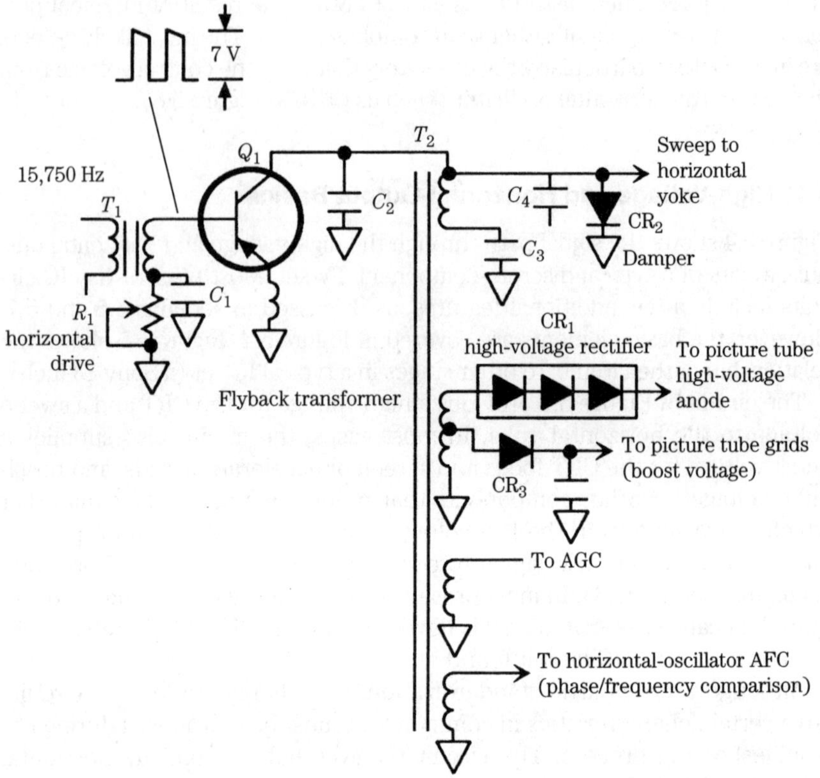

A

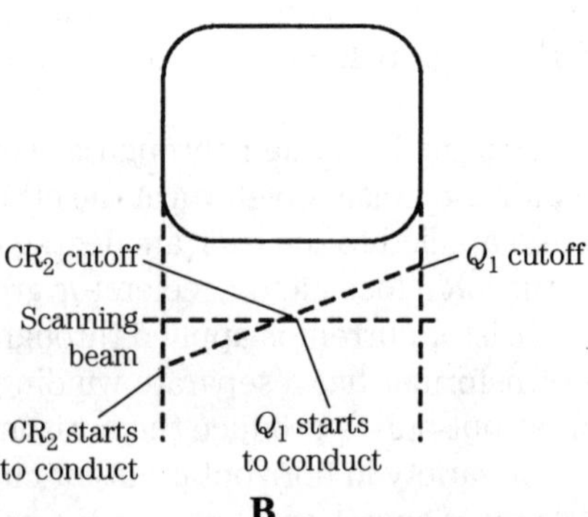

B

Figure 5.4 Classic high-voltage and horizontal-output circuit.

During the horizontal forward scan (when the picture is displayed), CR2 conducts and Q1 is cut off from the start of the sweep to about midpoint. Then CR2 is cut off, and Q1 conducts for the remaining half of the sweep. This sequence is shown in Figure 5.4B.

The scan sequence is important in troubleshooting some sets, because any problems in the right-hand side of the picture are probably the result of defects in Q1 (or related components), whereas trouble in the left-hand side is probably the result of defective CR2. Of course, where CR2 is built into Q1, both are replaced simultaneously if either is defective.

When Q1 is cut off, the CRT is blanked, and current flows rapidly through the horizontal yoke in the opposite direction, pulling the electron beam back to the left side of the screen (known as the horizontal retrace or flyback). It is the pulse developed during the flyback interval that is used by the other windings on flyback transformer (FBT) T2.

5.4.1 Recommended troubleshooting approach

Many trouble symptoms caused by horizontal output circuits can also be caused by defects in other circuits. A dark screen (no raster) or insufficient width are two good examples.

If the low-voltage supply is completely inoperative (or almost zero), the screen can be dark. If the low-voltage supply is producing a low output, the picture width can be decreased. Of course, in either case, the sound is absent or abnormal. On the other hand, if the horizontal-drive circuits are defective (no drive signal to the horizontal output), there is no horizontal sweep or high voltage, even though the sound is probably normal. (In any service situation where the screen is dark but there is normal sound, never overlook a possibly defective CRT.)

The most logical troubleshooting approach is to analyze the symptoms and then isolate the trouble to the horizontal-output circuits by an input waveform check. That is, if the sound is normal (good low-voltage supply), measure the input waveform (from the horizontal drive) at the Q1 base. Generally, this is about 6 to 8 V and appears similar to that shown in Figure 5.4. Check the waveform against the service literature.

If the input waveform is normal, the trouble is in the horizontal output circuit (unless the CRT is bad). Of course, if the input signal is not normal, the next step is to check the horizontal-drive circuits as described in section 5.3. If the input is good, make the following checks.

Dark screen. In the case of a completely dark screen, the next obvious check is to measure the high voltage at the CRT anode. Then measure the accelerator grid and focus voltage from the boost circuit. If the voltages are present and normal, the CRT is defective. (Of course, if you have not already thought of it, check that the CRT filament or heater is on!)

High voltage or boost voltage absent. If either the high voltage or boost voltage is absent or abnormal, this isolates the trouble to the corresponding circuit. If both voltages are absent (with the drive signal good), the problem is in Q1 or related parts.

High voltage only absent. If only the high voltage is absent, check for alternating current at the anode side of CR1 (unless this is not recommended in the service literature). When measuring the high voltage, always use a meter with a high-voltage probe. Observe all of the usual precautions when measuring high voltages. Do not make an arc test with any solid-state or IC set! Some notes on high-voltage measurement are included in chapter 11.

Measuring voltages. Measure the boost voltage with a meter and low-capacity probe. With the exception of high voltage, most of the voltages in a solid-state or IC set can be measured using a low-capacity probe and a meter (or scope). Observe any and all precautions given in the service literature. Typically, there are warning notes on the schematic such as "Do not measure high ac voltages."

Transistor Q1 measurements. It is often quite helpful if you can measure the emitter-collector current of Q1. Usually, this is not practical, so you must use dc voltage measurements to supplement the waveform check. If the dc voltages and waveforms at Q1 are correct, it is reasonable to assume that the circuit is good up to the FBT T2.

Flyback and yoke checkers. I have no recommendations regarding flyback and yoke checkers. If you use such checkers, make certain to follow the instructions, and be sure that the checker is suitable for the set you are servicing. With a few exceptions, it is possible to service the horizontal circuits of most sets using waveform and voltage measurements.

Common horizontal output troubles. The most common causes of trouble in the horizontal-output and high-voltage circuits are capacitors, horizontal-output transistor (because of the high currents), diodes, and transformers.

5.4.2 Typical troubles

The following sections discuss symptoms that could be caused by defects in the high-voltage supply and horizontal-output circuits.

Dark screen. If the screen is dark but sound is normal, suspect the horizontal circuits. Of course, the CRT can be defective, or the problem can be something simple (such as a lack of filament voltage for the CRT).

Before checking individual parts, there are some circuit tests that help isolate the problem. Check the drive-voltage waveform at the Q1 base and the sweep output at the Q1 collector; check for high voltage at the CRT anode (observing all precautions); check for boost voltage and any auxiliary voltages (focus, accelerator or screen grid, AFC, etc.).

The first check to be made depends on whatever is the most convenient. For example, it is logical to check the Q1 waveforms before checking the high voltage. In some sets, Q1 might be at a very inaccessible location, so the best bet is to check the high voltage first.

If the drive waveform at the base of Q1 is absent or abnormal, the problem is ahead of the horizontal-output circuits. If the base waveform is good but the Q1 collector waveform is bad, Q1 is the first suspect. If the collector waveform is good but one or more of the output voltages is bad, check individual parts in the related circuit. Also check voltages at each of the Q1 elements.

Shorted capacitors (a common problem in the case of a dark screen) show up when dc voltages are measured. Open or leaking capacitors are not located easily. Generally, an open or leaking capacitor produces an abnormal waveform.

Leaking transistors show up when waveforms are measured. The problem is confirmed further when the dc voltages are measured. Collector-base leaking in Q1 is a common problem. If leakage is bad enough to cause complete failure, which results in a dark screen, the dc voltages at the Q1 elements are incorrect.

Remember that Q1 operates at or near zero bias or possibly with reverse bias. If there is any substantial forward bias on Q1, it is probably the result of leakage. Any collector-base leakage forward-biases a transistor. In the case of a normally cut off Q1, the undesired forward bias attenuates to the collector waveform.

If the capacitors and Q1 appear to be in order, check the diodes. If the sweep-output waveform (Q1 collector) is abnormal, check the damper CR2 (which is often built into Q1 on many present-day sets). If the output voltages are abnormal (with a good sweep output), check the corresponding diodes (CR1 and CR3).

The high-voltage rectifier CR1 can be a series of diodes. If any of these diodes are shorted or develop excessive leakage, the remaining diodes can break down. Usually, the service literature recommends replacing all high-voltage diodes simultaneously.

A shorted CR2 is usually easy to pinpoint, because the Q1 collector waveform and dc voltage are abnormal. An open CR2 usually does not produce a dark screen. If any of the other diodes are defective, this shows up as absent or abnormal output voltages.

If the flyback transformer T2 has an open winding or if a winding is shorted, the problem is usually self evident. However, if there is only a partial short, leakage between winding, or a high-voltage arc, it might be difficult to check. Substitution is the only sure check, unless you have a flyback-transformer tester suitable for the circuits. Unfortunately, replacement of the flyback transformer is not an easy job. So do not try substitution except as a last resort (when all other parts are good).

Picture overscan. This symptom occurs when the picture becomes dim, even with the brightness control fully on, and there is an enlargement of the picture or raster. Usually, the enlargement is uniform, but there might be some defocusing.

In some circuits, the brightness control operates in reverse. That is, rotating the control for an increase in brightness produces a decrease, after reaching a critical point of control. If picture overscan is not accompanied by insufficient width, suspect the high-voltage circuit only.

The first obvious test is to measure the high voltage (observing all precautions). If the high voltage is normal (not likely), try a new CRT. Also, there might be a corona problem (leakage from the high-voltage lead), especially in large-screen TV sets. The only practical cure is to replace the lead. If the high voltage is low, check any high-voltage filter capacitor for leakage.

Narrow picture. The most logical cause of a narrow picture (that cannot be corrected by adjustment of width controls) is insufficient horizontal drive. This is usually accompanied by other symptoms, such as decreased brightness, picture distortion, and the like. Very often, a narrow picture is the result of a marginal breakdown, rather than a complete breakdown. For example, if Q1 has some collector-base leakage, Q1 is forward-biased, and the sweep circuit is decreased.

The most likely defects depend on symptoms that accompany the narrow picture. For example, if there is distortion on the left-hand side of the picture, look for an open or leaking CR2. If there is right-hand distortion, check for a defective Q1.

The first step is to measure the Q1 collector waveform as well as the sweep waveform to the horizontal yoke. These waveforms rarely, if ever, are normal when the picture is narrow. Waveform measurements should be followed by voltage measurements at the Q1 elements.

Do not overlook insufficient horizontal drive, especially when waveform and voltage measurements appear normal. Pay particular attention to drive pulses at the base of Q1. Even a slight drop in amplitude can reduce sweep output to the horizontal yoke.

Foldback or foldover. Horizontal foldback usually occurs only on one side of the picture screen. A portion of the picture is folded back on one edge of the display. In some rare cases, there is a fold in the center. If there is a foldover (with a good drive signal), all of the horizontal-sweep components are suspect. Again, look for Q1 problems if foldback is on the right, and CR2 problems if the foldback is on the left.

The first test is to measure both the base (drive) and collector (output) of Q1. Then measure the horizontal yoke waveform (if different from the Q1 collector waveform). Invariably, the yoke waveform is distorted. Remember that the horizontal-sweep system is essentially a resonant circuit. The inductance of T2 combines with C4 to resonate at about 50 kHz (typical). If the resonant frequency is not correct (generally low), the waveform is distorted, resulting in foldback.

Nonlinear horizontal display. Any nonlinearity in the horizontal display is almost always found with at least one other problem (such as narrow picture, overscan, or blooming, lack of brightness, foldback, etc.). Thus, the most likely causes of nonlinearity are the same as for other symptoms. Also, the troubleshooting/test sequence should be the same.

Horizontal nonlinearity should not be confused with the keystone effect (where the picture is wider at the top than at the bottom, or vice versa), and is almost always caused by a problem in the horizontal yoke. (One set of horizontal-deflection coils has shorted turns and is unbalanced with the other set of coils.)

5.5 Basic IC Vertical and Horizontal Sweep

Figure 5.5 shows the circuits necessary to generate the vertical and horizontal drive pulses (synchronized to the transmitted sync signals) that produce a raster on the picture-tube (CRT) screen. In this particular IC set (Sony 13 inch), many of the functions are performed in IC501. Notice that the vertical-sweep signals applied to the vertical CRT yoke are output from IC502, while the horizontal yoke receives sweep signals from discrete components (drive Q501, transformer T501, and output Q502).

5.5.1 Vertical oscillator

Figure 5.6 shows the vertical-oscillator circuits for a typical TV set (the Sony 13 inch). The main purpose of the circuit is to provide drive pulses for the vertical-output stages that produce sweep signals to the vertical yoke. In addition, the circuits separate the vertical and horizontal sync signals present in the video-detector output.

The composite video (via the RGB interface, chapter 6) is applied to pin 22 of IC501 through C509, R506, and C508, which act as a low-pass filter to remove video/chroma information (leaving only sync). The sync pulses at pin 22 are amplified and applied to pin 20. (Although the circuit between these two pins is called a separator, the function is more like that of an amplifier, because both vertical and horizontal sync appears at pin 20.)

The sync information is applied to the tuning circuits (Figure 2.5) as described in chapter 2, and to sync separator Q506. The output from Q506 is used to lock the vertical oscillator in IC501 with the transmitted sync. The ver-

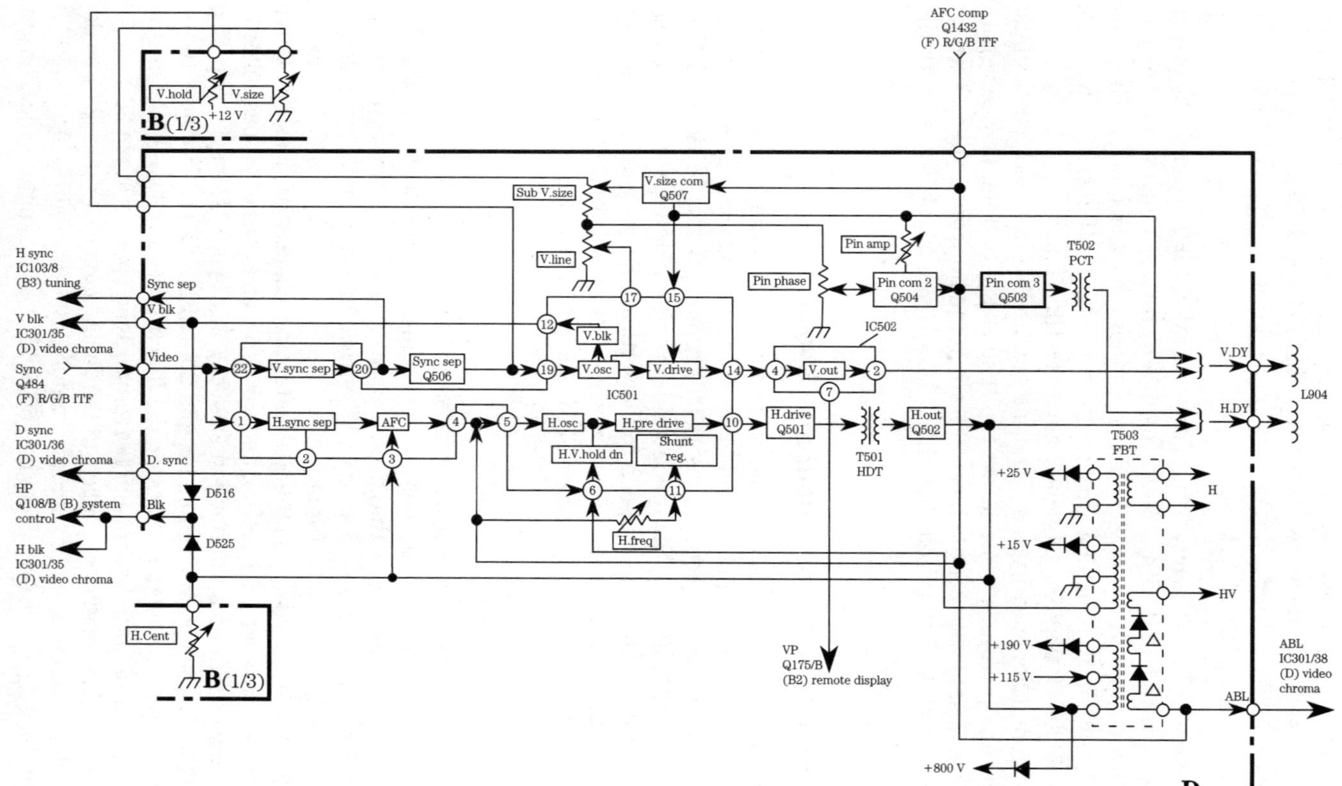

Figure 5.5 Horizontal and vertical sweep (E).

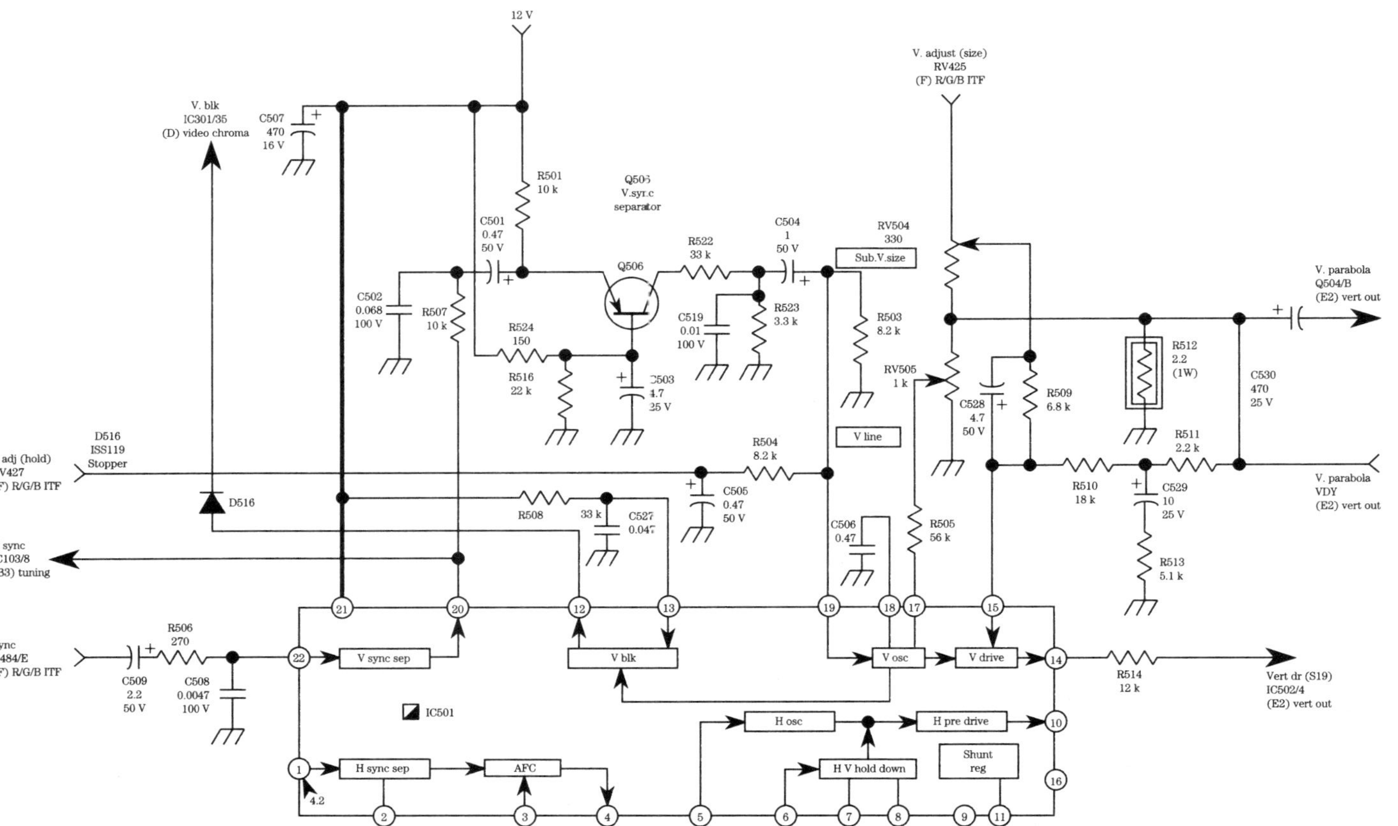

Figure 5.6 Vertical oscillator circuits (E1).

tical oscillator frequency can be adjusted by RV505. The amplitude of the vertical-drive signal to the vertical-output circuits (and thus the vertical height of the CRT display) is adjusted by RV504 and RV425 on the RGB interface board.

A correction signal from the vertical yoke (Figure 5.7) is applied to the vertical-drive circuit at pin 15 of IC501. This feedback signal (known as the parabola or "S" correction signal) serves to maintain linearity of the drive signal at pin 14. The same feedback signal is applied to the pincushion-correction circuit as described next.

5.5.2 Vertical output

Figure 5.7 shows the vertical-output circuits for a typical TV set (the Sony 13 inch). The main purpose of the circuit is to provide sweep signals for the vertical yoke. In addition, the circuits provide the signals necessary for pincushion correction in the horizontal sweep.

The vertical-drive signals from the vertical oscillator (Figure 5.6) are applied to pin 4 of IC502. The deflection-output circuits within IC502 are powered by the rectified output from flyback transformer T503. The +25 V at D514 is doubled by the D505/C532 combination to produce a vertical-output supply voltage of about 40–45 V. It is important to note that the vertical-output stages of IC502 operate solely from the 25 V, taken from FBT T503 in the horizontal output circuits (Figure 5.9). This is done so that any variation in horizontal sweep will have a corresponding variation in the vertical sweep, thus maintaining a constant horizontal/vertical aspect ratio of 4 to 3.

The vertical-pulse output from pin 7 of IC502 is applied to the remote/display circuits as described in chapter 9. The vertical-sweep output from pin 2 of IC502 is centered by S501. Switch 501 is mounted on a circuit board and has three positions. Each position applies a different dc level to the vertical yoke, and thus moves the vertical display up or down.

A sample of the vertical-output signal (the vertical parabola) is returned to the vertical-drive circuits to maintain linearity as described. The same signal is also applied to the horizontal yoke (Figure 5.9) through Q504/Q503 and T502, along with a distortion-compensation signal from the RGB interface (chapter 6). The combined signals minimize the pincushion effect (where the CRT display appears to bend outward). RV507 and RV508 provide for pincushion adjustments (as described in chapter 11).

5.5.3 Horizontal oscillator

Figure 5.8 shows the horizontal-oscillator circuits for a typical TV set (the Sony 13 inch). The main purpose of the circuit is to provide drive pulses for the horizontal-output stages that produce sweep signals to the horizontal yoke. In addition, the circuits provide for shut-off of the horizontal output and high-voltage stages in the event of excessive CRT voltages.

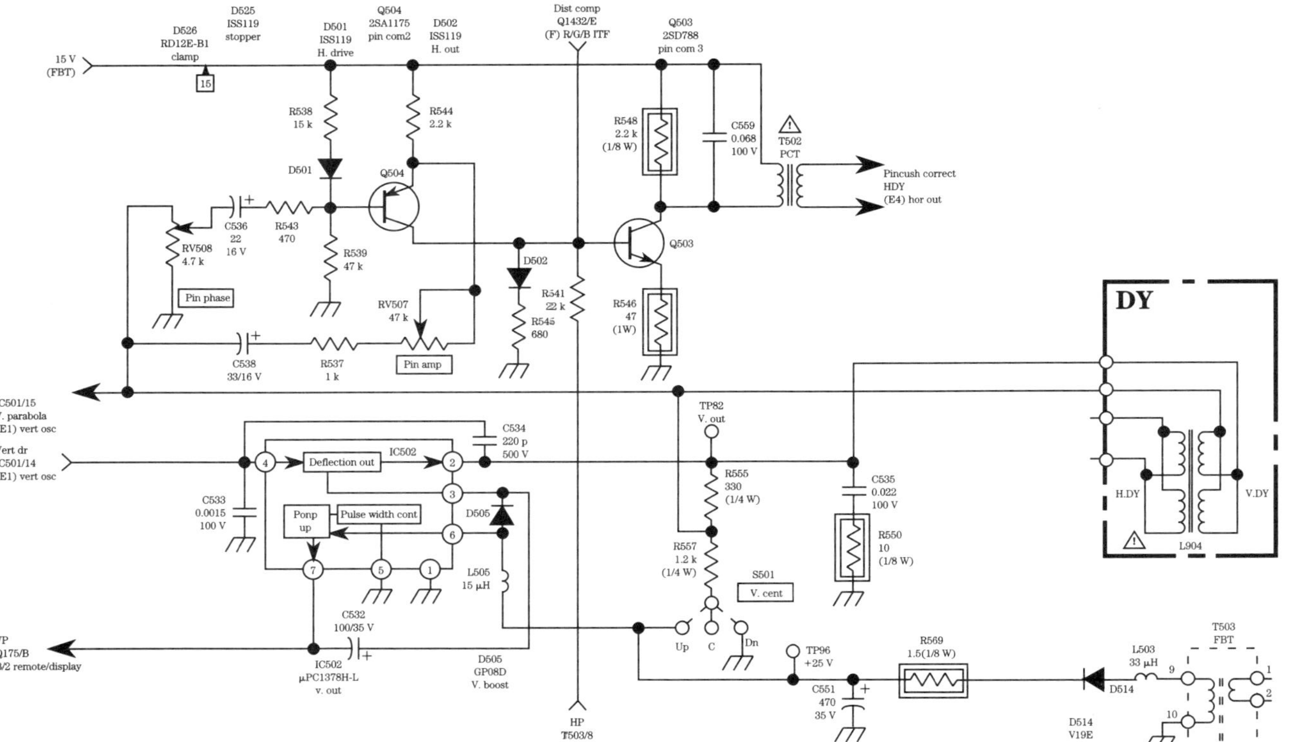

Figure 5.7 Vertical output circuits (E2).

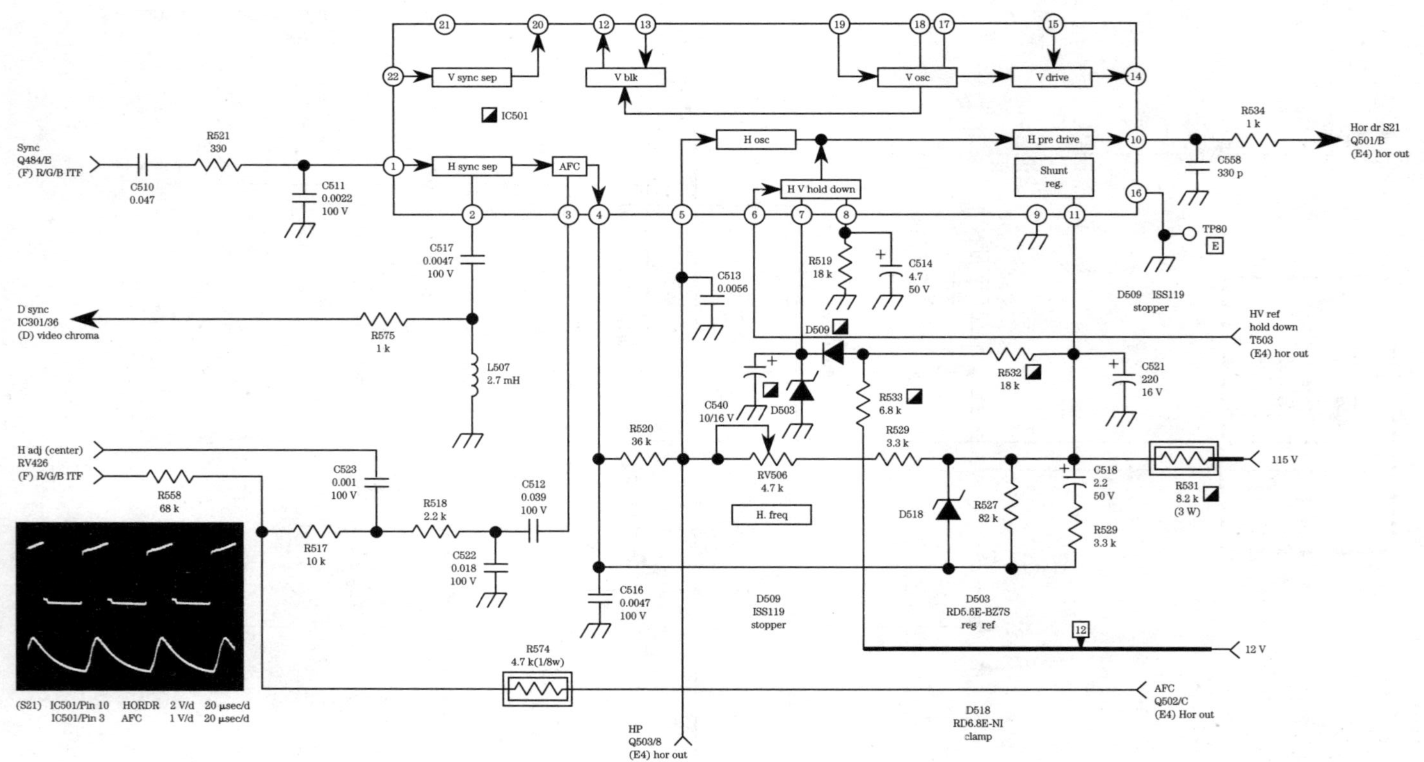

Figure 5.8 Horizontal oscillator circuits (E3).

The composite video (via the RGB interface, chapter 6) is applied to pin 1 of IC501 through C510, R521, and C511, which act as a low-pass filter to remove video/chroma information (leaving only sync). The sync pulses at pin 1 are amplified and applied to the AFC circuits within IC501. (Again, the circuit at pin 1 is essentially an amplifier, although it is called a separator.) A delayed sync output at pin 2 is applied to the video/chroma circuits (Figure 4.6) for burst gating, as described in chapter 4.

The sync information from the separator is applied to the AFC, and compared with a reference horizontal signal at pin 3. The AFC reference signal is taken from the horizontal-output transistor Q502 (Figure 5.9). The AFC compares both signals and produces a correction signal at pin 4. The correction signal locks the horizontal oscillator in IC501 to the transmitted sync. The horizontal oscillator frequency can be adjusted by RV506.

5.5.4 Horizontal output and high voltage

Figure 5.9 shows the horizontal-output and high-voltage circuits for a typical TV set (the Sony 13 inch). The main purpose of the circuit is to provide signals for the horizontal yoke. In addition, the circuits provide a number of sync and control signals to other stages.

The horizontal-drive signals from the horizontal oscillator/predriver (Figure 5.8) are applied to the flyback transformer T503 and the horizontal yoke through horizontal drive Q501, horizontal pulse transformer T501, and horizontal output Q502. (Notice that Q502 has a built-in damper diode.)

The output from Q502 applied to the yoke produces the horizontal sweep. This same output applied to one winding of T503 induces various voltages and signals at the other windings, as follows.

The ac output at pins 1 and 2 is used for the CRT heater, while the dc output at pin 3 is the CRT high voltage. The ac output at pin 4 is rectified by D511 and used as the CRT screen or accelerator-grid voltage. The ac output at pin 6 is rectified by D515 and used as the CRT focus voltage.

The ac output at pin 9 is rectified by D514 and used as the +25-V supply for the vertical output IC502 (Figure 5.7). The ac winding at pin 7 is rectified by D510/D513 and used as the +12-V supply (Figure 5.8).

The output at pin 8 is used by several circuits. First, the output is applied to pin 38 of IC301 (Figure 4.6), and is used as the ABL or automatic brightness limiter discussed in chapter 4. The same output is applied to the base of Q503 as part of the pincushion-correction function (Figure 5.7) discussed in section 5.5.2. The output is also applied to the horizontal oscillator (Figure 5.8) at pin 5 of IC501 as a feedback signal to hold the oscillator on frequency.

The output at pin 11 (rectified by D512) is used as the high-voltage hold-down signal applied to pin 6 of IC501 (Figure 5.8). The rectified output from D512 is compared to a fixed reference voltage at pin 7 of IC501. If the

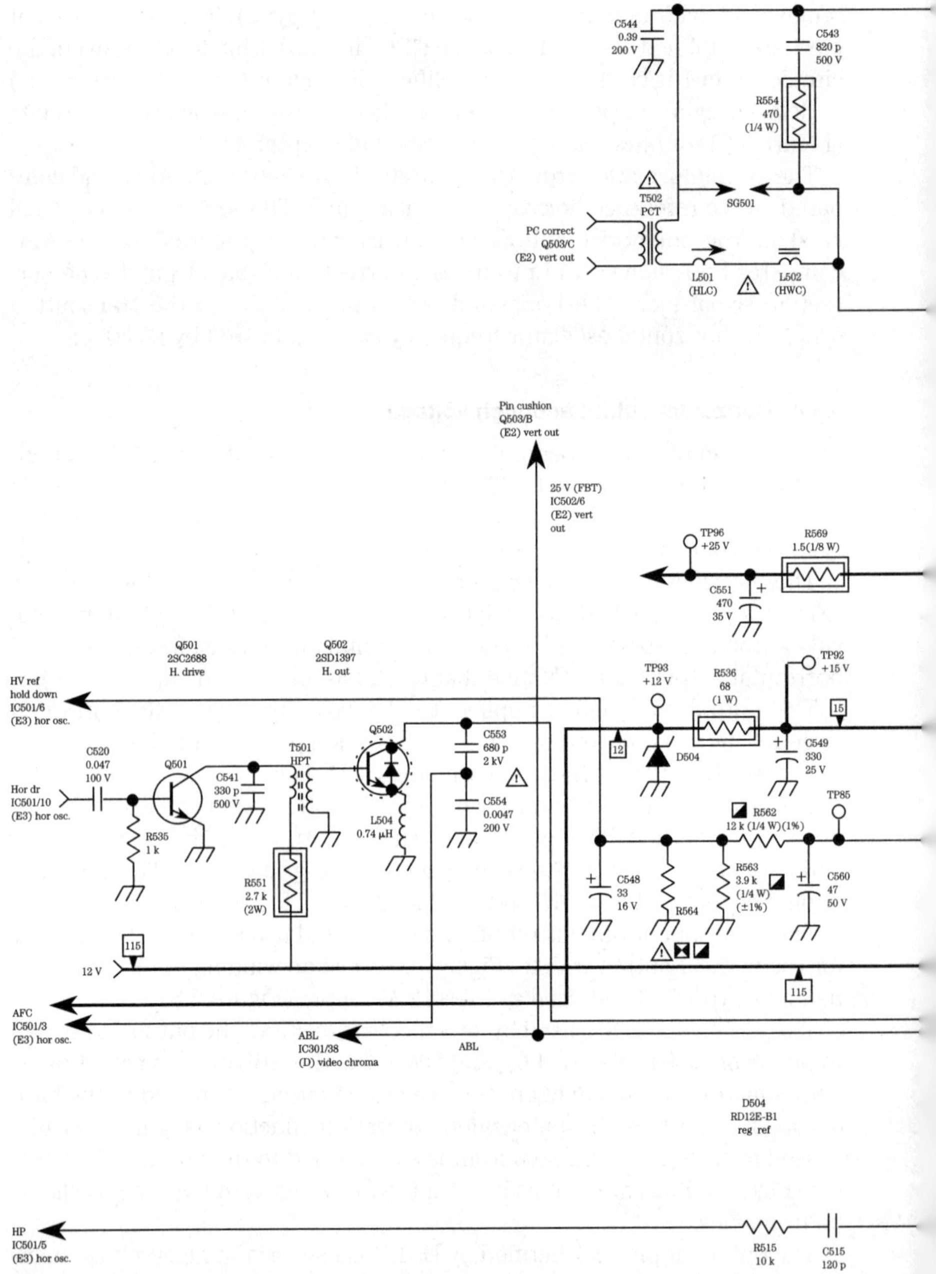

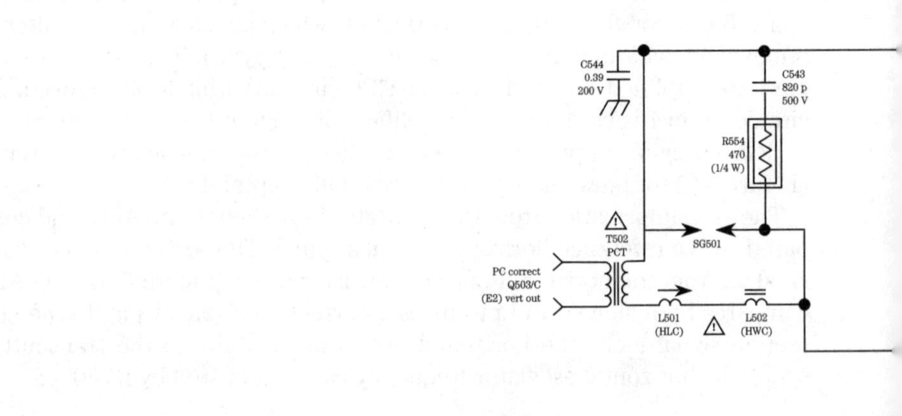

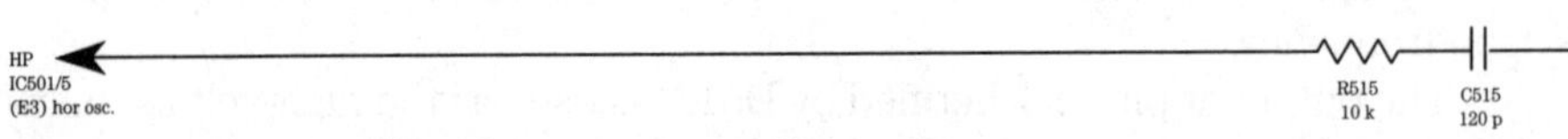

Figure 5.9 Horizontal-output and high-voltage circuits (E4).

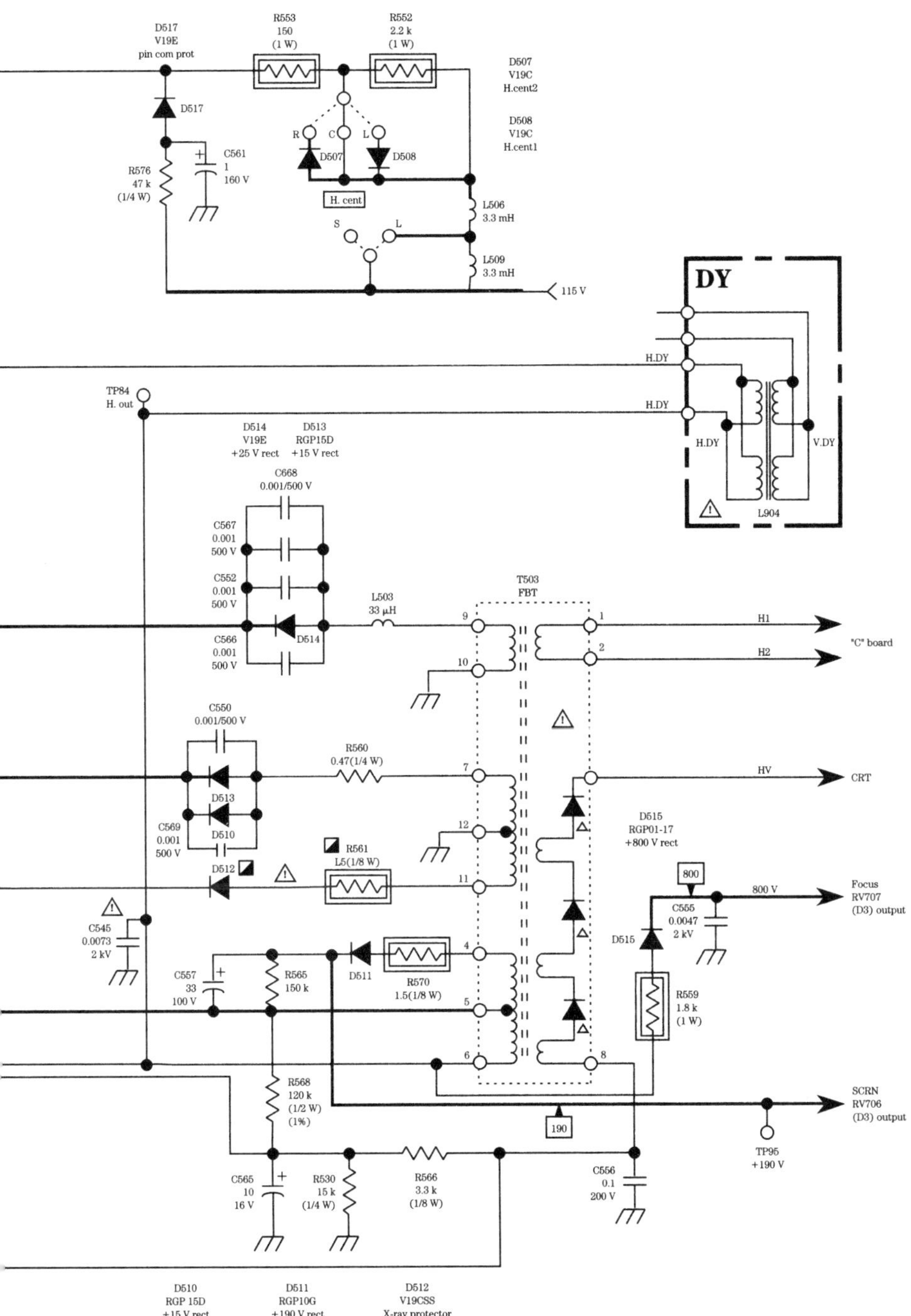
D517
V19E
pin com prot
R553
150
(1 W)
R552
2.2 k
(1 W)
D507
V19C
H.cent2
D508
V19C
H.cent1
D517
R576
47 k
(1/4 W)
C561
1
160 V
R
C
L
D507
D508
H. cent
S
L
L506
3.3 mH
L509
3.3 mH
115 V
DY
H.DY
H.DY
H.DY
V.DY
L904
TP84
H. out
D514
V19E
+25 V rect
D513
RGP15D
+15 V rect
C668
0.001/500 V
C567
0.001
500 V
C552
0.001
500 V
C566
0.001
500 V
D514
L503
33 μH
T503
FBT
H1
H2
"C" board
C550
0.001/500 V
R560
0.47(1/4 W)
D513
C569
0.001
500 V
D510
HV
CRT
D515
RGP01-17
+800 V rect
R561
L5(1/8 W)
D512
800
800 V
Focus
RV707
(D3) output
C555
0.0047
2 kV
D515
C545
0.0073
2 kV
C557
33
100 V
R565
150 k
D511
R570
1.5(1/8 W)
R559
1.8 k
(1 W)
R568
120 k
(1/2 W)
(1%)
190
SCRN
RV706
(D3) output
TP95
+190 V
C565
10
16 V
R530
15 k
(1/4 W)
R566
3.3 k
(1/8 W)
C556
0.1
200 V
D510
RGP 15D
+15 V rect
D511
RGP10G
+190 V rect
D512
V19CSS
X-ray protector

voltage developed by T503 exceeds a safe level (where dangerous X-rays might be emitted), the voltage at pin 6 of IC501 varies sufficiently from the reference at pin 7, and the hold-down circuit in IC501 disables the horizontal oscillator. This removes all power from IC503 and the CRT.

Once the high-voltage hold-down circuit is operated, it is necessary to remove power from the circuit and then restore power. If the malfunction is cleared, the circuits will operate normally when power is reapplied. If not, the hold-down circuit will again turn off the horizontal oscillator.

The horizontal-sweep output is centered by solder-bridge adjustments that select corresponding diodes D507 or D508. Selection of the diode applies a different dc level to the horizontal yoke, and thus moves the horizontal display left or right.

5.6 Vertical and Horizontal Sweep with a Jungle Chip

This section describes vertical/horizontal sweep circuits that use a so-called jungle chip, such as shown in Figure 4.8. As discussed in chapter 4, such chips are multipin and incorporate many of the functions found in discrete form on other sets. The chip shown in Figure 4.8 is designated as IC501 in a Sony 19 inch. The following sections describe operation of the vertical/horizontal sweep circuits associated with IC301.

5.6.1 Vertical oscillator

Figure 5.10 shows the vertical oscillator/drive circuits within IC301. Composite video from the video-processing stages (Figure 4.9) is applied to IC301 at pin 42. After amplification, both horizontal and vertical sync signals are separated. The vertical sync is applied to the vertical oscillator in the normal manner. The vertical-oscillator signal is applied to the vertical output (Figure 5.11) through the vertical-drive stages in IC301, at pin 17.

The feedback or correction signal from the vertical yoke is applied to the vertical-drive at pin 20 of IC301, and serves to maintain linearity of the drive signal pin 17. The vertical-sweep frequency and vertical size are adjusted by RV503 and RV501, respectively. Notice that vertical-size control RV501 receives +25 V from the flyback transformer so that any variation in horizontal sweep will have a corresponding variation in vertical sweep, thus maintaining the 4/3 aspect ratio.

5.6.2 Vertical output

Figure 5.11 shows the vertical-output circuits for a typical TV set with jungle IC (the Sony 19 inch). (Notice that none of the circuits are part of the jungle IC). The main purpose of the circuits is to provide sweep signals for the vertical yoke. In addition, the circuits provide control for the vertical oscillator and on-screen display (chapter 9).

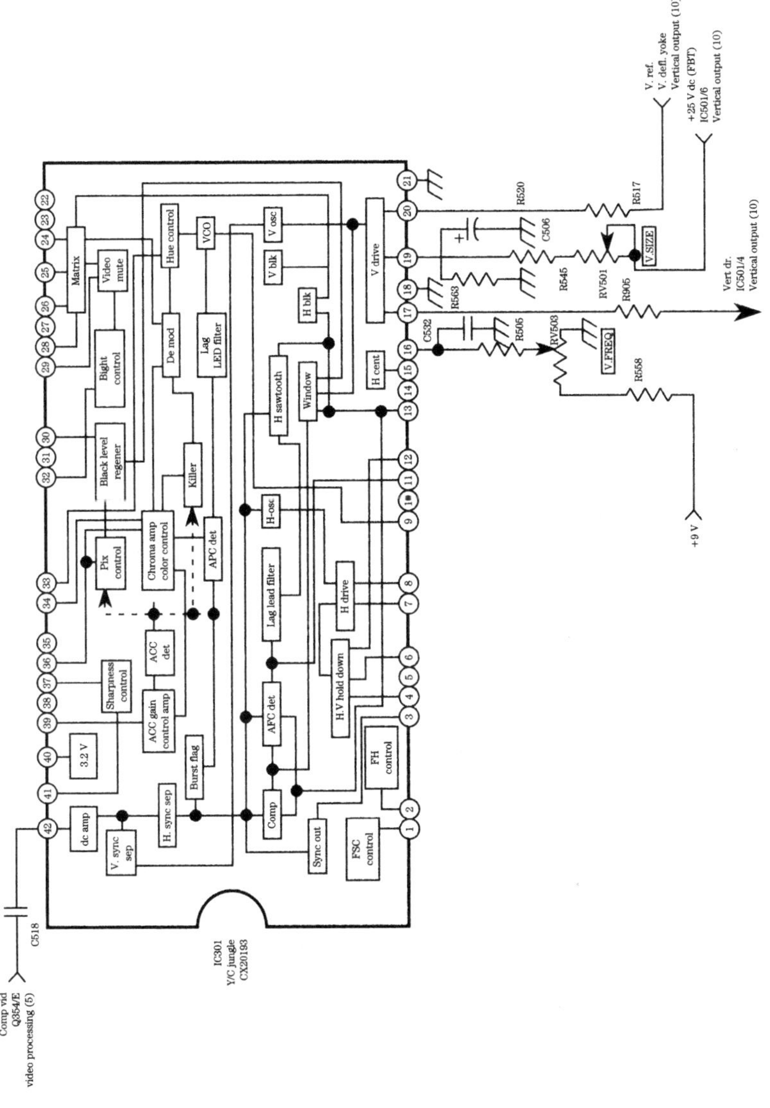

Figure 5.10 Vertical oscillator in jungle-chip (9).

The vertical-drive signal from pin 17 of IC301 is applied to pin 4 of IC501. The drive signal is amplified by IC501 and applied to the vertical coil of the yoke. Pin 3 of IC501 receives a boosted voltage from the pump-up circuit at pin 7. During the vertical trace period, C515 is charged, or pumped up, through the pump-up circuit. During the retrace period, C515 discharges into pin 3, increasing the vertical output retrace signal. The vertical output signal is also used to drive inverter Q171, and provides a vertical pulse to the on-screen display (chapter 9). As discussed, the linearity-correction feedback signal from the yoke is applied to the vertical oscillator at pin 20 of the IC301 (Figure 5.10). The vertical-bias control RV505 generally does not need to be readjusted unless vertical components are replaced.

The vertical signal is also applied to Q503 along with a horizontal-reference signal from the horizontal-output stage (Figure 5.13). These signals are combined in Q503 to generate the pincushion-correction signal, which is coupled to the horizontal coil of the yoke through T502. The pincushion-correction signal can be adjusted by RV504. Vertical centering is accomplished by S501, which applies and removes voltages and grounds to the vertical yoke, thus moving the vertical display up or down.

The +25 V from the flyback-transformer is unregulated, with the exact voltage level depending on the horizontal-oscillator output. This maintains the 4/3 horizontal/vertical aspect ratio, because the amplitude of the vertical output to the yoke varies directly with the horizontal output. (If the horizontal output increases, the vertical increases by the same amount, and vice versa.)

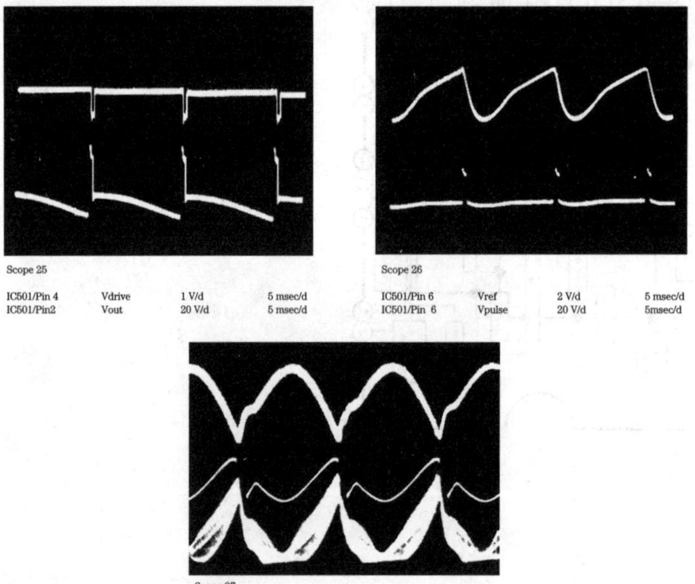

Figure 5.11 Vertical output with jungle-chip (10).

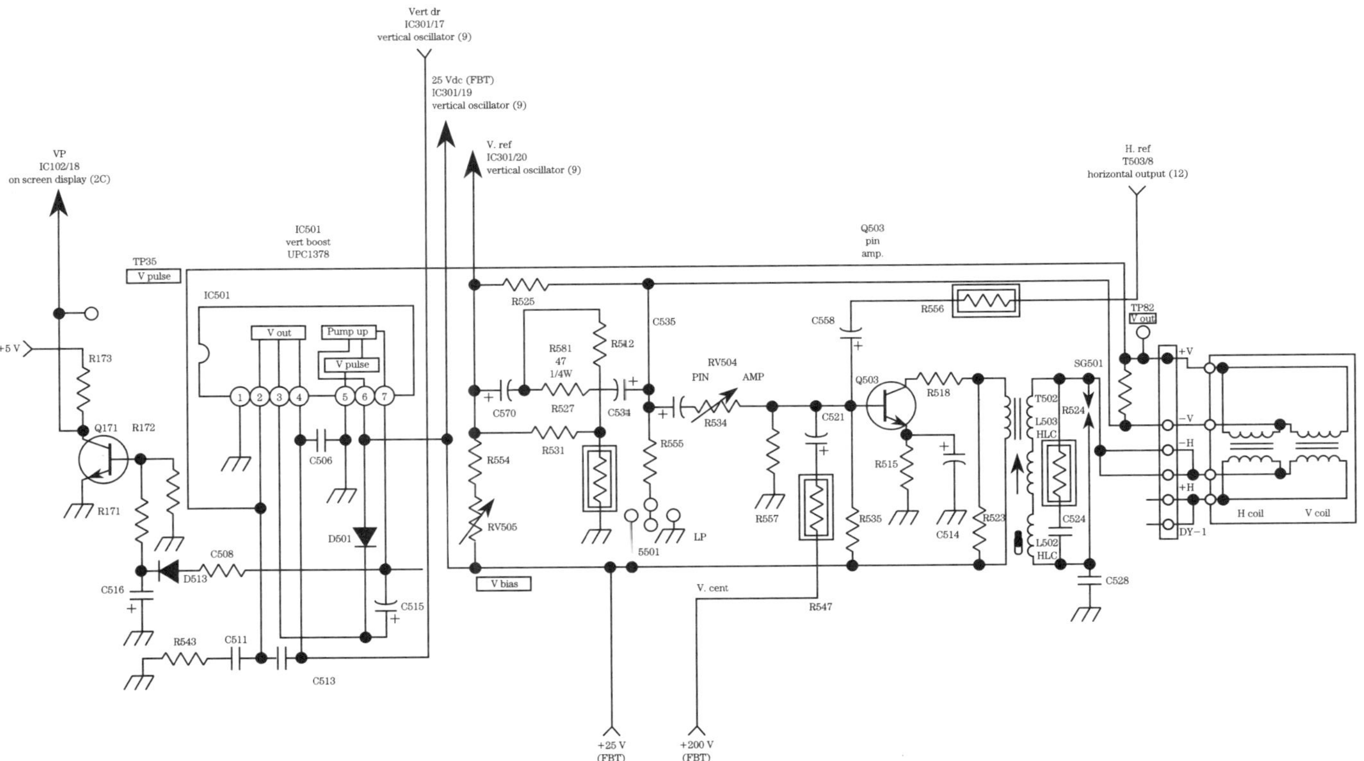

Figure 5.11 Continued.

5.6.3 Horizontal oscillator

Figure 5.12 shows the horizontal-oscillator circuits for a typical TV set with jungle IC (Sony 19 inch). As shown, many of the circuits are within IC301. The main purpose of the circuit is to provide drive pulses for the horizontal-output stages that produce sweep signals to the horizontal yoke (Figure 5.13). In addition, the circuits provide for shut-off of the horizontal output and high-voltage stages in the event of excessive CRT voltage, as well as control of the CRT brightness during normal operation.

The composite video is applied to pin 42 of IC301 through C518. After amplification, the sync pulses are separated and applied to the AFC circuits within IC301. The sync information is compared with a reference horizontal signal at pin 13. The AFC reference signal is taken from the horizontal-output transistor Q502 (Figure 5.13). The AFC compares both signals and produces a correction signal that locks the horizontal oscillator in IC301 to the transmitted sync. The horizontal-oscillator frequency can be adjusted by RV503 and monitored at TP7 (pin 11 of IC301).

Transistors Q504/Q505 provide the high-voltage hold-down function (also known as the high-voltage protection or X-ray protection function). No matter what it is called, a reference from the horizontal-output transformer T502 (Figure 5.13) is applied to pin 6 of IC301, establishing a dc reference level at pin 6. In the event that the output high voltage becomes excessive (as monitored at pin 7 of T503, Figure 5.13), Q504 turns Q505 on, making pin 12 of IC301 high. If the high at pin 12 exceeds the reference level at pin 6, the high-voltage hold-down circuit operates, and the horizontal oscillator is disabled. This removes all power from T503 and the CRT.

With the high-voltage hold-down circuit on, the level at the collector of Q504 is applied through D307 to the video-processing circuits (Figure 4.10) as a brightness-control signal. Once hold down is activated, it is necessary to remove and then restore power. If the malfunction is cleared, the circuits will operate normally when power is reapplied. If not, the hold-down circuit will again turn off the horizontal oscillator. Notice that the ABL sensing signal from the horizontal output is applied to the ABL picture-control in the video-processing circuits (Figure 4.10), through R528/R530.

5.6.4 Horizontal output and high voltage

Figure 5.13 shows the horizontal-output and high-voltage circuits for a typical TV set (Sony 19 inch). The main purpose of the circuit is to provide sweep signals for the horizontal yoke. In addition, the circuits provide a number of sync and control signals to other stages.

The horizontal drive signals from the horizontal oscillator (Figure 5.12) are applied to the flyback transformer T503 and the horizontal yoke through horizontal drive Q501, horizontal pulse transformer T501 and horizontal output Q502. (Notice that Q502 has a built-in damper diode.)

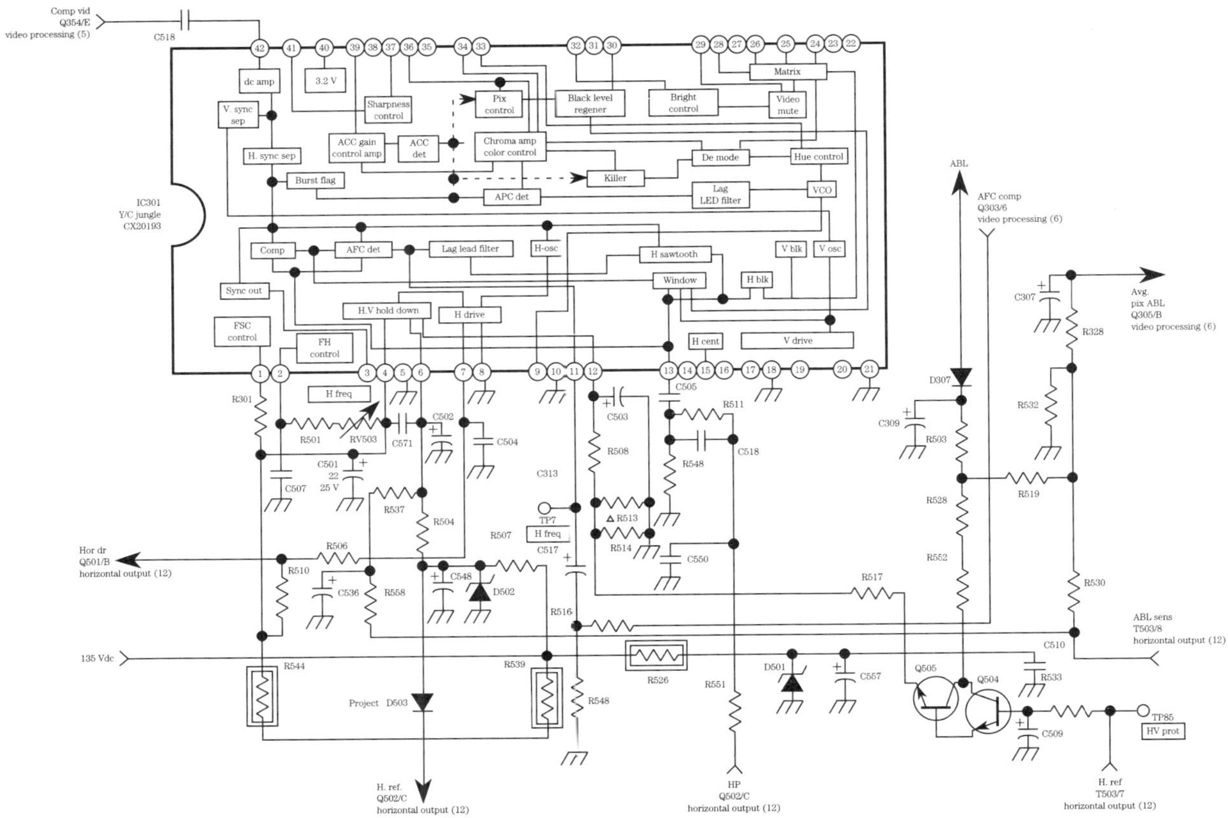

Figure 5.12 Horizontal oscillator in jungle-chip (11).

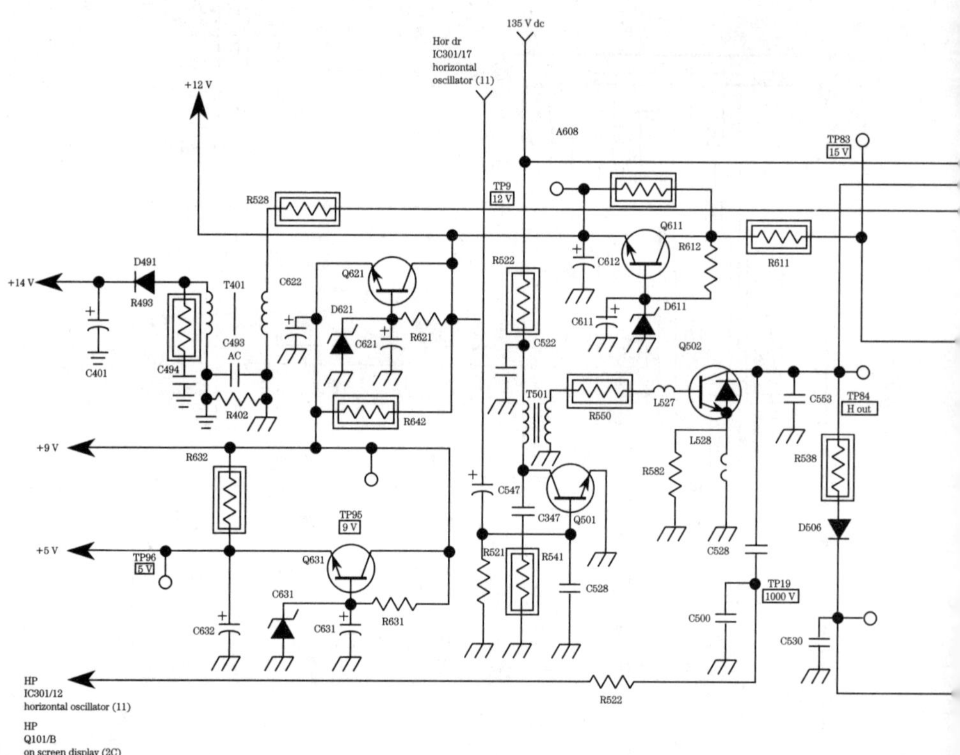

Figure 5.13 Horizontal oscillator with jungle-chip (12).

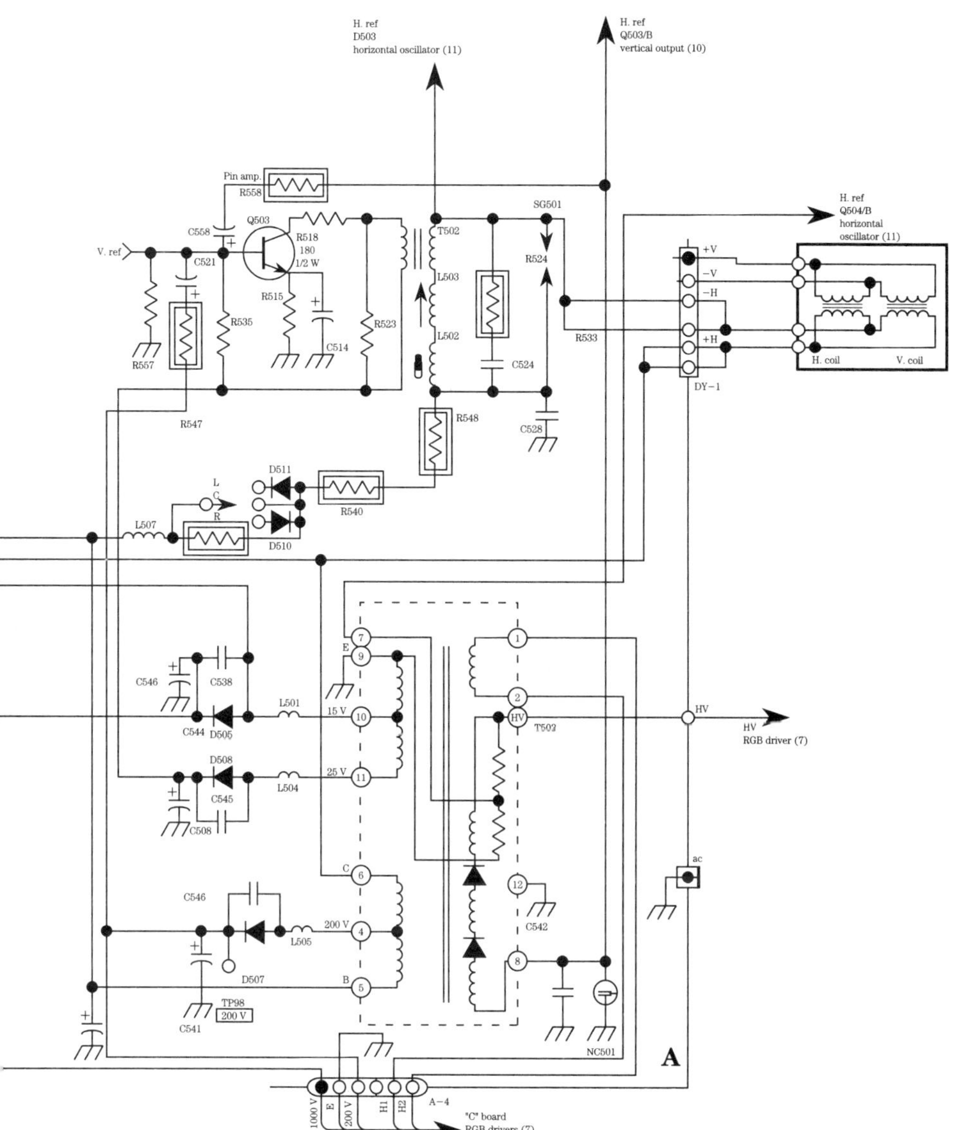

H. ref
D503
horizontal oscillator (11)
H. ref
Q503/B
vertical output (10)
H. ref
Q504/B
horizontal
oscillator (11)
Pin amp.
R558
C558
Q503
R518
180
1/2 W
V. ref
C521
R515
R535
R557
C514
R523
R547
T502
L503
L502
SG501
R524
C524
R533
C528
R548
+V
−V
−H
+H
DY−1
H. coil
V. coil
D511
L
C
R
L507
R540
D510
C546
C538
L501
15 V
C544
D505
D508
L504
25 V
C545
C508
HV
T502
HV
HV
RGB driver (7)
ac
C542
C546
200 V
L505
D507
TP98
200 V
C541
NC501
A
1000 V
E
200 V
H1
H2
A−4
"C" board
RGB drivers (7)

The output from Q502 applied to the yoke produces the horizontal sweep. This same output applied to one winding of T503 induces various voltages and signals at the other windings as follows.

The ac output at pins 1 and 2 is used for the CRT heater, while the dc output at the HV pin is the CRT high voltage. (Notice that this CRT voltage is about 25 kV.) The output of D505 is 15 V, which is regulated by Q611 to 12 V. The ac at pin 10 of T503 is also applied through T401 and rectified by D491 to 14 V. The 12 V regulated by Q621 is also reduced to 9 V, which is further regulated by Q631 to 5 V. The output at pin 7 of T503 is the horizontal-reference signal for the high-voltage hold-down circuit, while the output at pin 8 is the ABL reference, both previously discussed.

The horizontal output at Q502 is rectified by D506, producing a 1000 V for the CRT focus and screen (Figure 4.12). The ac at pin 4 of T503 is rectified by D507 to provide 200 V for the RGB drivers and CRT grid. The ac at pin 11 of T503 is rectified by D508 to provide 25 V used by various circuits.

The horizontal-sweep output is centered by a switch that selects corresponding diodes D510 or D511 (or neither). Selection of the diodes applies a different dc level to the horizontal yoke, and thus moves the horizontal display left or right.

5.7 Horizontal and Vertical Sweep (25 Inch)

This section describes the horizontal and vertical processing circuits where the majority of the stages are in a single IC (found in a Sony 25 inch). Figure 5.14 shows the principle circuits in block form. Such circuits include sync separators, oscillators, AFC, blanking, high-voltage hold-down (X-ray protection), and predrivers. Notice that the drivers and output stages are discrete, as is the true sync separator.

5.7.1 Vertical processing

Composite sync from the RGB interface (chapter 6) is applied to pins 1 and 22 of IC1501 through buffer Q1510. The sync signals (pulses) are negative at a level of about 1.5 Vp-p. The amplified sync signals are applied to system control (chapter 7) as a vertical reference, and to vertical-sync separator Q1501, which separates the vertical sync from the composite sync.

The vertical-sync pulses (at about 3 Vp-p) are applied to the vertical oscillator through pin 19. The vertical-oscillator output is applied to system control through a blanking circuit at pin 12. These pulses are mixed with horizontal-blanking pulses from Q1504, and are applied to blanking and clamp circuits of the video processor. S-correction for the vertical oscillator is provided by a 10 Vp-p parabolic vertical signal from the vertical-deflection yoke, processed by Q1511 and input at pin 17 of IC1501.

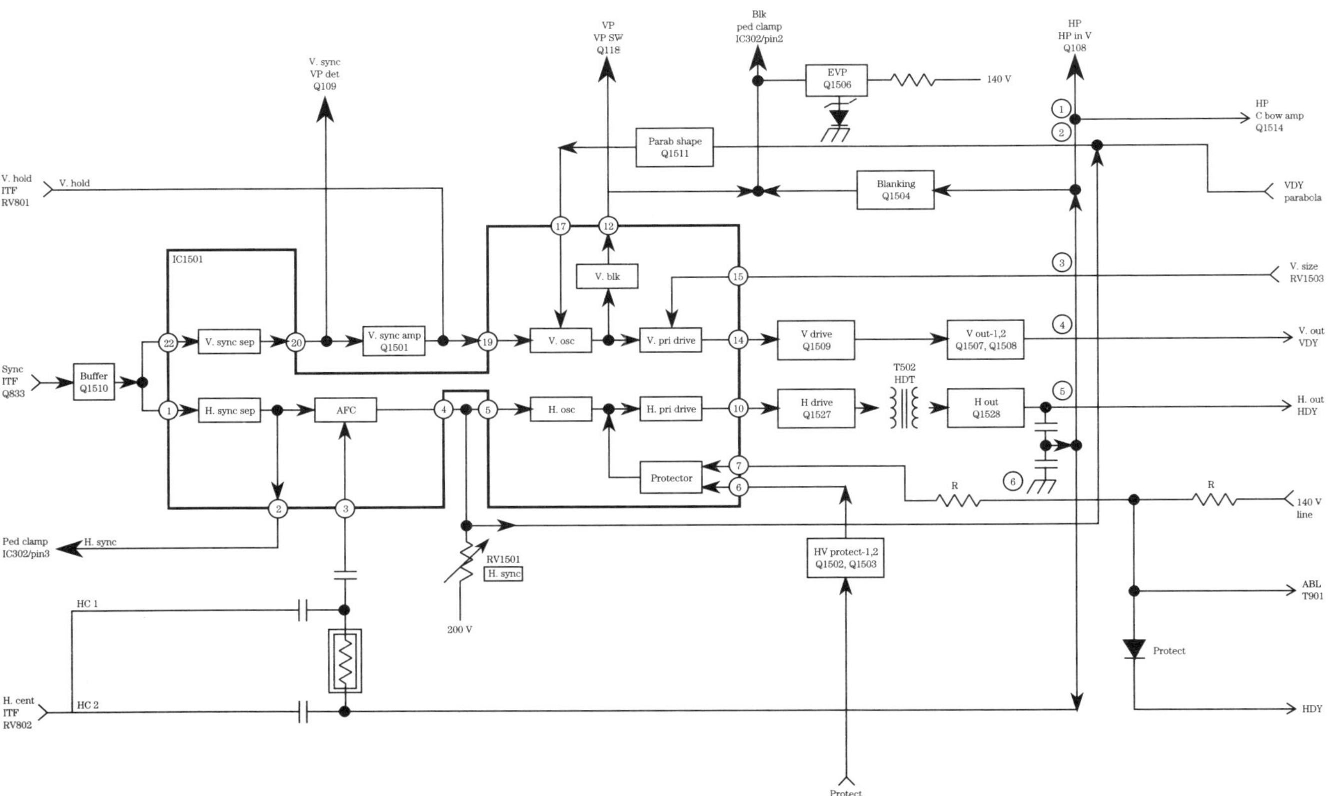

Figure 5.14 Horizontal and vertical sweep circuits (G).

5.7.2 Horizontal processing

Composite sync is applied to the horizontal-sync separator through buffer Q1510 and pin 1 of IC1510. The separated horizontal-sync output is applied to the AFC circuits in IC1501, and to a pedestal clamp in the video processing circuits through pin 2. The level of the negative pulses at pin 2 are about 12 Vp-p.

A sample of the horizontal-output signal from Q1528 is fed back to pin 3 of IC1501 as the AFC reference pulse. The output of the AFC circuits, at pins 4 and 5, is used to synchronize the horizontal oscillator. This signal is a sawtooth at a level of about 5 Vp-p. The horizontal oscillator output is adjusted in frequency by RV1501, and is applied to the horizontal driver Q1527 through a horizontal predriver in IC1501. The signal at pin 10 is at a level of about 3 Vp-p.

The reference voltage for the protector circuit at pin 7 of IC1501 is derived from the 140-V line across a series of dropping resistors R. The high-voltage protect transistors Q1502/Q1503 monitor the high-voltage sample. Should the high voltage rise above a safe level, conduction in Q1502/Q1503 increases to a level where pin 6 is above the level at pin 7. When this occurs, the protector circuit in IC1501 latches, shutting down the horizontal operation. To restore operation, power must be turned off, then back on. Keep this in mind when troubleshooting a dark screen or no raster symptom.

The connections to the ABL winding of the flyback transformer, and through the protect diode to the horizontal-deflection yoke, can also cause hold-down operation. Any excessive load in the flyback transformer or the horizontal-deflection yoke will lower the reference at pin 7 below the normal level at pin 6, activating the protector circuit in IC1501.

The excessive-voltage protection (EVP) circuit Q1506 monitors a sample of the 140-V line, and compares the line to a zener reference. Should the 140-V line rise about 10%, Q1506 turns on, and blanks the video processing stages.

5.7.3 Pincushion correction and convergence

Because of the large, flat CRT used in this 25 inch set, and the intended application as a monitor, the convergence and pincushion-correction circuit requirements are more stringent than those of conventional sets. Figure 5.15 shows the circuits involved in block form. The following sections describe circuit operation.

5.7.4 Pincushion correction circuits

The pincushion effect occurs because of the horizontal deflection either overshooting or undershooting the extreme left and right edge of the CRT. This effect is more pronounced in the middle of the picture, which coincides with the middle of the vertical sweep. By mixing a sample of the vertical sweep with the horizontal sweep, in controlled amounts, the horizontal sweep can be modified at specific points or areas of the vertical plane. This is the purpose of the pincushion-correction circuits.

A sample of the vertical sweep (a vertical parabola) is taken from the vertical-deflection yoke (VDY) and applied to the PIN (pincushion correction) IC at pin 11. The level of this sample is set by RV1512 to about 1 Vp-p. Feedback from the horizontal-deflection yoke (HDY) is provided at pin 13, with a level of about 2.5 Vp-p. The output signal at pin 7 is fed back to pin 10 across the PIN CORRECT adjustment RV1510. The signal level at pin 7 is also about 2.5 Vp-p.

The resulting pincushion-correction signal at pin 15 is at a level of about 4 Vp-p. This signal is applied to the horizontal-deflection yoke through PIN predrive Q1523 and PIN drive Q1524.

The following precautions should be observed when troubleshooting or adjusting to correct a pincushion problem in this or any set. Follow the service-manual procedures exactly. Usually, the procedures are very specific about the areas of the picture affected by each adjustment. During the adjustment procedure, keep in mind the interactive nature of the adjustments. The procedures should be repeated for best results. Finally, the waveform shapes and amplitude shown in the service manual will vary from set to set, and are to be used for reference only. The effectiveness of the circuit must be judged by the results observed.

5.7.5 Convergence circuits

The convergence circuits, including the convergence yoke (CY), are like the pincushion circuits in that convergence correction requires samples of both the horizontal and vertical sweeps. The horizontal sample is provided to the CBOW amplifier Q1514 at a level of about 10 Vp-p. The vertical sample is provided to the YBOW amplifier Q1513 at the same level.

As with the pincushion circuits, the service manual waveforms and amplitudes will vary from set to set, and operation must be judged from observing the effect of controls on the CRT. All of the adjustments are interactive and must be repeated. From a troubleshooting standpoint, operation of each stage is easily confirmed by observing the effect of each adjustment on the related circuit, and the overall effect on the CRT. Again follow the service manual procedures.

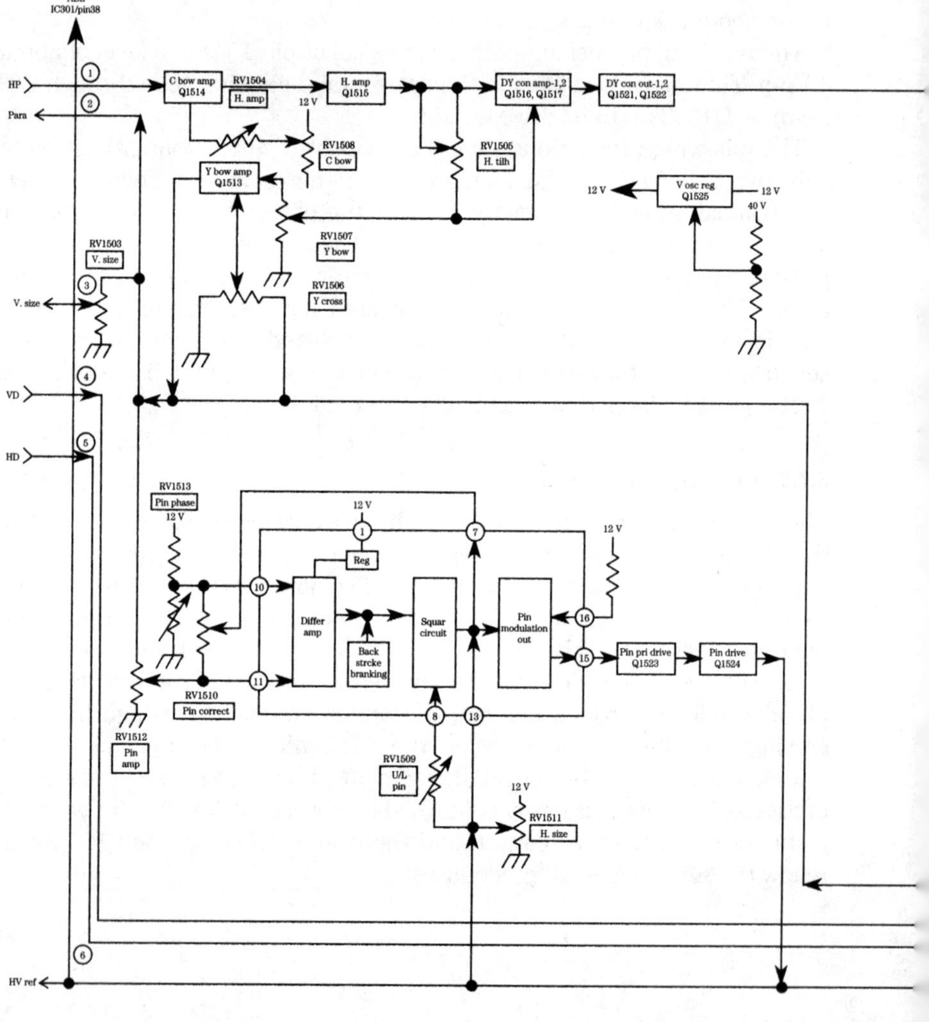

Figure 5.15 Pincushion correction and convergence circuits (H).

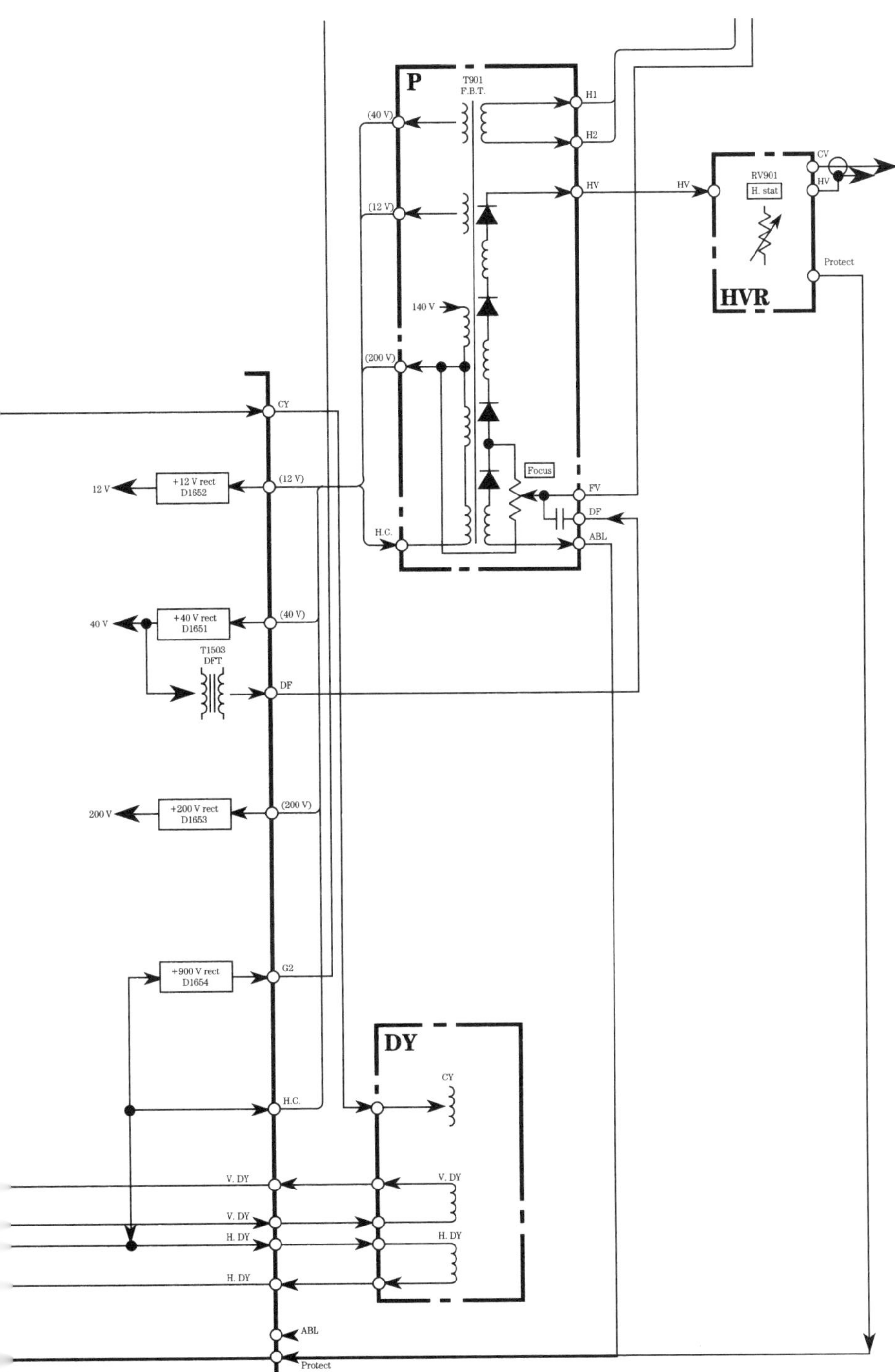
P
T901
F.B.T.
(40 V)
(12 V)
140 V
(200 V)
H1
H2
HV
Focus
FV
DF
ABL
H.C.
HV
RV901
H. stat
CV
HV
Protect
HVR
CY
12 V
+12 V rect
D1652
(12 V)
40 V
+40 V rect
D1651
(40 V)
T1503
DFT
DF
200 V
+200 V rect
D1653
(200 V)
+900 V rect
D1654
G2
DY
CY
H.C.
V. DY
V. DY
H. DY
H. DY
V. DY
H. DY
ABL
Protect

Chapter

6

RGB Interface Circuits

This chapter is devoted to TV-set circuits that provide interconnection or interface between the CRT/speakers and video/audio inputs. In some sets, these circuits are on a separate board, while on other sets, the circuits are integrated with circuits on various boards. Either way, the purpose of the circuits is to interface the set with VCRs, computers, video games, etc., thus making the set a monitor TV. Although called RGB, the circuits usually involve both color video and audio, possibly stereo/SAP audio (chapter 10).

6.1 RGB Basics

Figure 6.1 shows the RGB interface circuits for a typical TV set (Sony 13 inch) where the majority of interface circuits are on a single board. Because the board circuits are primarily simple amplifiers and switching ICs, the circuits are shown in block form. In this set, the RGB interface has three modes of operation: TV, video, and RGB. Also, in the RGB mode, the circuits have the capability of processing both digital and analog RGB signals.

6.1.1 TV mode

In the TV mode, the FM-demodulated audio signal from the VIF stage (Figure 3.4) is applied to buffer transistor Q864. A sample of the signal is applied to MPX OUT connector CNJ409 through Q861 for use by external multiplex MPX stereo/SAP decoders (chapter 10). The buffered signal from Q864 is applied to TV AUDIO OUT connector CNJ407, and to Q882 through IC881. In the TV mode, analog switch IC881 is enabled by highs at pins 12

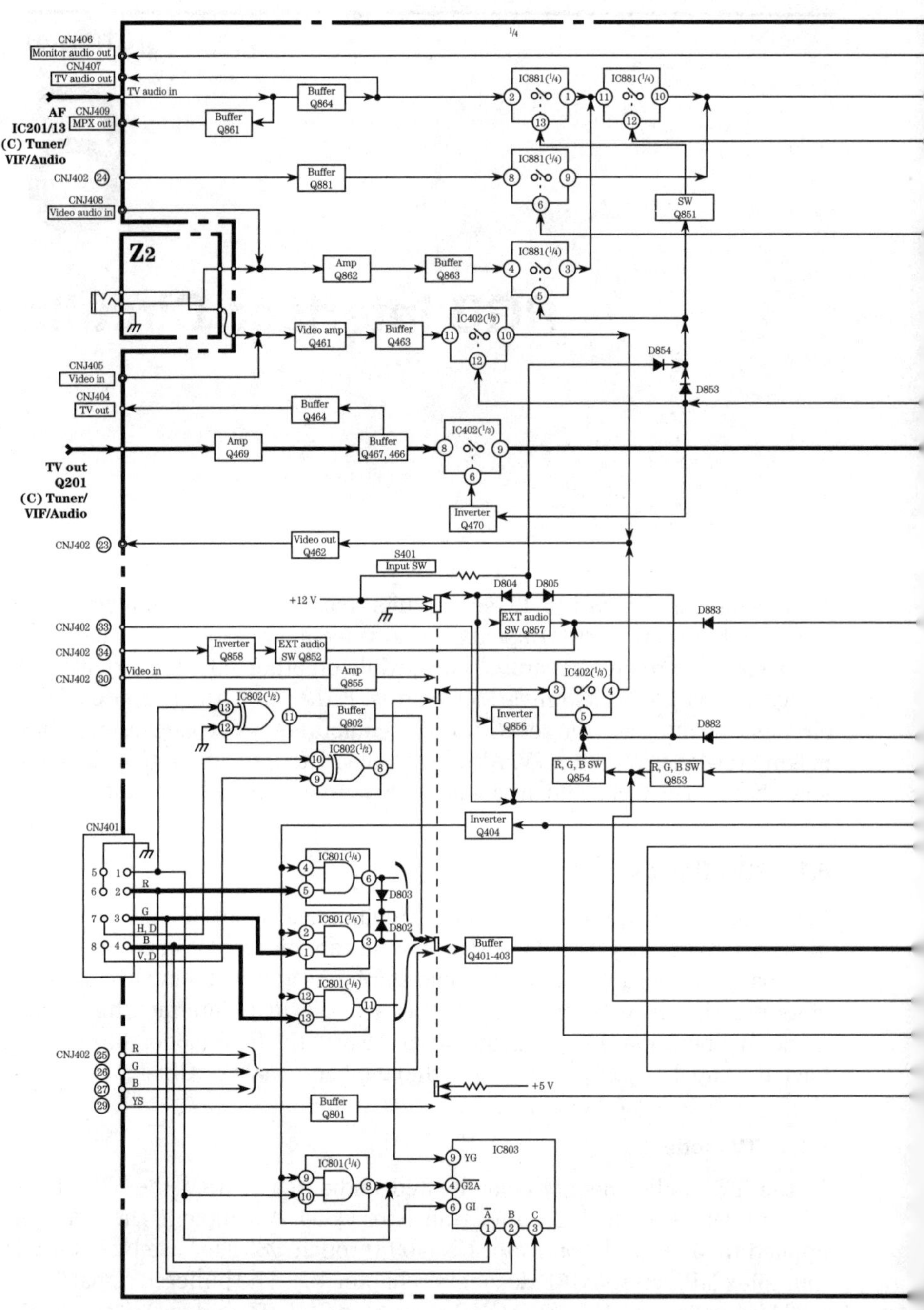

Figure 6.1 RGB interface circuits (F).

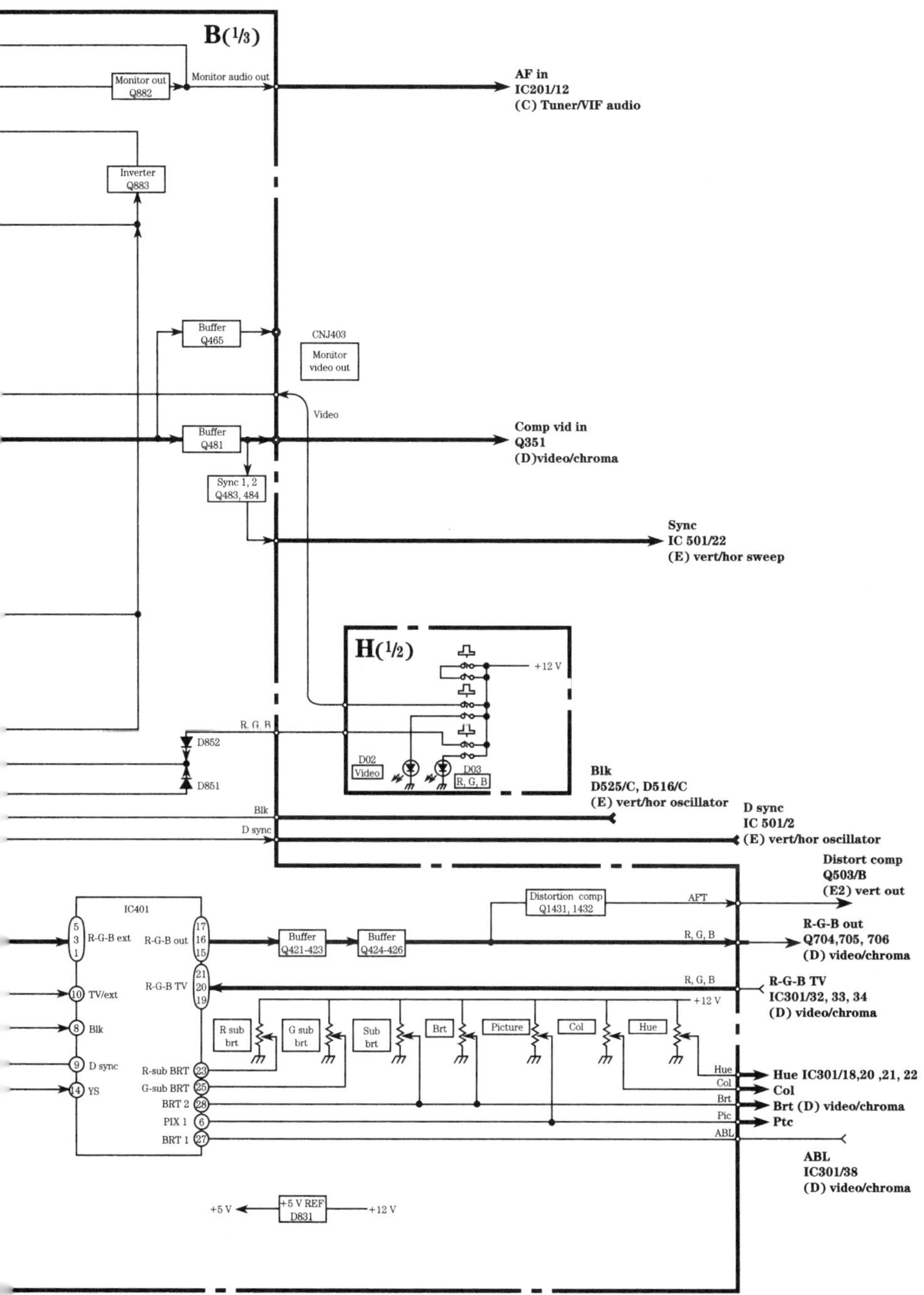
B(1/3)
Monitor out Q882
Monitor audio out
AF in
IC201/12
(C) Tuner/VIF audio
Inverter Q883
Buffer Q465
CNJ403
Monitor video out
Video
Buffer Q481
Comp vid in
Q351
(D)video/chroma
Sync 1, 2 Q483, 484
Sync
IC 501/22
(E) vert/hor sweep
H(1/2)
+12 V
R, G, B
D852
D851
D02 Video
D03 R, G, B
Blk
D525/C, D516/C
(E) vert/hor oscillator
Blk
D sync
D sync
IC 501/2
(E) vert/hor oscillator
Distort comp
Q503/B
(E2) vert out
IC401
R-G-B ext
R-G-B out
Distortion comp Q1431, 1432
AFT
Buffer Q421-423
Buffer Q424-426
R, G, B
R-G-B out
Q704,705, 706
(D) video/chroma
R-G-B TV
R, G, B
R-G-B TV
IC301/32, 33, 34
(D) video/chroma
+12 V
TV/ext
Blk
D sync
YS
R sub brt
G sub brt
Sub brt
Brt
Picture
Col
Hue
R-sub BRT
G-sub BRT
BRT 2
PIX 1
BRT 1
Hue
Col
Brt
Pic
ABL
Hue IC301/18,20 ,21, 22
Col
Brt (D) video/chroma
Ptc
ABL
IC301/38
(D) video/chroma
+5 V
+5 V REF D831
+12 V

and 13, as selected by INPUT switch S401. The output of Q882 is applied to MONITOR AUDIO OUT connector CNJ406, and is returned to the VIF stage at pin 12 of IC201 (Figure 3.4).

The composite video signal from the VIF stage (Figure 3.4) is applied to amplifier Q469 and buffers Q466/Q467. This video signal is applied to TV OUT connector CNJ404, and to buffers Q465/Q481 through switch IC402 (also enabled by S401). Buffer Q465 supplies composite video to MONITOR VIDEO OUT connector CNJ403. Buffer Q481 supplies composite video to Q483/Q484 and to the video/chroma stages (Figure 4.5). Transistors Q483/Q484 are sync separators and provide composite sync to the vertical/horizontal sync sweep (Figures 5.5 and 5.8). Notice that the composite video is also applied to pin 23 of CNJ402 through Q462.

6.1.2 Video mode

The video mode is selected by pushing the front-panel VIDEO switch. This applies 12 V to front-panel VIDEO indicator D02, and to pin 12 of switch IC402 through the video connector, enabling the external composite-video input switching. The 12-V high is inverted by Q470 to a low, removing the TV composite-video path between pins 8/9 of IC402. The 12 V is also applied to Q851 through D853. This turns off Q851 and removes the TV audio path between pins 1 and 2 of IC881. However, the external audio path between pins 10 and 11 of IC831 remains intact.

The external audio signal is applied at VIDEO AUDIO IN CNJ408, or from a combination audio/video input jack located on the front panel. Either way, the audio is amplified by Q862, buffered by Q863, and applied to the monitor amplifier Q882 through the enabled section (pins 3/4 and 10/11) of IC881. The output from Q882 is applied to the VIF stage at pin 12 of IC201 (Figure 3.4), and to MONITOR AUDIO OUT connector CNJ406.

The external composite video (say from a VCR) is applied at VIDEO IN CNJ405, or from the front-panel jack. The video is amplified by Q461, buffered by Q463, and applied to pin 23 of CNJ403 through the enabled section (pins 10/11) of IC402 and video amplifier Q462. The video is also applied to MONITOR VIDEO OUT connector CNJ403 through buffer Q465, to the video/chroma stages (Figure 4.5) and sync separators Q483 and Q484, as described in section 6.1.1 for the TV mode.

6.1.3 RGB mode

The RGB circuits are used in both the TV mode and RGB mode. In the TV mode, pin 10 of IC401 is high, the RGB signals developed by the video/chroma processing stages (Figure 4.6) are applied to pins 19–21 of IC401, and are returned to video/chroma output drivers through pins 15–17 and buffers Q421–Q426.

The RGB mode is selected by pushing the front-panel RGB switch. This applies 12 V to front-panel RGB indicator D03, and to RGB switch Q853 through D852. The low output from Q853 is applied to pin 10 of IC401. This enables the IC401 RGB mode, where the external RGB inputs at pins 1–3 are used. The low from Q853 also turns Q854 off, driving pin 5 of IC402 high and completing the path between pins 3 and 4.

Digital RGB signals from an external computer or video are applied at pins 2–4 of CNJ401, and buffered by gates in IC801. The RGB gate outputs are applied to pins 1–3 of IC401 through buffers Q401–Q403. With a low at pin 10 of IC401, the digital RGB signals are output from IC401 at pins 15–17, as is the case of the TV mode.

6.2 RGB Digital and Analog

Figure 6.2 shows the RGB interface circuit for another TV set (Sony 25 inch). Again, the circuits are shown in block form. These RGB circuits have only two basic modes of operation. However, the switching required is somewhat elaborate. For that reason, Table 6.1 is included to facilitate tracing the signal from in the various modes.

6.2.1 Digital RGB mode

The digital RGB mode is selected by a low from pin 34 of the system controller IC301 (chapter 7), or by a high from a computer at pin 33 of the RGB multiconnector. Either input causes pin 13 of IC806 to go low, and pin 12 to go high. This sets pin 11 of IC804 and pin 10 of IC801 high.

In the digital RGB mode, the A/D (analog/digital) input at pin 21 of the multiconnector is low. This makes pin 10 of IC804 low, connecting pins 2–15, and pin 11 high, connecting pins 13 and 14. The low A/D input also enables the A/D switches Q818–Q821 for the digital mode. The A/D low is inverted by IC806, turning on Q807, and providing pull-up voltages for the RGB AND-gate outputs at IC803. With all of the switching set, the digital RGB signals at the multiconnector are passed by the IC803 AND gates and the RGB digital buffers Q812, Q814, Q816, to the RGB switching in the Y/C processing stages.

If audio-select pin 34 of the multiconnector is made high, pin 9 of IC801 goes high, and the paths between pins 3 and 4 and 11 and 13 are completed. This allows right and left external audio at pins 20/24 of the multiconnector to be applied at the AUDIO MONITOR OUTPUT jacks, as well as to the audio-switching circuits, after amplification by IC807. In the digital RGB mode, vertical and horizontal sync from the multiconnector (pin 31) is buffered by IC803 and applied through pins 2/15 and 13/14 of IC804 to sync-output buffer Q833.

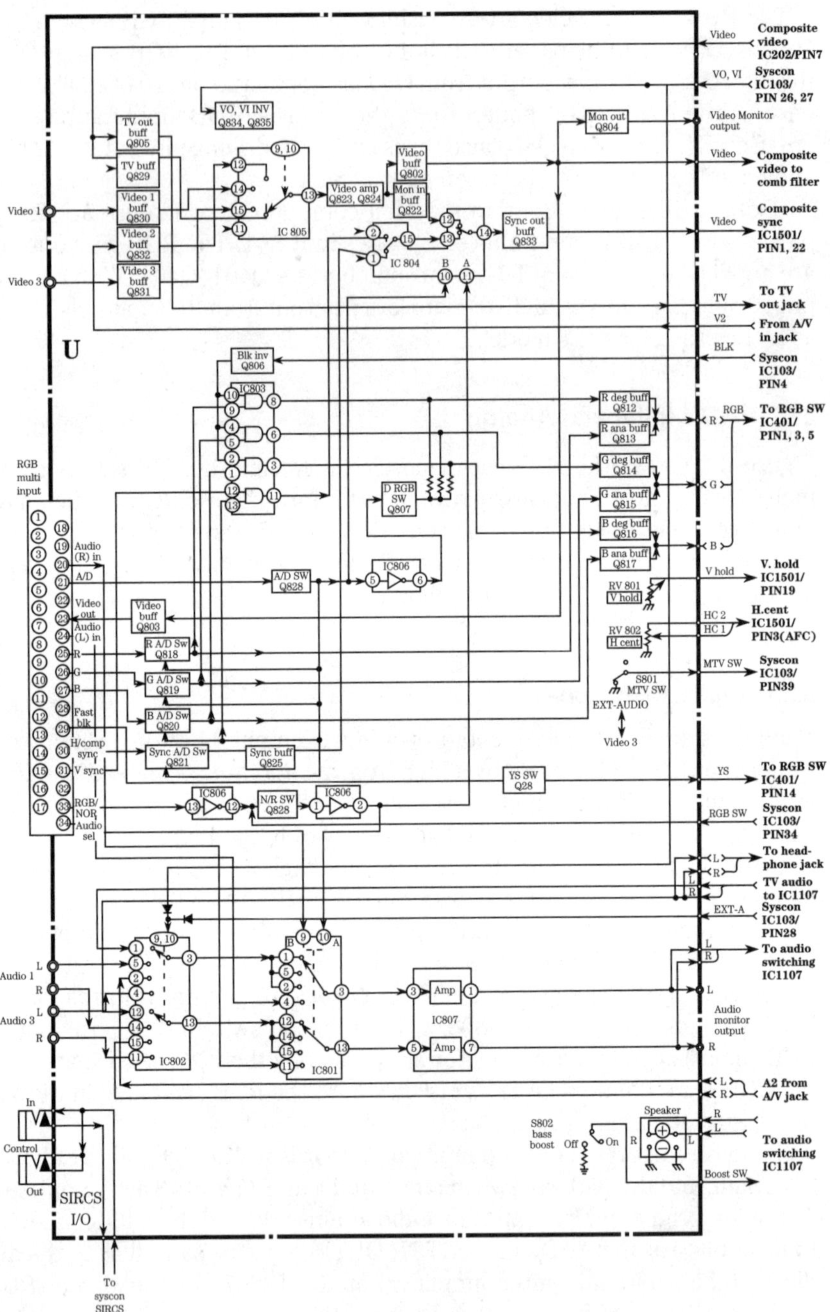

Figure 6.2 RGB digital/analog interface circuits (B).

TABLE 6.1 RGB Digital/Analog Logic

Sig	IC/Pin	TV	V1	V2	V3	Ext. A
V0	103/26	0	1	0	1	0
V1	103/27	0	0	1	1	0
V0	805/96	1	0	1	0	1
V1	805/10	1	1	0	0	1
V0	802/10	0	0	1	1	1
V1	802/9	0	1	0	1	1
		TV	Dig RGB	Ana RGB		
B	804/10	1	0	1		
A	804/11	0	1	1		

6.2.2 Analog RGB mode

The RGB switching for the analog mode is similar to that of the digital mode, except that multiconnector pin 21 (A/D) goes high. This turns off Q807, inhibiting the IC803 AND-gate outputs. Pin 10 of IC804 is made high, connecting pins 1/15, and setting A/D switches Q818–Q821 to the analog mode. The analog RGB signals are then processed by the analog RGB buffers Q813, Q815, and Q817. The composite sync input at pin 30 passes Q821, Q825, IC304 pins 1/15 and 13/14 to the sync-output buffer Q833.

6.3 Recommended Troubleshooting Approach

Because of the circuits involved (amplifier, buffers, and IC switches) the obvious troubleshooting approach is signal tracing. This also involves making sure that the IC switches are closed to pass the corresponding signals.

As an example, assume that the circuit of Figure 6.1 is operated in the TV mode, and the FM-demodulated audio from pin 15 of IC201 (Figure 3.4) is not returned to pin 12. First check for audio at MPX OUT connector CNJ409 and at TV AUDIO OUT connector CNJ407. If the audio is missing at CNJ409, suspect buffer Q861. If the audio is absent at CNJ407, suspect buffer Q864.

Next, check for audio at pins 1/2 and 10/11 of IC881. If the audio is available at pin 2, but not at pin 10, suspect IC881. Before pulling IC881, check that pins 12 and 13 are both high. If not, check back to the switch-control source. In the TV mode, the input at Q883 is inverted to a high and applied to pin 12 of IC881. Pin 13 of IC881 is made high switch-transistor Q851. Finally, check for audio at MONITOR AUDIO OUT connector CNJ406. If there is audio at pin 10 of IC881, but not at CNJ406 or at pin 12 of IC201, suspect buffer Q882.

Now assume that the video mode is selected by pushing the front-panel VIDEO switch. The FM-demodulated (TV) audio might still be present at CNJ407 and CNJ409, but TV audio should not be present in the signal path to pin 12 of IC201, beyond pin 2 of IC881. If TV audio appears beyond pin 2 of IC881, check for a low at pin 13 of IC881, and that Q851 is turned off by +12 V from the VIDEO switch through D853.

To check the video-mode audio signal path, apply audio to VIDEO AUDIO IN connector CNJ408, or at the combination audio/video input jack located on the front panel. If the video-mode audio is absent at pin 4 of IC881, suspect Q862 and Q863. If the audio is present at pin 4, but absent at pin 3, check for a high at pin 5. The high (through D853) that turns Q851 off is also applied to pin 5 of IC881 in the video mode. If pin 5 is high, but pins 3 and 4 are not connected, suspect IC881.

If the video-mode audio is available at pin 3 of IC881, but not at pin 10, check for a high at pin 12. If the high is absent, suspect Q883 and check back to the switch-control source. If the high is present, but there is no audio at pin 10, suspect IC881. Finally, if there is audio at pin 10 of IC881, but not at CNJ406, suspect buffer Q882.

Chapter

7

System-Control Circuits

This chapter is devoted to TV-set circuits that provide control of such functions as servicing the keyboard (front-panel controls), the remote-control (chapter 9) input processing, and tuning operations (chapter 2), including band selection, tuning memory, and phase/frequency lock. In most present-day sets, such functions are under control of a microprocessor (also called a controller, microcontroller, system controller or syscon, etc.). In addition to the functions just mentioned, the microprocessor usually handles the on-screen display (chapter 9), power on/off (chapter 8), audio volume control (chapter 3), and picture-level control (chapter 4).

System-control circuits have several functions, and there are many circuit configurations. Thus, it is very difficult to present a typical configuration. However, most system-control circuits have certain inputs and outputs that can be monitored. If the inputs are normal, but one or more of the outputs are abnormal, the problem can be localized to a specific part or IC. If you are not familiar with the basic control-microprocessor functions, read *Lenk's Digital Handbook* (McGraw-Hill, 1993).

7.1 System-Control Basics

Figure 7.1 shows the basic system-control circuits for a typical TV set (Sony 13 inch). Figure 7.2 shows some additional details. The following sections describe the basic system-control functions and troubleshooting approach.

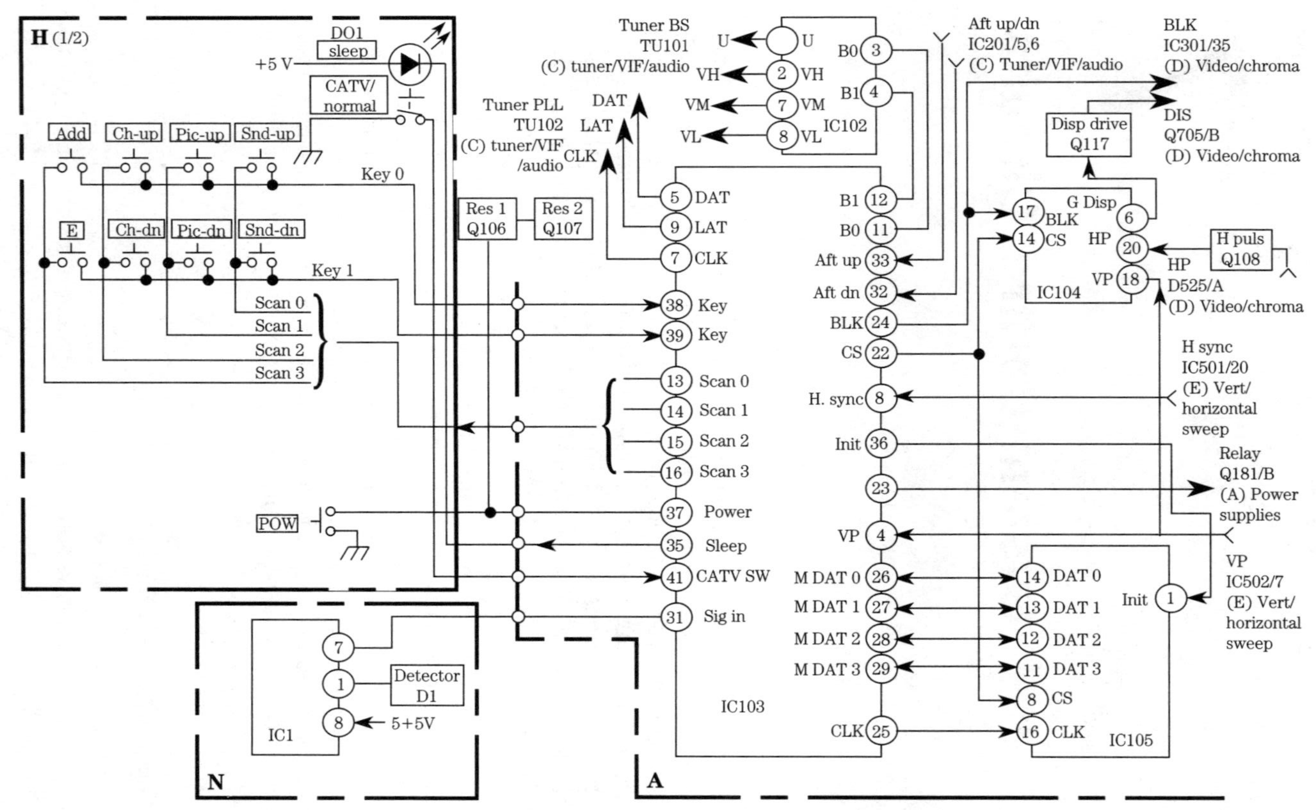

Figure 7.1 System-control circuits (B).

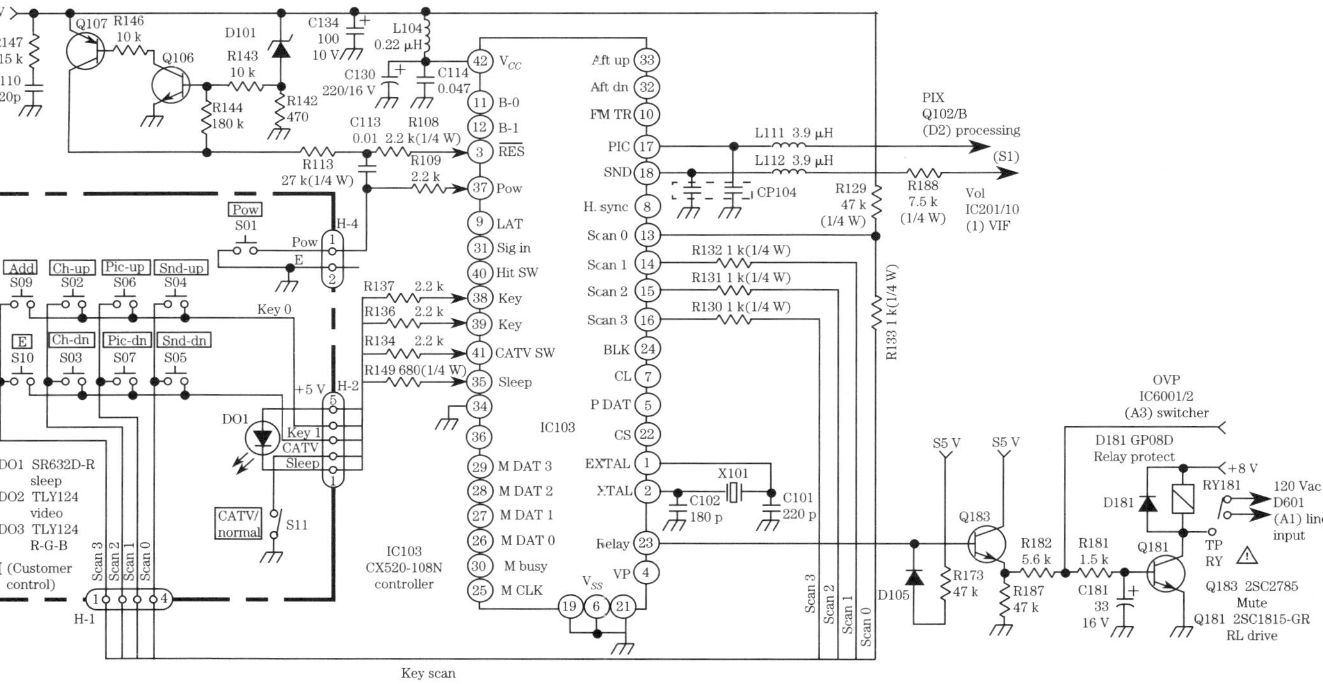

Figure 7.2 System-control circuit details (B1).

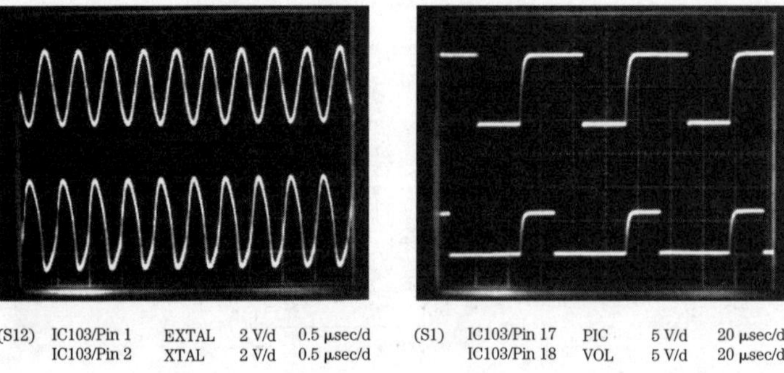

Figure 7.2 Continued.

7.1.1 Reset

The microprocessor or IC103 reset function is under control of Q106 and Q107. This circuit combination operates when power is initially applied, and each time the power is turned on and off.

When the set is first powered up or plugged in, Q106 and Q107 are normally off. As the standby 5-V or S5-V supply (chapter 8) comes up, IC103 is reset and (eventually) the breakover characteristics of a zener D101 are exceeded. This causes Q106/Q107 to turn on, and applies S5 V to the reset and input at pin 3 of IC103. The zener action of D101 ensures that the IC103 reset input remains low long enough to fully reset IC103.

During normal operation, the power on/off function is controlled by POWER switch S01 on the H board. When S01 is pressed, a low is applied to the power input at pin 37, instructing IC103 to energize the power-on relay (chapter 8). Pin 23 of IC103 goes high. At the same time, the low at S01 grounds one end of C113. This pulls reset pin 3 low for the charge time of C113, resetting IC103.

7.1.2 Master clock

The master clock signal is generated within IC103. Crystal X101 at pins 1 and 2 determine the operating frequency. For this set, the frequency is 2 MHz, with a sine wave of about 4 Vp-p at pins 1 and 2. The master-clock signal should be available once power is applied to IC103, and IC103 has been reset. This master-clock or system-clock signal synchronizes or coordinates all of the IC103 microprocessor functions. It is essential that the frequency, amplitude, and waveform be correct for the set to operate properly.

7.1.3 Keyboard service

The keyboard (or customer control) inputs shown in Figures 7.1 and 7.2 are serviced using the scan 0–3 output lines at pins 13–16 of IC103, and the key

0–1 input lines at pins 38 and 39. These input/output or I/O lines work with the keyboard matrix components located on the H board.

When any key is pressed, an oscillator within IC103 is turned on, generating the appropriate scanner signals (positive-going pulses). The scanner outputs at pins 13–16 of IC103 are connected to one contact of the specific keyboard switches. The remaining contact of each switch is connected to a specific input at pin 38 or 39.

When a given key is pressed, a scanner signal is applied to the corresponding input through the closed switch contacts. A specific scanner signal is applied to a specific input for each switch closure. Similarly, each keyboard switch is represented by a unique combination of scanner output and signal input. The scanner signal applied to a particular input is decoded by IC103, and produces a specific output representing the function of the corresponding keyboard switch.

As an example, assume that the CH UP (channel up) switch S02 is pressed. This applies the scan-2 signal to the key-0 input at pin 38, and causes IC103 to apply channel-up commands to the tuner band-select IC102 (from pins 11 and 12 of IC103) as described in chapter 2. When the tuner has reached the selected channel and is properly tuned to the channel, the AFT signals at pins 32 and 33 are removed, and IC103 stops sending channel commands to IC102. If the CH UP switch is held down, the channels will be scanned continuously, going from low to high.

Front-panel CATV/NORMAL switch S11 is not scanned. Instead, S11 provides a high or low (open or ground) directly to pin 41 of IC103. In some cases, the channel commands to the tuner band-select IC102 are different when cable is used (to select the additional channels available on cable).

When S11 is set CATV, pin 41 goes low, instructing IC103 to issue the corresponding channel commands to IC102. During NORMAL TV operation, S11 is open, and IC103 produces the corresponding band-select commands.

The sleep function is not operated by a front-panel switch, even though there is a SLEEP indicator D01. Instead, the sleep function is a fixed-time (one hour) shutoff operated by the remote control (chapter 9). When sleep is selected on the remote transmitter, pin 35 of IC103 goes low, turning on SLEEP LED D01. This indicates that the sleep function is in operation. One hour after the sleep function is selected, pin 23 of IC103 goes low, turning off the power-on relay RY181 (chapter 8).

The picture-level (PIC UP/PIC DN) and sound level (SND UP/SND DN) commands at pins 17 and 18 of IC103 are positive-going PWM (pulse width modulated) signals at a frequency of 15 kHz. These signals are filtered by the capacitor/coil combination, producing a corresponding dc voltage for picture control (at Q102, Figure 4.6) and sound control (at pin 10 of IC201, Figure 3.4).

The level of the dc control voltages varies, depending on the width of the

15-kHz PWM signals. As pulse width increases, the dc voltages increase, and vice versa. In turn, the width of the PWM signals is determined by the scanned picture and sound switches S04-S07. For example, to decrease the picture level, hold S07 until the desired level is reached.

7.1.4 Power on/off sequence

The power on/off functions are controlled by Q181/Q183 and relay RY181, as selected by the signal at pin 23 of IC103. As discussed in chapter 8, +8 V from the standby supply is applied to RT181. When a power-on command is received by IC103 (from front-panel switch S01, or the remote), pin 23 of IC103 goes high. This turns on Q181/Q183 and energizes RY181. The normally open contacts of RT181 close, applying 120 V to the line-input circuit at D601 (chapter 8). This turns the set on. If the set is on, a low pin at 37 of IC103 causes pin 23 to go low. This removes power and turns the set off.

The OVP (over-voltage protection) input from the switching supply (chapter 8) is applied to the base of Q181. Should an over-voltage condition occur, the OVP input goes low, turning Q181 off, and removing power to RY181. This removes power and turns the set off.

7.1.5 Recommended troubleshooting approach

Troubleshooting for the system-control circuits starts with the microprocessor. Although the system-control functions are often complex, troubleshooting is not necessarily difficult, because you are dealing with basic input/output and on/off functions. This is especially true when only one or two functions are absent or abnormal.

As an example, assume that sound or audio level does not increase when the front-panel SND UP switch S04 is pressed. First check if the sound level can be increased with the remote. If not, the problem is likely in IC103, or in the control circuits from pin 18 of IC103 to IC201. If the remote does have control over the sound level (both up and down), then the problem is between S04 and IC103. (The SND UP switch applies a scan-0 output to the key-0 input at pin 38.)

The problem becomes more complex when all of the system-control functions are absent or abnormal (or are erratic). Should this occur, again, start with the microprocessor. The following paragraphs describe some basic approaches.

The first step is to check all power and ground connections to the microprocessor (and any other IC in system control). Make certain to check all power and ground connections to each IC, because many ICs have more than one power and one ground connection. For example, as shown in Figure 7.2, IC103 requires standby 5 V at pin 42, and four grounds at pins 6, 19, 21, and 34.

With all of the power and ground connections confirmed, check that all ICs are properly reset. IC103 is reset when the power cord is first connected and standby power is applied. Pin 3 of IC103 should be low (near zero) when standby power is first applied, and then rise to about 5 V. The low causes circuits within IC103 to reset. After reset, when pin 3 rises to about 5 V, IC103 remains ready to perform the various microprocessor functions.

If the IC103 reset function is not as described, suspect Q106/Q107 and the associated parts. Notice that IC103 is not reset by temporary drops in the standby power (because of the charge on C110), but should be reset once each time the standby power is removed and reapplied.

As in the case of any microprocessor, if the reset pin is open or shorted to ground or to power, the IC will not reset (or will remain reset, low) when the power is switched on and off. So if you find a reset pin (or line) that is always high, always low, or apparently connected to nothing (floating), check the pin connections and reset circuits.

The next function to check is the clock signal at pins 1/2 of IC103. It is possible to measure the presence of a clock signal with a scope. However, a frequency counter provides the most accurate measurement. Obviously, if the clock is not operating, the microprocessor cannot function. On the other hand, if the clock is off frequency, the IC might appear to have a good clock, but the microprocessor functions might be impaired. (Notice that crystal-controlled clocks do not usually drift off frequency, but can go into some overtone frequency, typically a third overtone.)

Once you are certain that the system-control microprocessor has proper ground and power connections, and that all reset and clock signals are available, the final step is to monitor all input and output signals at the microprocessor IC103. Use the procedure previously described for sound- or audio-level control. That is, if the scan outputs from IC103 are available to the switches, but the scan signals do not appear at the appropriate key inputs, suspect the switches. If the correct key inputs are available, but the corresponding control outputs from IC103 are absent or abnormal, suspect IC103. If the control outputs from IC103 are good, but the function is not good, the problem is in the circuits external to IC103.

7.2 System Control (19 Inch)

Figure 7.3 shows the system-control circuits for a Sony 19-inch TV set. The following describes the control functions on a pin for pin basis. Notice that the front-panel key inputs are omitted, because such circuits are shown in Figure 2.6 and discussed in section 2.4.

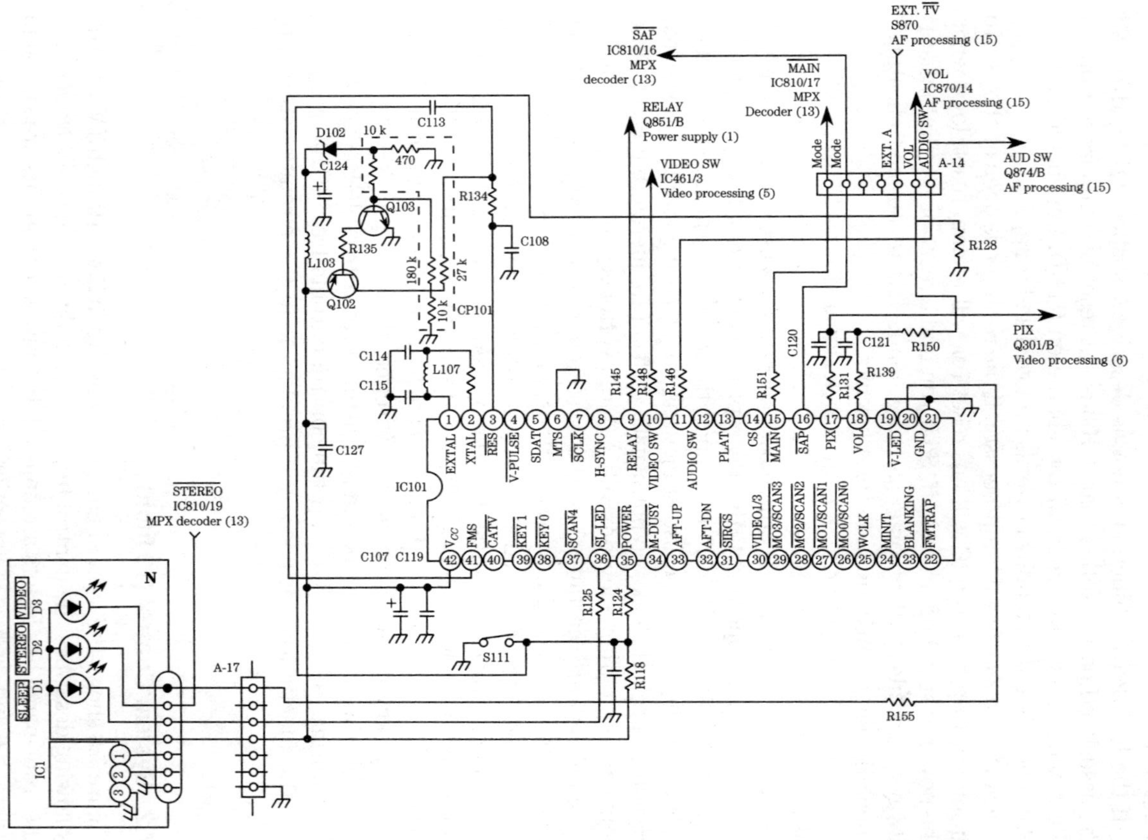

Figure 7.3 System-control circuits with remote input (2A).

PIN	SIGNAL	FUNCTION
1 and 2	Sysclk (System clock)	The system clock is generated within IC101, with the operating frequency set by the component at pins 1/2. The operating frequency is 2 MHz at 4 Vp-p.
3	Reset $\overline{\text{RES}}$	IC101 is reset on a negative transition at pin 3. Reset occurs in two modes. When the power switch S111 is closed, pin 3 is momentarily pulled low through the charging action of C113. This creates a sharp negative pulse at pin 3, and resets IC101. The other reset mode occurs when the set is initially plugged in. As the S5-V line comes up, pin 3 is held low long enough for IC101 to be reset. C124 introduces a short delay in the turnon of Q102/Q103 to ensure that the S5-V line is completely up before IC101 starts to operate. Zener D102 sets the turn-on threshold for Q103.
9	RELAY	The main power relay for this set is controlled by this output. Pin 9 goes high on command to energize the power relay (chapter 8).
10	VIDEO SW	If the front-panel TV/VIDEO switch is pushed, pin 10 goes high to select external video (if pin 41 is high) and to perform switching as described in chapter 4 (Figure 4.9).
11	AUDIO SW	This output goes high to select external audio when the front-panel TV/VIDEO switch is pushed (if pin 41 is low) and to perform switching as described in chapter 10.
15	$\overline{\text{MAIN}}$	The audio has three modes of operation selected by repeated pressing of the MTS switch. Pin 15 outputs a low for MAIN or BOTH audio modes to the MPX decoder (chapter 10).
16	$\overline{\text{SAP}}$	The third audio mode SAP is also selected by pressing MTS. Pin 16 outputs a low for SAP or BOTH audio modes to the MPX decoder (chapter 10).
17	PIX	This output is a PWM signal of 15 kHz, filtered by the RC network to produce a picture-control dc voltage for the video-processing circuits (Figure 4.10).
18	VOL	This output is a PWM signal of 15 kHz, filtered by the RC network to produce a volume-control dc voltage for the audio-processing (chapter 10).

20	$\overline{\text{V-LED}}$	The front-panel VIDEO indicator D3 is turned on by a low at pin 20 when external video is selected (if pin 41 is high).
35	$\overline{\text{POWER}}$	Pin 35 goes low when power switch S111 is closed. This instructs IC101 to turn the set on (if it is off) through operation of the main power relay (chapter 8). If the set is on, a low at pin 35 turns the set off (opens the main power relay).
36	$\overline{\text{S-LED}}$	The front-panel SLEEP indicator D1 is turned on by a low at pin 20 if the sleep mode is selected. Sleep is a fixed-timed (one hour) turnoff mode selected by the remote.
	STEREO	The front-panel STEREO indicator D2 is turned on by a low from the MPX decoder when a stereo pilot signal is detected (chapter 10).
41	FMS	This input pin is controlled by the front-panel EXT/TV switch S870 (chapter 10). Pin 41 is high for external, low for TV, and controls the operating state of pins 10, 11, and 20.

7.3 System Control (25 Inch)

Figure 7.4 shows the system-control circuits for a Sony 25-inch TV set. The following describes the control functions on a pin for pin basis. Notice that there is a display-control microprocessor IC101 in addition to the system-control IC103. These two microprocessors have separate clocks, as well as separate power, ground, and reset connections. The recommended troubleshooting approach described in section 7.1 applies to both IC101 and IC103.

PIN	SIGNAL	FUNCTION
1 and 2	Clock XTAL EXTAL	The system clock is generated within IC103 at a frequency of 2 MHz as set by the crystal.
3	RESET	IC103 is reset on a negative transition at pin 3 (as described in section 7.2) by reset circuits Q106/Q107.
4	V-PULSE	All tuning and display-data transfers occur during the vertical-retrace period. IC103 monitors V-pulse (which originates from the vertical oscillator Figure 5.14) to determine the retrace period.

5	SDAT	This is the serial data bus for sending data to the PLL control unit and to the display controller IC101.
7	SCLK	This is the synchronizing signal for the serial data transmitted on the data line at pin 5.
8	HSYNC	This signal is developed from the composite video to inform IC103 that a tuning operation is successful, and a station is being received (similar to pin 34 of IC1 in Figure 2.4). IC103 responds by monitoring AFT UP/DOWN at pins 32/33 to fine-tune the station.
9	PLAT	This is the tuning PLL latch signal that latches data from pin 5 into the PLL. PLAT occurs immediately after valid PLL data bits are transmitted on pin 5.
10	CS	This chip-select signal enables IC101 to accept data from pin 5 of IC103, and occurs only when valid data bits are on the SDAT line at pin 5.
11 and 12	DCOL 1, 2	The logic state of pins 11 and 12 determines the operating state of the Trinitone circuits (chapter 4, Figure 4.13). The truth table shows the logic state required at pins 11 and 12 to activate white, blue, or red dynamic color operation.
13 and 14	BAND 0,1	The logic state at pins 13 and 14 selects the tuning band (similar to pins 1 and 2 of IC1 in Figure 2.4), as shown by the truth table.
15	FM TRAP	This output switches in the FM trap when cable channels 14, 15, 16, and 17 are being tuned. These channels are close to the FM broadcast band, and the FM trap prevents interference. The output at pin 15 is present only when there is an input at pin 40, indicating that cable has been selected.
16	RELAY	This output controls the power relay to turn the set on and off, and occurs in response to a power input at pin 41.
24	BLANK	This blanking output occurs during channel change to mute the audio and video between channels, producing a black screen during change.
25	LED	This output turns on the SLEEP indicator LED when the sleep function is selected.

26 and 27	VIDEO 0, 1	The logic state of pins 26 and 27 determines the external video input selected, as shown by the truth table. When the back-panel VID3/EXT A switch is set to EXT A, then pins 26 and 27 assume the TV state, and pin 28 (external audio) selects the external audio inputs for simulcast programs.
28	EXTA	This output to the interface circuits selects external audio instead of normal TV audio.
29	ANT	The antenna system has an AUX input, controlled by pin 29, that allows selection of a cable converter box for premium channels.
31	SIRCS	The SIRCS signal is input here, either from the remote or the key encoder (Figure 7.5), and then decoded by IC103 to produce the corresponding commands and function.
32 and 33	AFT UP/DN	The AFT UP/DN signals from the VIF stages are monitored, and produce corresponding fine-tune commands to the tuning PLL (similar to pins 35 and 36 of IC1 in Figure 2.4).
34	RGB	This output is to the RGB interface circuits (chapter 6), and selects external or internal RGB.
36 and 37	MS 0, 1	The logic state of these pins sets the mode of operation for the stereo decoder (chapter 10) in accordance with the truth table.
39	MTV	Back-panel AUD/VID3 switch sets this input for selecting external audio, through pin 28, for simulcast programs.
40	CATV	This input is set low by a front-panel customer switch when cable is in use. The input instructs the tuner PLL controller to tune cable channels, and to activate pin 15 (FM trap) if necessary.
41	POWER	The front-panel power switch sets pin 41 low to turn the set on or off, through operation of the relay output at pin 16.

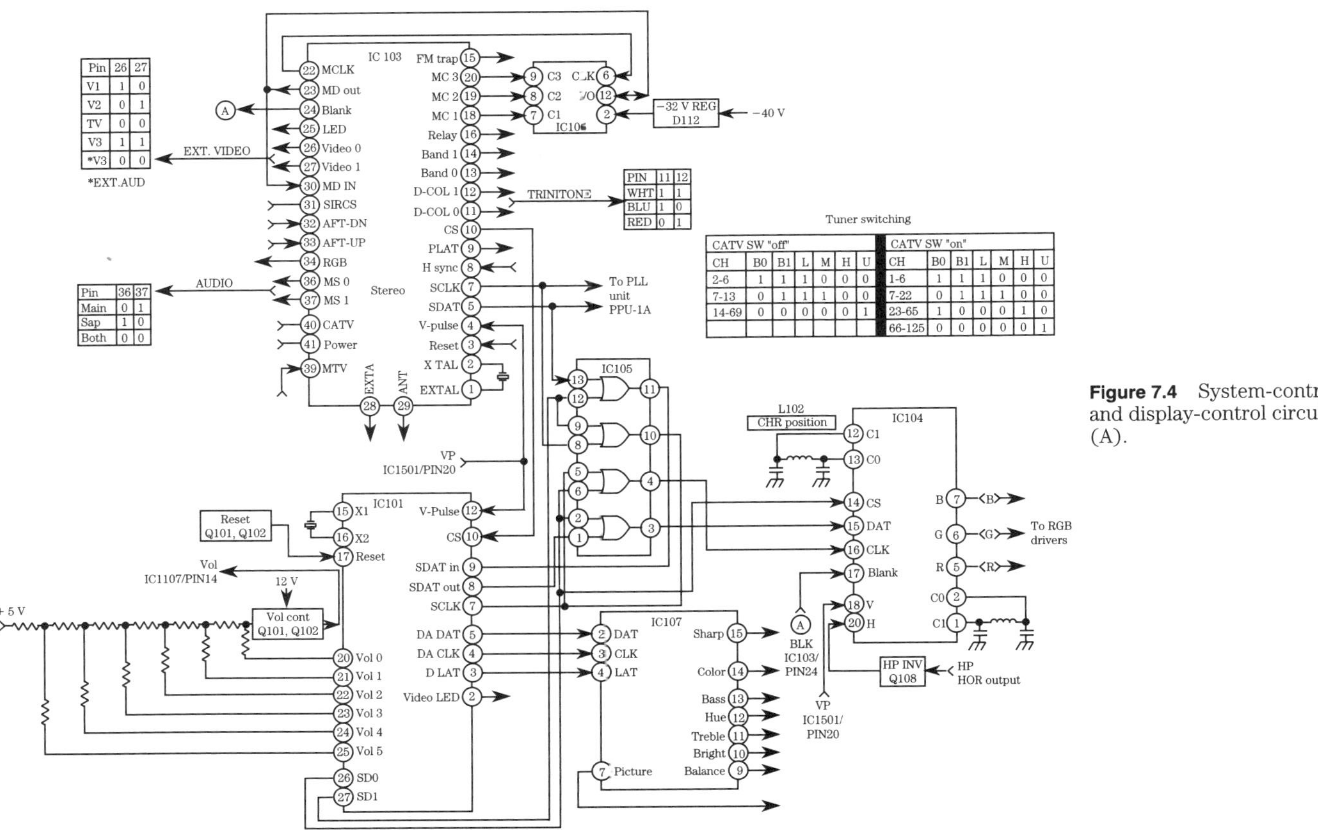

Figure 7.4 System-control and display-control circuits (A).

7.4 Key Encoder

Figure 7.5 shows the key-encoder circuits for a Sony 25-inch TV set. These circuits convert key closures to the standard Sony SIRCS signal for application to system-control microprocessor IC103 (at pin 31, Figure 7.4). The circuits are essentially the same as those used in the hand-held remote transmitters (or commanders as Sony calls them) described in chapter 9. The SIRCS switch transistor Q1301 allows IC103 to access both SIRCS signal sources (remote or front-panel key encoder).

In remote operation, the transmitted SIRCS signal is detected and processed by preamp IC1. The detected SIRCS signal is then applied to pin 31 of IC103 through Q1302/Q1303, an external control jack, and Q1301.

In front-panel operation, the standby 5-V power is dropped to 3 V for the key-encoder circuits. The clock at pins 7 and 8 of key-encoder microprocessor IC301 operates at a frequency of 450 kHz. IC301 has a standby mode in that the clock runs only when a key is closed.

The keyscan outputs K03–K06 are positive-going pulses. Key closures are detected through isolation diodes at the keyscan inputs K11–K18. When a key closure is detected and encoded to the appropriate SIRCS command, the indicator output at pin 6 of IC301 goes low, enabling the SIRCS switch S1301 to pass the SIRCS signal to IC103. When this occurs, the remote SIRCS signal from IC1 is inhibited through action of Q1302 and Q1303.

Notice that the K03 keyscan output (pin 11) is also applied to the connector on the T4 board. Key closures from foot-switches on an optional TV stand are applied on the key input lines K11–K14 and K16. These inputs provide for foot-switch control of power on/off, channel up, channel down, volume up, and volume down functions.

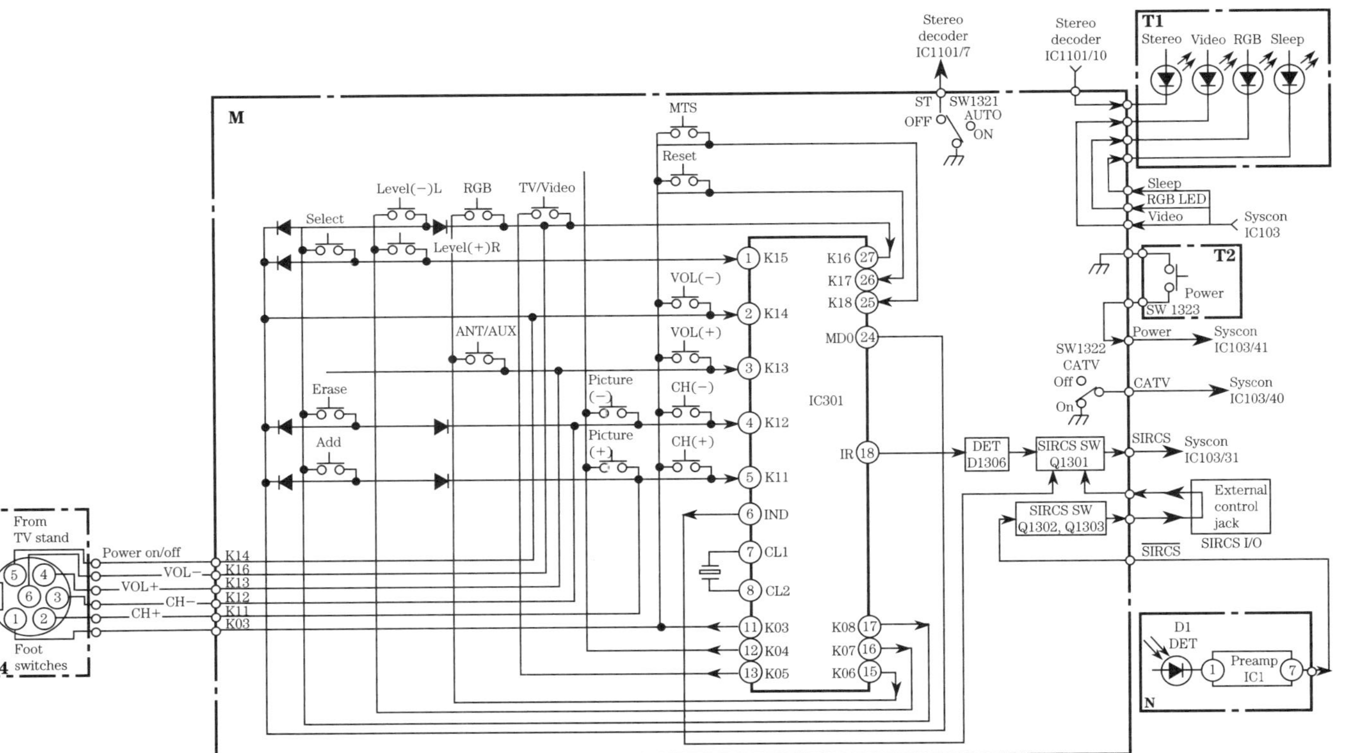

Figure 7.5 Key encoder circuits (C).

Chapter

8

Low-Voltage Power-Supply Circuits

This chapter is devoted to TV-set circuits that provide low-voltage power to the remaining circuits in the set. High voltages for operating the picture tube or CRT are discussed in chapter 5.

8.1 Low-Voltage Power-Supply Basics

Figure 8.1 shows the power-supply circuits for a portable black-and-white TV. The circuit consists essentially of a full-wave bridge rectifier CR1, followed by a discrete (zener CR2 and transistor Q1 through Q3) regulator. As with many portables, the set can be operated with self-contained rechargeable batteries. The set can also be operated from an auto battery (12 V) using an adapter plug connector to the cigarette lighter.

The circuit provides direct current at about 10 to 12 V. This voltage is sufficient for all circuits in the set except for the high voltage that is required by the CRT. As discussed in chapter 5, high voltages are supplied by the FBT and horizontal circuits.

The line voltage is dropped to about 12 V by T1 and is rectified by CR1. The output from CR1 is regulated by Q1–Q3, as well as CR2, and is distributed to three separate circuit branches, each at a different voltage level. Operation of the regulator is standard. The emitter of Q1 is held constant by CR2, whereas the base of Q1 depends on the output voltage. Any variations in output voltage (resulting from changes in input voltage or variations in load) change the base voltage in relation to the emitter voltage. These

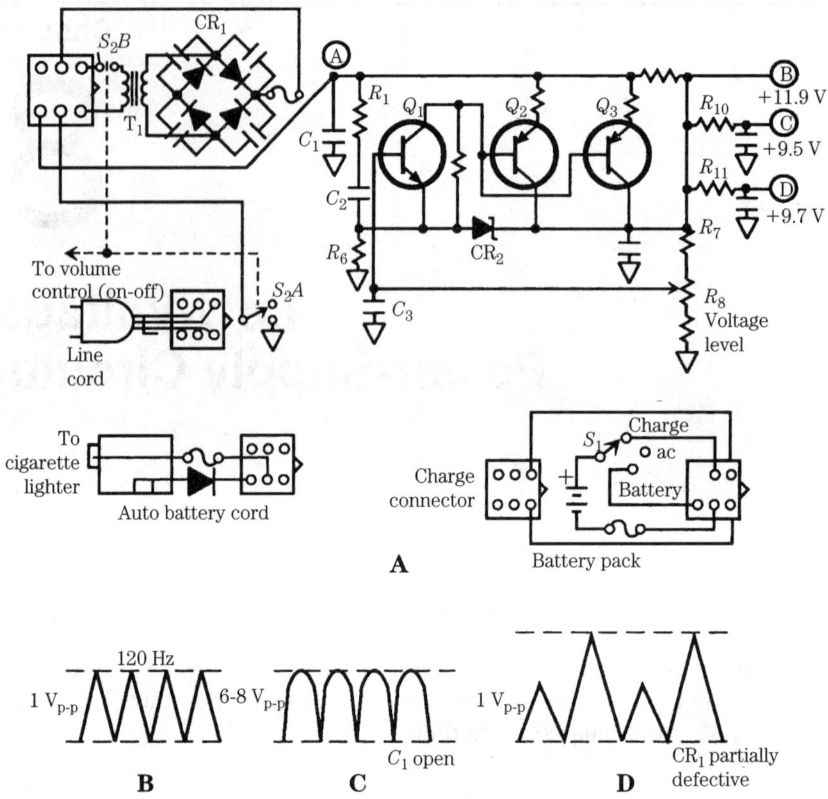

Figure 8.1 Low-voltage power supply for portable TV.

changes appear as a variation in Q1 collector voltage, and consequently in Q2 and Q3 base voltage.

Transistors Q2 and Q3 are connected in parallel with the output from CR1 and the load. Thus, Q2 and Q3 act as variable resistors to offset any changes in output voltage. The level of output voltage is set by R8.

When the set is to be operated on battery power, switch S1 is set to battery, the ac power plug is removed, and the battery-pack cord is plugged in. When the battery is to be recharged, S1 is set to charge, and the power cord is plugged into a special connector on the battery pack. Although there is no standardization, rechargeable batteries provide about 4 to 6 hours of operation and require about 8 to 12 hours of recharging. When the set is to be operated from an auto cigarette lighter, the ac power plug is removed and the auto battery cord is plugged in.

8.1.1 Recommended troubleshooting approach

If the symptoms indicate a possible defect in the low-voltage supply, the obvious approach is to measure the dc voltage. If there are many branches (three branches in Figure 8.1), measure the voltage on each branch. If any of the branches are open (say an open R10 or R11), the voltage on the other branches might or might not be affected. However, if any of the branches are shorted, the remaining branches are probably affected (the output voltage is lowered).

Current measurements. It is not practical to measure currents in present-day TV circuits where virtually all parts are mounted on PC boards. Likewise it is not practical to disconnect each output branch, in turn, until a short or other defect is found. Resistance measurements are far more practical.

Resistance measurements. If one or more output voltages appear to be abnormal, and it is not practical to measure the corresponding current, remove the power and measure resistance in each branch. Compare actual resistance against those in the service literature, if available. If you have no idea of the correct resistance, look for obvious low resistance (a complete short or resistance of only a few ohms).

Isolation transformer and substitute supply. Always use an isolation transformer when checking any solid-state circuit, discrete or IC. As a convenience, have a 12-Vdc supply to substitute for the low-voltage supply circuit in the set. If practical have a variable dc supply to match the various voltages found in present-day TV sets. Ideally, the substitute supply should have a voltmeter and ammeter to monitor power applied to the set.

Battery operation. If the set can be operated on batteries, switch to battery operation and see if the trouble is cleared. If so, the problem is definitely localized to the low-voltage supply. Likewise, if the set can be operated from an auto battery or similar arrangement, switch to that mode and check operation.

Zener replacement. If it becomes necessary to replace a zener, or any IC power package (linear or switching), always use an exact replacement. Some technicians replace zeners and power ICs with a slightly different voltage value or type, and then attempt to compensate by adjusting the regulator circuit. This might work.

Scopes. Do not overlook the use of a scope in troubleshooting power supplies. Even though you are dealing with dc voltages, there is always some ripple present. The ripple frequency and waveform produced by the ripple can help in localizing troubles, as discussed next.

8.1.2 Typical troubles

The following sections discuss symptoms that could be caused by defects in the low-voltage supply.

No sound and no picture raster. When there is no raster on the CRT, and no sound, it is likely that the low-voltage supply is totally inoperative or that it is producing a very low voltage. If the circuits are producing a voltage about 50 percent of normal (say 6 V in a 12-V system), there is some sound, even though the raster might be absent. The most likely defects are filter-capacitor leakage, a defect in the regulator (completely off), or a short in the output line.

Start the troubleshooting process by making voltage measurements at test points A, B, C, and D (or their equivalents in the set you are servicing). If the voltage is absent at all test points, check all fuses and switches as well as CR1 and T1 (right after you make sure the set is plugged in). If the voltage at B, C, and D is absent or very low but the voltage at A is high, suspect the regulator. If all of the voltages are low, look for a short in one or more of the output lines.

To check the regulator, first make sure that all three transistors are forward biased. Typically, the base of Q1 is about 2 V, with an emitter voltage 1.5 to 1.8 V. The bases of Q2/Q3 are about 11 V, with the emitters at 11.5 V. In any event, all three transistors must be forward-biased for the regulator to operate. Notice that it is possible for Q1 to cut off while Q2 and Q3 are forward biased. However, such a problem quickly points to a fault in Q1 or the associated parts.

A common fault in discrete-component regulators (which still exist even in this day of IC regulator packages) is a base-emitter short in the current-carrying transistors Q2 and Q3. Because these transistors are in parallel, it might be difficult to tell which transistor is at fault. If necessary, disconnect each transistor and check it separately. If either Q2 or Q3 has a base-emitter short, the regulator IC shuts off, and both the base and emitter are high.

No sound, no picture raster, and transformer buzzing. When there is no sound or picture, with transformer buzzing, there is probably an excessive current being drawn in the supply. The most likely defects are a short circuit ahead of the regulator (test point A, for example), shorted rectifiers, or shorted turns in the transformer.

Again, start by making voltage measurements at points A, B, C, and D. Notice that prolonged excessive current causes one or more fuses to blow. Thus, with these symptoms, the assumption must be an excessive current that is still below the rating of the fuses. This normally results in a very low voltage (but not an absent voltage).

Unlike the previous symptoms, the regulator is probably operating, so concentrate on shorts (particularly the rectifiers) and possibly shorted transformer turns. Both T1 and CR1 can be checked by substitution or a resistance test, whichever is convenient. If this does not localize the problem, check for short circuits in each branch of the output. Also look for overheated parts, such as the transformer, or burned PC wiring.

Distorted sound and no raster. These symptoms are similar to those previously described, except that there is some sound (often with intercarrier buzz). Start by making voltage measurements at test points A, B, C, and D. Then measure the amplitude and frequency of the waveforms at the same test points, particularly at the bridge output (test point A) and line output (test point B). An analysis of the waveforms can often pinpoint trouble immediately.

The normal waveform at the bridge output is a near sawtooth at twice the line frequency (typically 120 Hz), as shown in Figure 8.1B. If the filter and regulator are operating properly, the 120-Hz signal is suppressed at the final output. However, there might be some line-frequency (60-Hz) ripple at the regulator output (all three branches).

If there is a strong 120-Hz waveform at the regulator output, this indicates that the filter and/or regulator are not suppressing the 120-Hz bridge output. The most likely cause is an open C1 or a regulator defect. C1 could be leaking, but excessive leakage or a short will blow the fuse.

The waveform at point A also indicates the condition of C1. If the waveform is a 120-Hz half sinewave similar to Figure 8.1C, rather than the sawtooth of Figure 8.1B, C1 is probably open. This is confirmed further if the waveform amplitude at test point A increases from a typical 1 V (or less) to several volts, as shown in Figure 8.1C.

The condition of CR1 is also indicated by the waveform at A. If the waveform is 60 Hz (line frequency), one-half of CR1 is defective (most likely shorted or open). If the waveform is not symmetrical (Figure 8.1D), one diode of CR1 is probably defective (most likely leaking).

Notice that if any of the diodes in CR1 are shorted, T1 will probably run very hot. Thus, a hot T1 does not necessarily indicate a bad T1. However, if the waveform at test point A is good (indicating a good CR1 and C1) but T1 is hot, the most likely problem is a defective T1 (possibly shorted turns).

Also notice that all four diodes in CR1 are paralleled with capacitors to protect the diodes in case of sudden changes that might exceed the breakdown voltage. If one of these capacitors is shorted, it can give the appearance of a shorted diode. If CR1 is a sealed package with self-contained capacitors, the entire package must be replaced.

If the capacitors can be replaced separately from the diodes, make sure that both the capacitor and diode are good before replacing either. For example, if the capacitor is open, the corresponding diode might be damaged. If the diode is replaced, the trouble is repeated unless the capacitor is also replaced.

Remember that an increase in load (say, because of a short or partial short between PC traces) causes an increase in ripple amplitude, even with a good filter and regulator, because a larger load (lower load resistance) causes faster discharge of the filter capacitor between peaks of the sine-wave pulses from the rectifier.

Picture pulling and excessive vertical height. Thus far, we have discussed symptoms and troubles that result from a low power supply output because of defective parts or shorts. It is possible that the power supply can produce a high output voltage, resulting in picture pulling (the raster is stretched vertically and is bent). Hum bars across the CRT screen are usually present with this condition.

Assuming that the trouble is definitely in the low-voltage supply, the most likely cause is in the regulator. A defect in the rectifier, filter, or transformer usually results in a low supply output. It is possible for the regulator to be cut off, causing the output to increase in voltage. The logical approach is to start the troubleshooting process by making voltage and waveform measurements at all test points.

If the voltage at test point B is nearly the same as at test point A (within about 0.5 V), the regulator is probably cut off. The voltage at test points C and D will also be high. The condition can be confirmed further if the waveform at test points B, C, and D is at 120 Hz (indicating that the regulator is not suppressing the 120-Hz bridge output). With the trouble definitely pinned down to the regulator, test all of the transistors (Q1–Q3) and CR2.

Low brightness and insufficient height. When the low-voltage supply drops below normal but is not really low, one of the first symptoms is low picture brightness. Picture height and width are also reduced, but height is generally reduced more. There are several possible causes for a small reduction in supply voltage. One of the bridge-rectifier diodes (or the related protective capacitor) can be shorted. Or the diode can be open, but this is less likely. Capacitor C1 can be leaking. There might be a partial short in one of the output lines, thus dropping the voltage at all outputs by a small amount.

Start by checking the voltages at test points B, C, and D. Try to correct a low-voltage condition at all outputs by adjustment of R8. It is possible that the low output is caused by component aging. Look for trouble in CR1 or C1 if R8 must be set to an extreme for correct voltage output.

Use the previously discussed method to check GR1 and C1. For example, a waveform similar to Figure 8.1C at test point A indicates an open C1. The waveform in Figure 8.1D indicates a partial defect in CR1. Remember that a poor solder connection or a break in PC wiring can simulate a defective (open) C1.

Insufficient height and width, weak sound, and snowy picture. These are also symptoms of below-normal voltage from the low-voltage supply. Again, start by checking voltages at test points B, C, and D, and try adjusting R8.

Also measure the waveform and voltage at test point A. If the indications are correct at A, but the voltages at B, C, and D are low (and cannot be adjusted), the regulator is at fault. This means that you must dig into the circuit, looking for such problems as short circuits (from solder splashes) and breaks in PC wiring. Let us consider a few examples.

When troubleshooting any circuit such as the regulator in Figure 8.1, voltage and resistance checks are generally more useful than scope checks. Likewise, capacitor failures are more likely than resistor failures, which is usually true in most TV circuits.

Assume that R7 is open (broken PC wiring, cracked resistor, etc.). This shows up in several ways. First, R8 has little or no control of the regulator circuit (little effect on output voltages). The base voltage of Q1 is abnormal, as is the resistance at test point B. However, the waveforms throughout the regulator circuit can remain normal (or so close to normal as to be unnoticed).

Now assume that C3 is shorted. This cuts off Q1 and provides a partial short across the output lines. Such a defect is generally easy to locate because the voltage at the base of Q1 (normally 2 V) is zero. Likewise the resistance from the base of Q1 to ground is zero.

When tracking troubles by means of voltage and resistance measurements, it is often necessary to check several components for one abnormal reading. For example, assume that the emitter of Q1 shows a high reading (say 3 V instead of 1.5 V). There are three logical suspects. First, C2 can be leaking. This applies a large voltage to the emitter of Q1 through R1. Next on the list is a defective R6 where the resistance has increased (say because of a partial break in the composition resistance material). Of course, CR2 is supposed to overcome minor changes in R6 resistance. However, a large change in R6 can prevent CR2 from maintaining the desired 1.5 V at the emitter of Q1. Finally, CR2 can be at fault, even though zeners are generally rugged.

8.2 Low-Voltage Supply (13 Inch)

Figure 8.2 shows the low-voltage supply circuits for a Sony 13-inch TV set, as well as the FBT high-voltage circuits, in basic block form. Figures 8-3 through 8-5 show the circuits in greater detail. The high-voltage circuits are discussed in chapter 5.

The relay-control signal from system control (chapter 7) is applied to the power relay RY131 through relay drivers. In turn, RY181 controls line ac to the main supply, rectifier D601, and to the degaussing coils. The ac power is applied to a standby transformer T901 (to generate +5 V and –36 V) at all times.

The rectified +135-V output from D601 is applied to a switching supply that generates +14-V and +115-V outputs. The switching supply also produces the OVP (over-voltage protection) signal that cuts off the supplies in the event of an over-voltage condition.

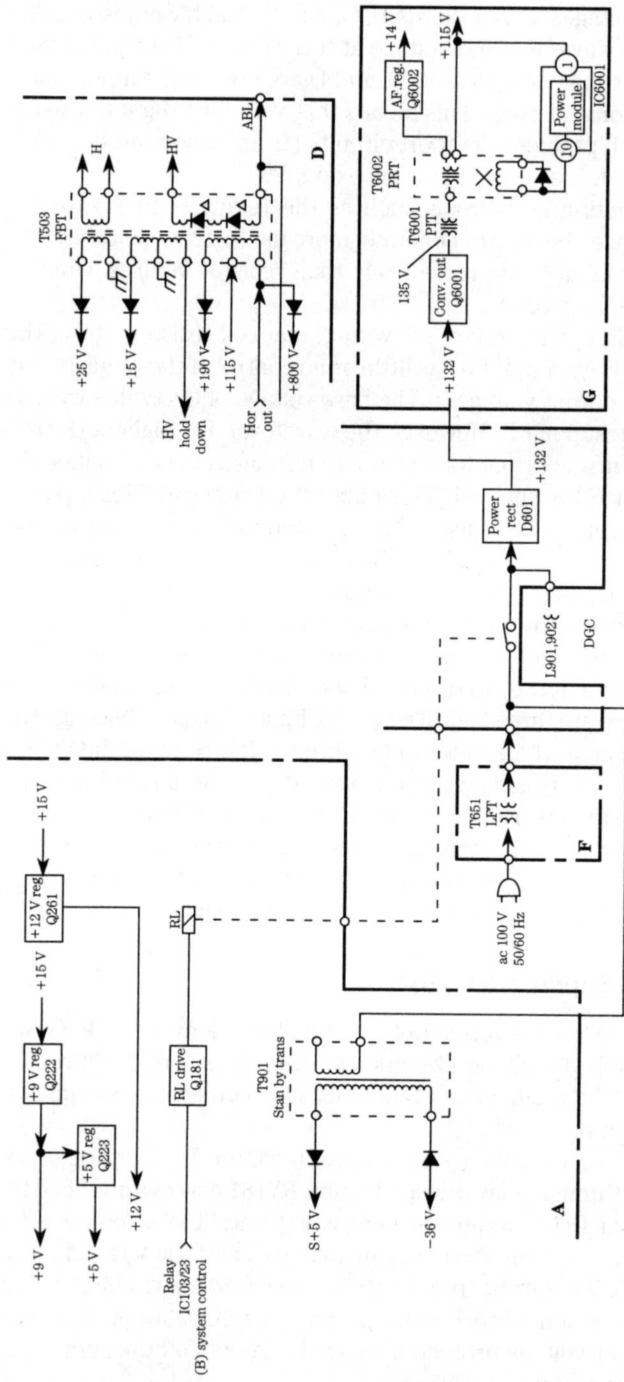

Figure 8.2 Power supply circuits (A).

8.2.1 Line input circuits

As shown in Figure 8.3, ac line power is applied to the standby supply (Figure 8.4) at all times when the line cord is plugged in. The ac power is applied to the degaussing coils L901/L902 through thermal switch THP601, and to main power rectifier D601, when RY181 is actuated by a power turn-on command from system control. The thermal characteristics of THP601 provide momentary degaussing of the CRT each time the set is turned on.

8.2.2 Auxiliary/standby power circuits

As shown in Figure 8.4, ac line power is applied to standby transformer T901. The output from one secondary of T901 is rectified by D182 and produces the standby –36 V for the tuning memory (chapter 2). The output from the other secondary of T901 is rectified by D183 and D184 to produce the standby 5-V (S5-V) supply. The 8 V is also used to power relay RY181.

Transistors Q222 and Q223 receive 15 V from the FBT and horizontal stages (chapter 5), and regulate the 15 V down to 5 V and 9 V. Transistor Q221 also receives 15 V from the FBT, and regulates the 15 V to 12 V.

The power-on command from system control is applied to relay RY181 through Q183/Q181. When actuated, the normally-open contacts of RY181 close, applying line power to the degaussing coils and main power supply (Figure 8.3), as described. The pulse produced by the power-on command is applied to Q251 through C259. This mutes the audio stages (Figure 3.5) for one-half to one second during power-on.

Over-voltage protection is provided by the switching supply (Figure 8.5) in the form of the OVP input to Q181. Should an over-voltage condition occur, the OVP signal goes low, turning Q181 off. As a result, relay RY181 is deenergized, and the contacts open to remove power from all but the standby circuits.

8.2.3 Switching supply

As shown in Figure 8.5, 135 V from the main supply is applied to power-input transformer T6001 and transistor Q6001. One secondary winding of T6001 and Q6001 form a 71-kHz oscillator. The oscillator signal is induced across T6001 windings and applied to power-regulator transformer T6002. The oscillator signal at R6007 is rectified by D6003, producing 17 V at series regulator Q6002. In turn, Q6002 regulates the 17 V down to 14 V. The output from the other winding of T6002 is rectified by D6002/D6007, producing 115 V, which is the primary B+ for the horizontal-output stages (Figure 5.9).

Regulation of the 14-V and 115-V lines is performed by the IC6001 module. Reference voltages are set across R6006, R6005, and R6004 at pins 8, 6, and 1, respectively. These reference voltages control the conduction of two transistors within IC6001. In turn, this controls the current through the

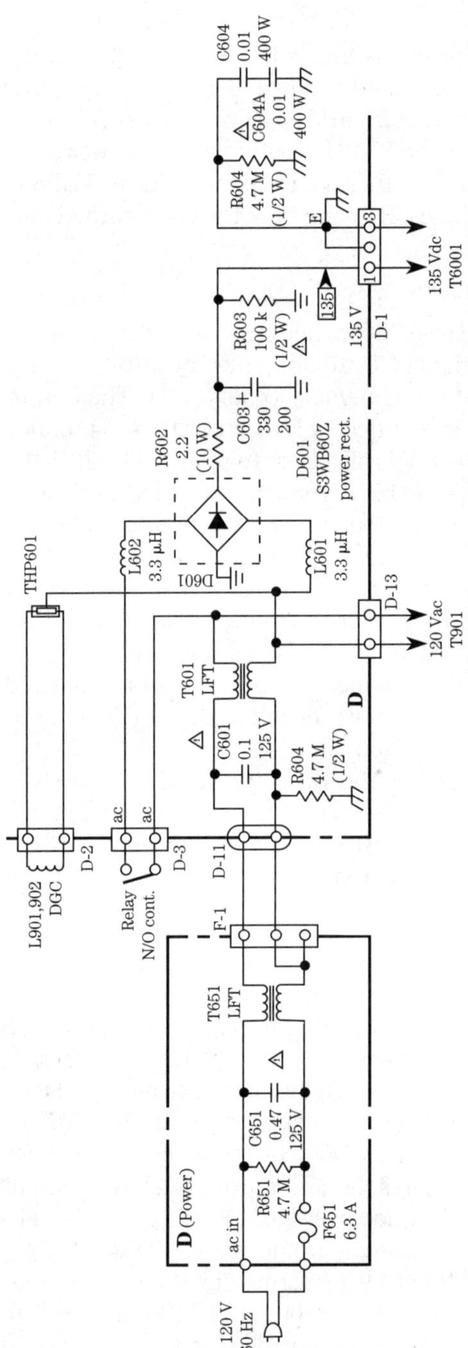

Figure 8.3 Line input circuits (A1).

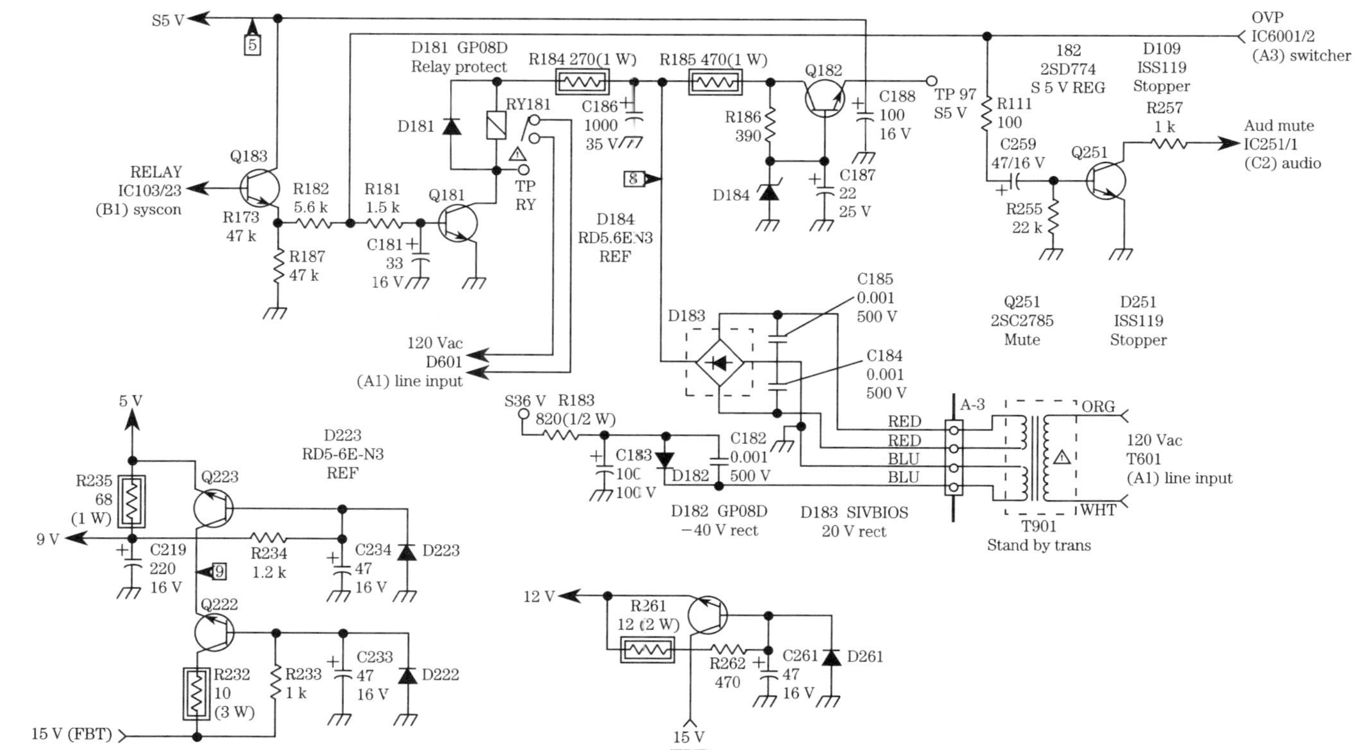

Figure 8.4 Auxiliary and standby circuits (A2).

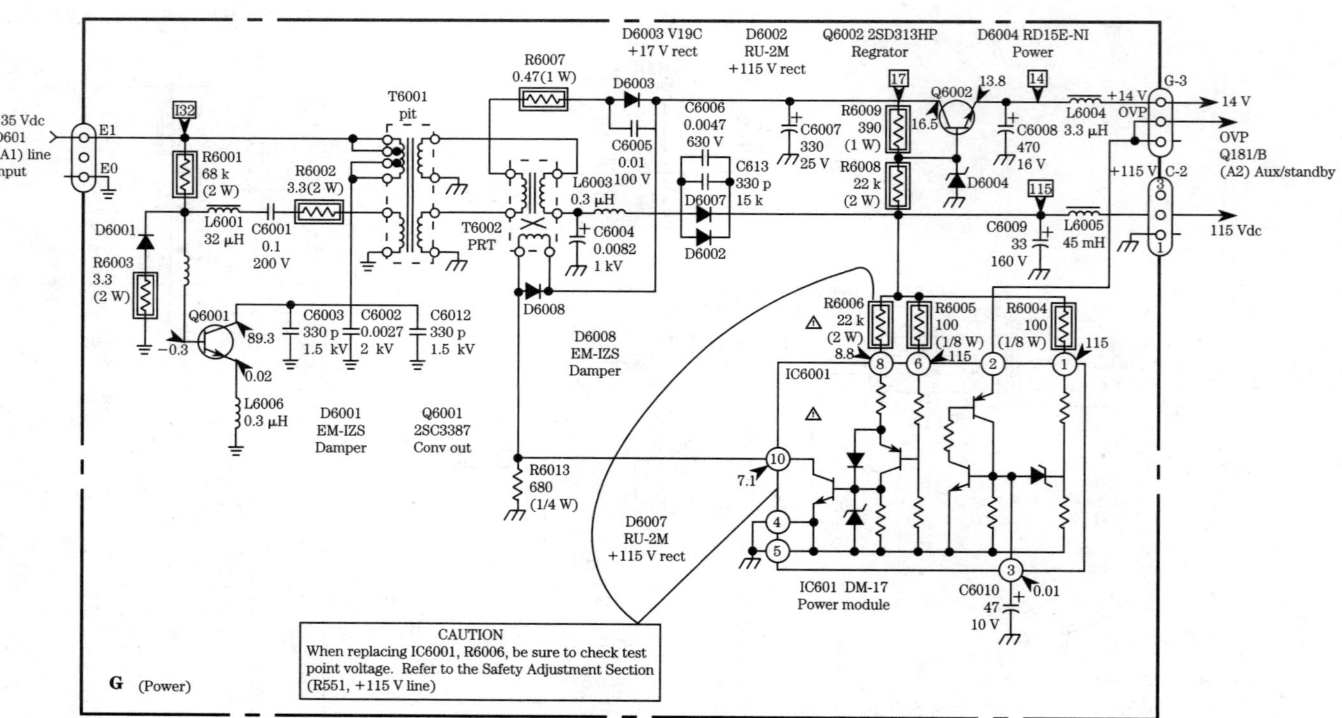

Figure 8.5 Switching supply circuits (A3).

control winding of T6002. If the output current of the 14-V or 115-V lines tends to increase (lowering the line voltage), the internal transistors of IC6001 conduct more. The current through pin 10 of IC6001 and the control winding of T6002 increases, lowering the winding inductance, and increasing the level of the line voltages to offset the initial drop in line voltage. The opposite occurs if the line voltages tend to increase.

The 115 V at pin 1 is applied across a zener within IC6001, setting the reference level of the remaining two internal transistors. Should an over-voltage condition occur, such that the breakdown voltage of the zener is exceeded, the internal transistors turn on, making pin 2 of IC6001 low. This low output is the OVP signal applied to Q181 to turn off the main power by opening RY181.

8.3 Low-Voltage Supply (19 Inch)

Figure 8-6 shows the low-voltage supply circuits for a Sony 19-inch TV set. The ac line power is applied to the standby supply transformer T901 at all times when the line cord is plugged in. The ac power is applied to the degaussing coils through thermal switch THP601, and to main-power rectifier D601, when power relay RL651 is actuated by a power turn-on command from system control through Q651. The thermal characteristics of THP601 provide momentary degaussing of the CRT each time the set is turned on.

The output from one secondary of T901 is half-wave rectified by D671, and produces the standby –38 V for the tuning memory (chapter 2). The output from the other secondary of T901 is full-wave rectified by D651, and regulated by D682, to produce the standby 5-V supply (used primarily by the system-control microprocessor and tuning memory). The power-control relay RL651 is powered by +20 V taken from the junction of R681/R682. The +20 V is also applied to other circuits within the set.

The rectified output from D601 is about 160 V. This is regulated down to 135 V by power regulator IC601. (Notice that the remaining circuits in IC601 are for an audio output stage not used in this particular model.) The reference for pin 1 of IC601 is set by R604, R605, and R608. The +135 V is taken from pin 6.

8.4 Low-Voltage Supply (25 Inch)

Figure 8.7 shows the low-voltage supply circuits for a Sony 25-inch TV set, in basic block form. Figure 8.8 shows some of the circuits in greater detail.

The relay-control signal from system control (chapter 7) is applied to the power relay RY601 through relay-drive transistor Q602. In turn, RY601 controls line ac to the main-power rectifiers D622/D623, and to the degaussing coils. The ac power is applied to a standby transformer T601 (to generate 18 V, RM5 V, and –40 V) at all times. The ac line voltage is

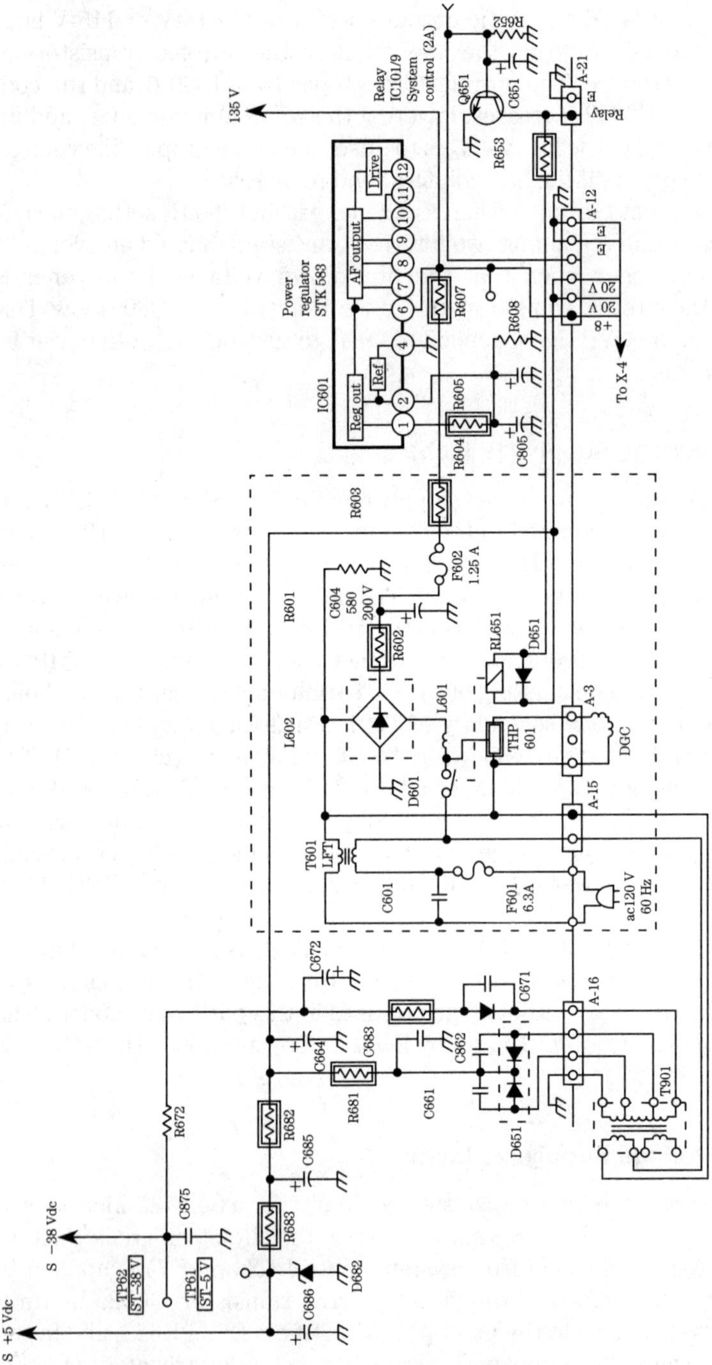

Figure 8.6 Low-voltage supply with regulator (1).

rectified by D622/D623, which act as a voltage doubler to develop 320 V for the switching supply.

The switching supply (Figure 8.8) consists of converter transistors Q651/Q652, the converter-drive transformer T651, the power-regulating transformer T652, power-interface transformer T653, and rectification circuits on the secondaries of T653 to provide +140 V, +16 V, +8 V, and +30 V audio B+.

Regulation is provided by part of IC651, which monitors the 140-V and 30-V lines. In turn, the output from IC601 controls the current through a control winding of T652 and thus maintains the output-voltage levels. The 140-V line is also monitored by another part of IC601. Should the 140-V line increase to about 150 V, IC601 produces an OVP output that removes drive to power relay RY601, and turns off the set.

8.4.1 Switching supply

As shown in Figure 8.8, 320 V from the main supply is applied to converter transistors Q651/Q652. The coupling of converter-drive transformer T651 is such that the base of Q651 and Q652 are 180° out of phase. The oscillator output of Q651/Q652 (at 40 kHz) is coupled through T651, power-regulator transformer T652 and the primary of power-input transformer T653.

The control winding of T652 is magnetically coupled to the primary winding. By controlling current in the control winding, the amount of converter signal reaching the primary of T653 can be controlled. This control function is used to maintain the dc output voltage levels, as well as to the three dc secondary winding of T653. The signal is then full-wave rectified to produce 140 V, and half-wave rectified to produce 16 V, 8 V, and 30 V.

The 140-V and 30-V lines are monitored by control module IC651 with reference voltages at pins 8 and 7, respectively. These reference voltages control conduction to transistors within IC651. In turn, this controls the current through the control winding of T652. If the output current of the 140-V or 30-V lines tends to increase (lowering the line voltage), the internal transistors of IC651 conduct more. The current through pin 10 of IC651 and the control winding of T651 increase the level of the line voltage to offset the initial drop in line voltage. The opposite occurs if the line voltages tend to increase.

The 140 V at pin 1 is applied across a zener within IC651, setting the reference level of the internal transistors. Should the 140-V line rise to about 150 V, such that the zener breakdown is exceeded, the internal transistors turn on, making pin 2 of IC651 low. This low is the OVP signal applied to relay-driver Q602 to turn off the main power by opening relay RY601.

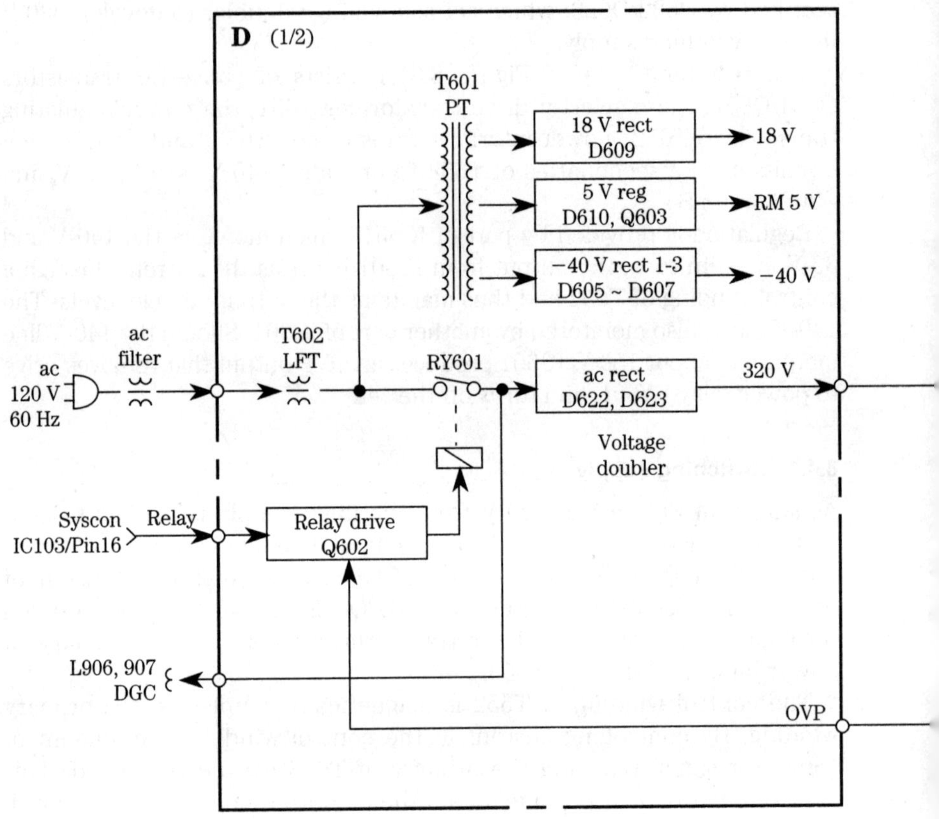

Figure 8.7 Power-supply circuits (D).

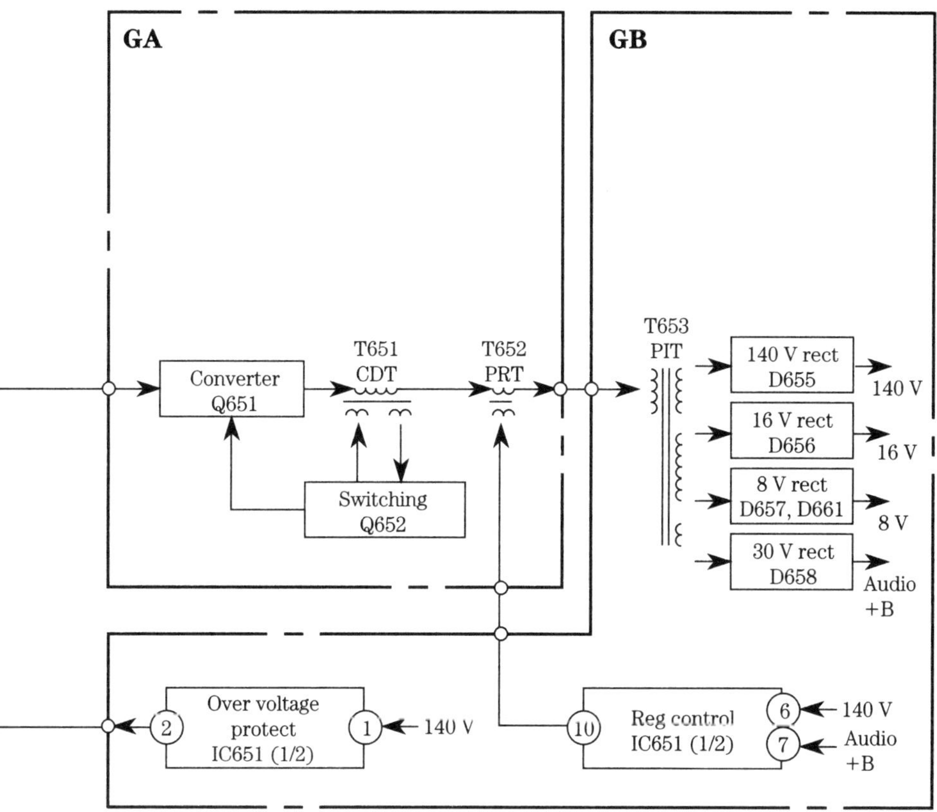
GA
GB
Converter
Q651
T651
CDT
T652
PRT
Switching
Q652
T653
PIT
140 V rect
D655
140 V
16 V rect
D656
16 V
8 V rect
D657, D661
8 V
30 V rect
D658
Audio
+B
Over voltage
protect
IC651 (1/2)
2
1
140 V
10
Reg control
IC651 (1/2)
6
7
140 V
Audio
+B

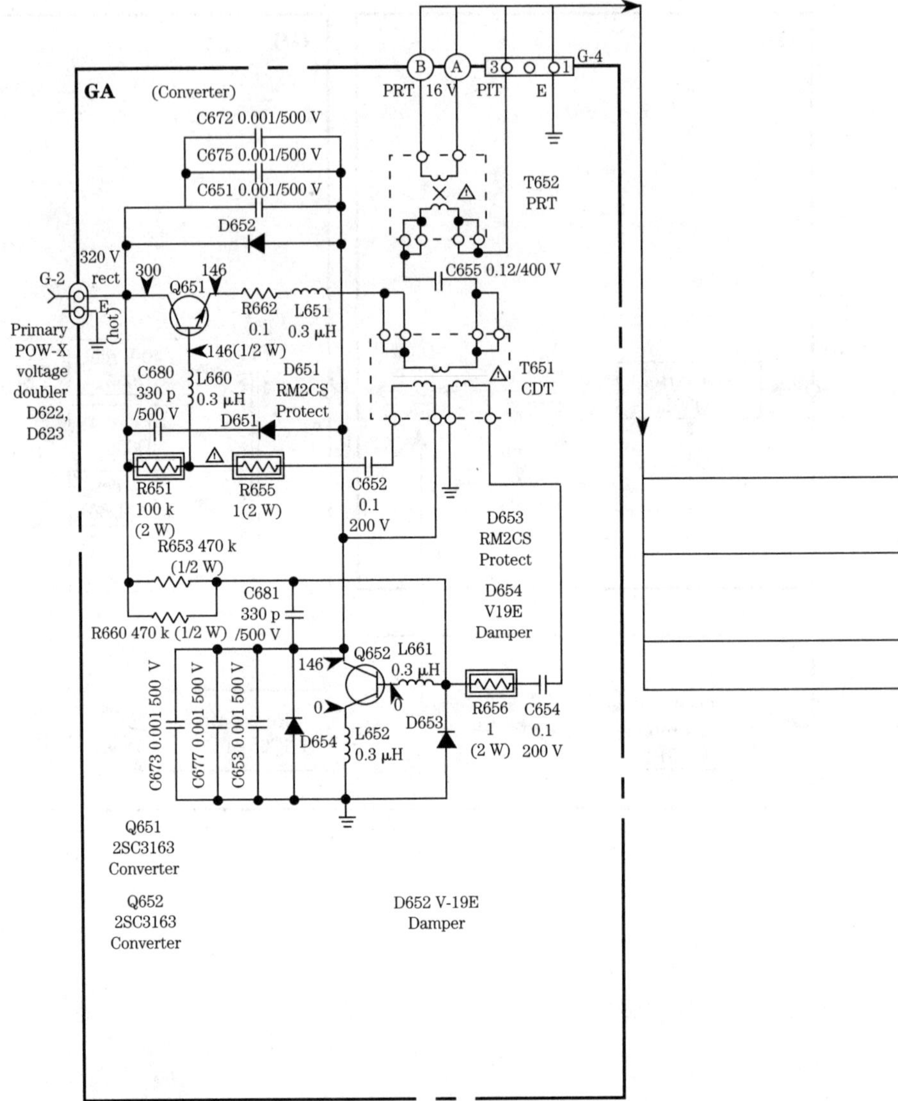

Figure 8.8 Power supply circuit details (E).

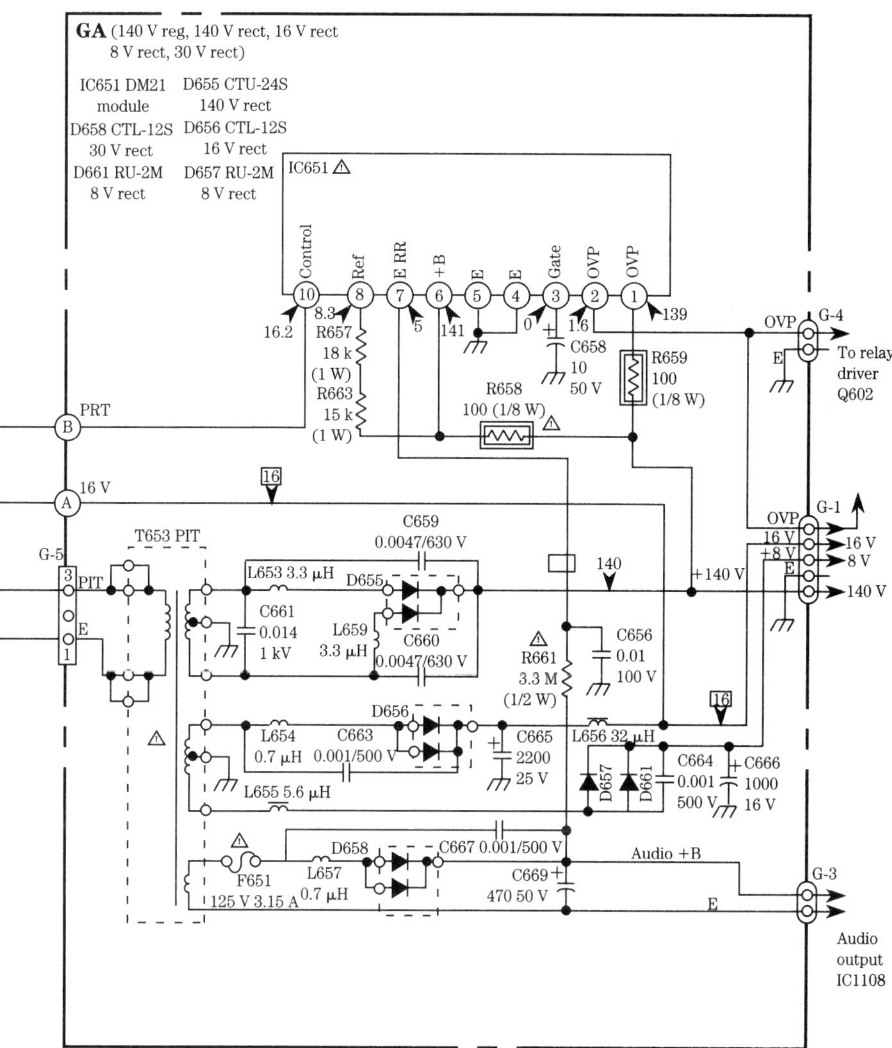
GA (140 V reg, 140 V rect, 16 V rect
8 V rect, 30 V rect)
IC651 DM21 module
D655 CTU-24S 140 V rect
D658 CTL-12S 30 V rect
D656 CTL-12S 16 V rect
D661 RU-2M 8 V rect
D657 RU-2M 8 V rect
IC651
Control
Ref
E RR
+B
E
E
Gate
OVP
OVP
16.2
8.3
5
141
0
1.6
139
R657 18 k (1 W)
R663 15 k (1 W)
R658 100 (1/8 W)
C658 10 50 V
R659 100 (1/8 W)
PRT
16 V
T653 PIT
G-5
PIT
E
C659 0.0047/630 V
L653 3.3 µH
D655
C661 0.014 1 kV
L659 3.3 µH
C660 0.0047/630 V
140
+140 V
R661 3.3 M (1/2 W)
C656 0.01 100 V
D656
L654 0.7 µH
C663 0.001/500 V
C665 2200 25 V
L656 32 µH
L655 5.6 µH
D657
D661
C664 0.001 500 V
C666 1000 16 V
F651 125 V 3.15 A
L657 0.7 µH
D658
C667 0.001/500 V
C669 470 50 V
Audio +B
E
OVP
G-4
To relay driver Q602
G-1
16 V
+8 V
140 V
8 V
G-3
Audio output IC1108

Chapter

9

On-Screen Display and Remote Circuits

This chapter is devoted to TV-set circuits that provide on-screen display (OSD) and remote-control operation. This includes circuits that provide for presetting of channel on/off times, blocking of channels, setting and display of day/time, and so on.

9.1 OSD Basics

Most present-day TV sets are provided with some form of OSD circuit. The on-screen channel display is a typical example. The OSD circuits use a character-generator IC to produce the desired numbers and letters. The IC receives sync signals from the same source as the CRT (so that the characters appear at a given position on the screen). Some OSD circuits include positioning controls. The characters to be displayed are determined by a microprocessor, which in turn receives commands from push-button controls. In some cases, the commands are hard-wired into the microprocessor.

9.1.1 Typical OSD circuit

Figure 9.1 shows the date and time OSD circuits for the electronic viewfinder (EVF) of a camcorder. These OSD circuits provide date and time information that appears on the EVF and that can be recorded on tape if desired. Although these OSD circuits are for a camcorder, the OSD circuits of TV sets are similar. The remaining sections of this chapter describe a cross-

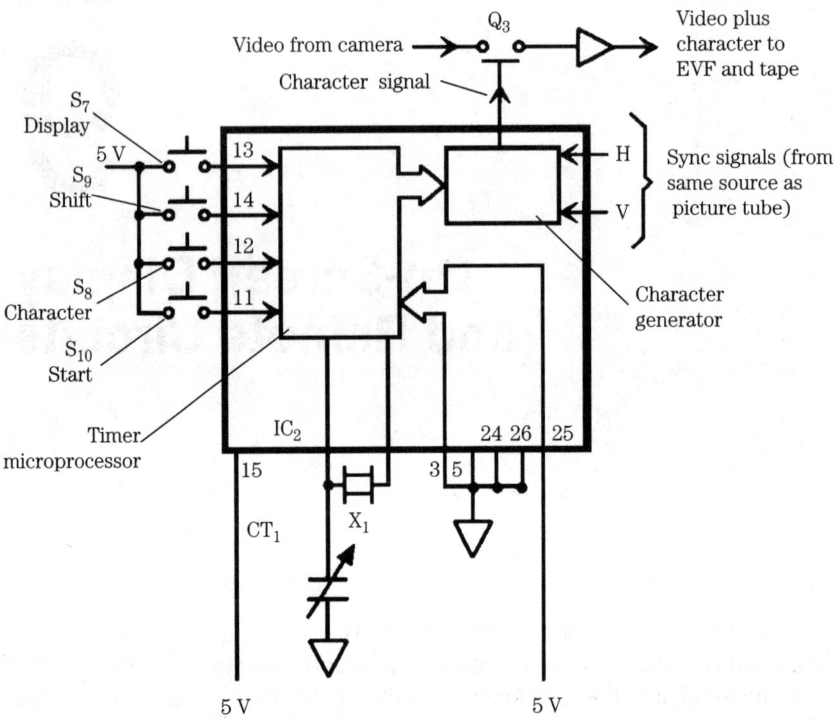

Figure 9.1 Typical OSD circuit.

section of TV-set OSD circuits. Compare the following brief description to that given in section 9.5.

Notice that most of the OSD circuits are contained within IC2. When 5-V power is applied to IC2 at pin 15, crystal X1 generates a 32.786-kHz clock pulse, and the timer-microprocessor in IC2 starts operation. CT1 provides a means of adjusting the clock frequency (to position the character display on the screen).

The character generator in IC2 receives date and time information from the timer and generates corresponding character signals, synchronized with the H- and V-signals from the sync generator (the same source of H- and V-sync applied to the EVF, or to the CRT of a TV set). The character signal is applied to the base of Q3. When the character signal is high, Q3 turns on, and the character is added to the video passing from the camera to the EVF and tape (or to the video appearing on the TV-set CRT).

The display mode for date and time is selected by the logic at pins 3, 4, 24, 25, and 26 of IC2. The date and time setting and display switches S7 through S10 determine the display generated by the character generator.

When display switch S7 is pressed, the character display appears in the EVF screen (or CRT screen). The display mode is changed to date, time, date and time, and display-off repeatedly, every time S7 is pressed.

When switch S9 is pressed, the display blinks. The blinking display is changed to AM or PM, hour, minute, year, month, and day repeatedly, every time S9 is pressed.

When character switch S8 is pressed, the blinking number is incremented. In the case of the AM or PM blinking display, AM is changed to PM, and vice versa, every time S8 is pressed.

Start switch S10 is used to start and stop the setting for time and date. When S10 is first pressed for setting date and time, the AM or PM blinks, and the setting can be made. When S10 is pressed after setting the date or time, blinking stops and date/time counting starts.

9.1.2 Troubleshooting OSD circuits

If the EVF or TV-set CRT display is good in all other respects, but there is no date and time display (or a channel display in the case of a TV), suspect IC2 and/or Q3. It is also possible that X1 is not oscillating. If there is a character display, but the display is not properly positioned, try adjusting CT1.

If there is a properly positioned display, but certain functions of the display are absent or abnormal, check the corresponding circuit. For example, if the blinking number is not incremented, suspect S8. Check that pin 12 of IC2 goes high (5 V) when S8 is pressed. If so, suspect IC2; if not, suspect S8.

9.2 Remote-Control Basics

Most TV sets use some form of IF (infrared) remote control. Also, most present-day remote-control systems use some form of digital position modulation, or P-M. So we concentrate on both IF and P-M remote control in this chapter.

9.2.1 Typical remote-control transmitter circuits

Figure 9.2 shows the circuits of a typical hand-held IF remote transmitter. Similar circuits can be found in the remote transmitters for TV sets, VCRs, etc. The circuits are unique in one respect. The reference frequency of the transmitter is 255 kHz (rather than the more common 455 kHz). This variation in transmission frequency prevents the transmitter from interfering with remote operation of other video products.

The digital code representing a given function appears as a series of "bursts" from the transmitter. Each burst contains 10 pulses, of 50-μs duration each, at a frequency of 20 kHz (rather than the more common 38 kHz). The duration of each burst is 500 μs.

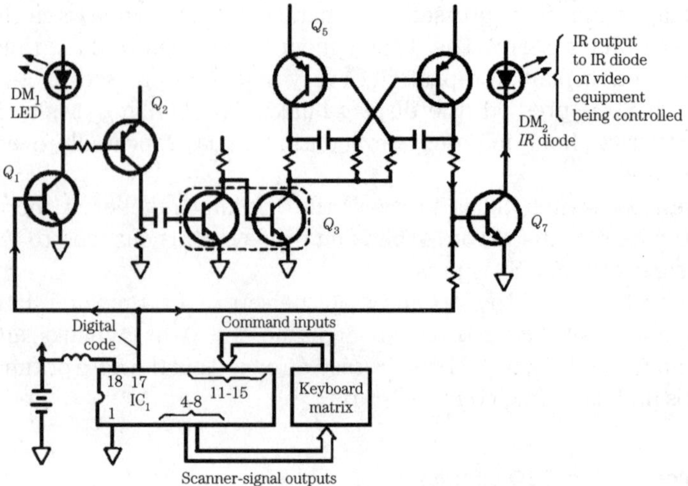

Figure 9.2 Typical hand-held IR remote transmitter.

The position or distance between recurring bursts identifies the digital 1s and 0s that make up a digital code. In this case, position or distance is determined by the amount of time between the rise of the first burst and the rise of the succeeding burst. A digital or logic 0 is 2 ms, while a logic 1 is 4 ms.

The function of IC1 is to generate an oscillator signal for the creation of scanner-signal outputs (at pins 4 through 8), which are applied to specific inputs (at pins 11 through 15) of IC1 (through the keyboard) to identify specific functions. A second function of IC1 is to decode the information it receives at the inputs (as selected by the keyboard) and produce a digital code (at pin 17) representing the selected function.

The code at pin 17 is applied to Q1/Q7, amplified, delayed for 2.2 ms, differentiated, and applied to Q3. Q3 conducts for 550 μs, enabling the 40-kHz multivibrator Q5. The output of Q5 is applied to the base of Q7, along with the output from pin 17 of IC1. The combination of both signals turns on Q7, activating the IR diode DM2. Because two signals are required to activate DM2, erroneous transmission is prevented.

Diode DM2 is pulsed and the IR output is sent to the light-sensitive IR diode (IR detector) on the TV set being controlled (section 9.2.3). The output from DM2 consists of a series of bursts, with the position or spacing between bursts determined by the digital code being transmitted.

9.2.2 Troubleshooting IR transmitter circuits

Once you have definitely pinpointed the problem to a remote-control transmitter by trying a different transmitter with the same equipment being controlled, look for weak or defective batteries. (Also look for any switch on the equipment being controlled that disables the remote-control function, as was the case on my TV set, which I forgot!)

Batteries are the most common causes of trouble in any remote-control transmitter. So start troubleshooting by putting in a known-good battery. The next most common problem is a defective switch contact. Defective switches usually show up when one or more functions are absent or abnormal, but other functions are normal. (Also defective switches often follow periods of operation by children or grandchildren with peanut-butter encrusted fingers.)

Note that when Q1 is turned on by the output at pin 17 of IC1, the remote-transmission LED DM1 is turned on. This indicates that the transmitter is sending commands to the TV set IR-sensitive diode. If DM1 does not turn on when a key is pressed, look for a defective battery. If the battery is definitely good, suspect IC1 or Q1. Not all remote transmitters have an indicator such as DM1.

If the batteries and switches appear to be good but you cannot transmit any command with the remote transmitter, press the keys while monitoring pin 17 of IC1. Although it is not practical to determine the actual digital code being transmitted, the presence of pulse bursts usually indicates that IC1 is good.

Next, try monitoring the pulse bursts through from Q1 to DM2. Also look for 40-kHz square waves at the collectors of Q5. If the pulse bursts appear at DM2, but the transmitter cannot transmit commands, suspect DM2.

9.2.3 Typical remote-control receiver circuits

Figure 9.3 shows the circuits of a typical IR remote receiver. IC1 has the dual function of decoding digital commands from the remote transmitter and converting them into signals (that are applied to system-control, tuner-control, and power circuits). These signals replace or supplement the commands applied by push buttons on the TV set being controlled.

The command in IC1 is taken from the signal transmitted by the remote-control transmitter. The transmitted IR signal is received by IR-sensitive diode PD1, and converted into an electrical signal applied to amplifier and detector IC91. The detected signal from IC91 is amplified and inverted by Q1, and applied to pin 12 of IC1.

The output at pin 36 of IC1 is the power-on signal, which turns on Q5 whenever the power button on the remote-control transmitter is pressed. Transistor Q5 is in parallel with the manual on/off push-button of the TV set. The output at pins 17, 29 through 32, 37, and 40 of IC1 are the channel-select commands. These commands are applied to a tuner PLL such as described in chapter 2.

9.2.4 Troubleshooting remote-control receiver circuits

If you do not get proper remote-control operation, substitute a known-good remote transmitter. Next, confirm the presence of the signal at pin 12 of

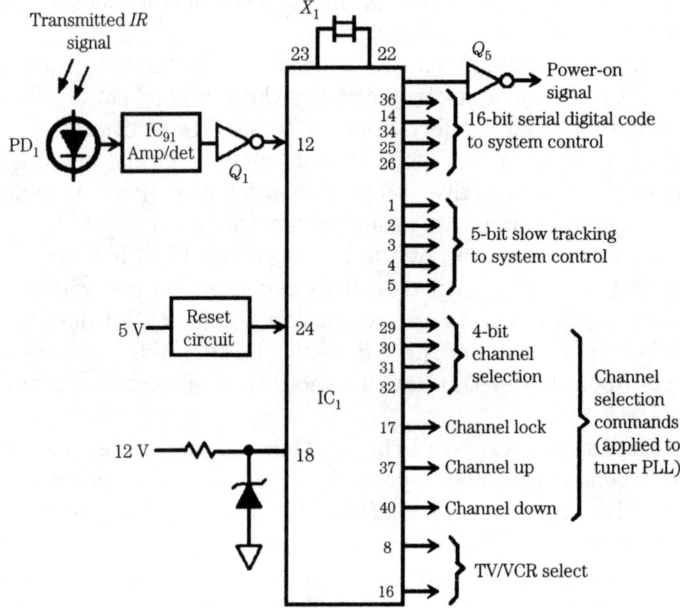

Figure 9.3 Typical IR remote receiver.

IC1. This signal is a serial-data stream that cannot be intelligently monitored with a scope. That is, there is no easy way to determine which digital code is being transmitted by monitoring the signal. However, it is reasonable to assume that if a signal is present, the code is correct.

If the signal is present, check for 5 V at pin 18 of IC1. If present, check for a clock signal at pin 23 of IC1, and for a reset high at pin 24. Pin 24 should initially go low when power is applied, and then return to high (about 5 V). If these inputs are correct but there is no remote-control operation (but manual operation is good), suspect IC1.

If you get remote-control operation for some but not all functions, check the corresponding output from IC1. For example, if you get channel-up operation but not channel-down (with a known-good remote transmitter), check for a channel-down output at pin 40 of IC1. If the output is absent or abnormal, suspect IC1. If the output is present, the problem is likely to be in the tuning system and should be checked as described in chapter 2.

9.3 OSD and Remote Circuits (13 Inch)

Figure 9.4 shows the OSD and remote-control circuits for a Sony 13-inch TV set.

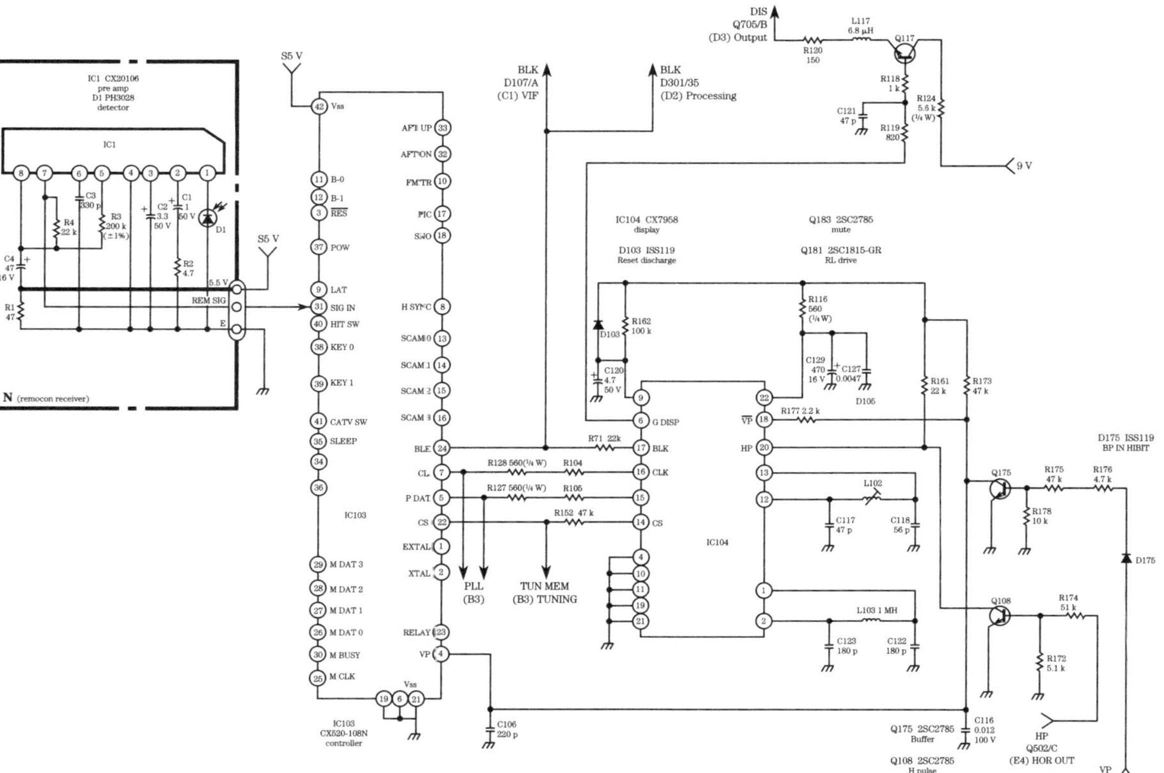

Figure 9.4 OSD and remote circuits (B2).

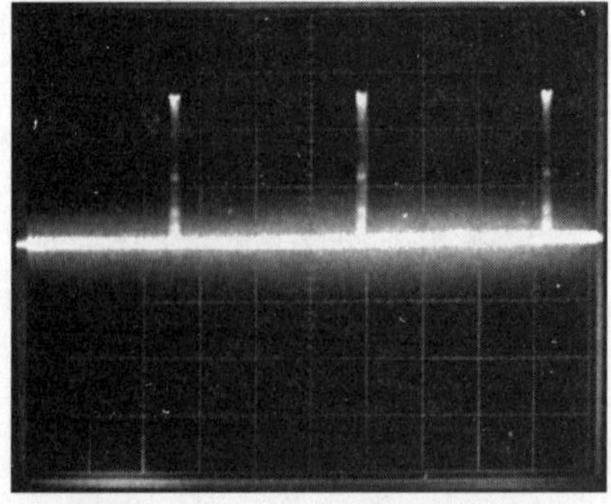

(S2) Q117/B DIS 2 V/d 5 msec/d

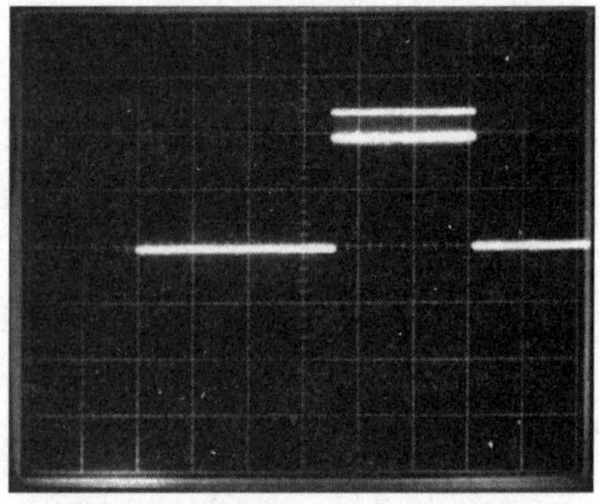

(S3) IC103/Pin 24 BLK 2 V/d 0.2 sec/d

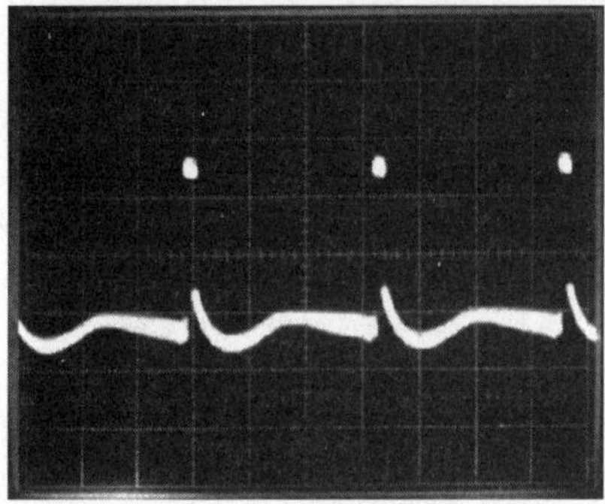

(S4) Q175/B VP 0.2 V/d 5 msec/d

Figure 9.4 Continued.

9.3.1 Remote-control receiver

The remote-control receiver, sometimes called the Head Amp, is composed of IC1 and IR detector D1. The transmitted IR signal is detected by D1 and applied to IC1 for processing. The 40-kHz pilot signal is removed, and the negative-going square-wave remote signal (digital data stream) is applied to system-control microprocessor IC103 at pin 31.

This system uses the standard Sony Infrared Remote Control System SIRCS for all remote-control operation. The SIRCS is characterized by the ability to selectively operate only pertinent equipment. For example, a TV will respond only to a TV remote transmitter and a VCR will respond only to a VCR remote. This is done by sending a product identifying code along with the control data.

9.3.2 OSD

The OSD functions are performed by IC104 (which corresponds to IC2 in Figure 9.1). However, display is provided by IC103 using the clock, data and chip-select lines on pins 7, 5, and 22, respectively. Timing for the display is generated with IC104 using the VP (vertical sync) and HP (horizontal sync) reference signals. These signals determine the starting points of the vertical and horizontal intervals, and originate from the same vertical/horizontal sweeps applied to the CRT.

The clock signals for IC104 are generated internally, and appear at pins 1 and 13. The components at these pins determine the exact clock frequencies. Notice that L102 is adjustable so that the clock at pin 13 can be adjusted in frequency. This permits the display to be positioned on the CRT screen.

Data bits for IC104 display operations are provided by IC103. The data bits are serially transmitted to pin 15 of IC104 in synchronization with the clock signal at pin 16. Notice that the clock and data lines are shared by the tuning PLL (Figure 2.5). Only the clock and data pulses occurring during the negative-going chip-select signal (pin 14 of IC104) apply to the IC104 display operation. Also notice that the chip-select line is shared by the tuning memory IC105 (Figure 2.5).

The signal developed in IC104, in accordance with the data transferred from IC103, is applied to the green gun of the CRT through pin 6 of IC104, Q117, and Q705 on the RGB output circuits (Figure 4.7). As a result, all OSD characters, numbers, etc., appear as solid green on the CRT screen.

9.3.3 Blanking output

During channel-change operation, IC103 outputs a positive-going pulse at pin 24. This blanking pulse mutes the audio by blanking the output to the VIF stage (Figure 3.4) and blanks the CRT with a pulse to pin 35 of IC301 (Figure 4.6). The blanking pulse is also applied at pin 17 of IC104 to remove the OSD signals from the green gun of the CRT. The result of this blanking action is muting of the audio and a totally blank CRT between channel changes.

9.3.4 Reference inputs

The internal clocks of IC104 are synchronized to the CRT vertical and horizontal rates by the VP and HP inputs applied through Q175 and Q108, respectively. The inverted VP pulse is also applied to pin 4 of IC103, telling IC103 when the vertical-blanking interval occurs. IC103 uses this information to update the OSD signals (and the PLL tuning signals) during the vertical-blanking period.

9.4 OSD Circuits (19 Inch)

Figure 9.5 shows the OSD circuits for a Sony 19-inch TV set. The OSD functions are performed by IC102 (which corresponds to IC2 in Figure 9.1). However, display data is provided by IC101 using the clock, data and chip-select lines on pins 7, 5, and 14, respectively. Timing for the display is generated within IC102 using the VP and HP sync signals. These signals determine the vertical/horizontal interval starting points, and originate from the same corresponding sweeps applied to the CRT.

The clock signals from IC102 are generated internally, and appear at pin 13. The clock frequency can be adjusted by L102 to position the OSD display on the CRT screen.

OSD data bits for IC102 operations are provided by IC101. The data bits are serially transmitted to pin 15 of IC102 in sync with the clock signal at pin 16. In turn, the OSD signals developed by IC102 are applied to the green gun of the CRT through pin 6 of IC102, Q104, and Q702 on the RGB output circuits (Figures 4.10 through 4.12). This results in solid-green OSD characters on the CRT.

During channel-change operations, IC101 outputs a blanking pulse at pin 23. This blanking pulse mutes the audio through Q282 (Figure 3.6) and blanks the CRT with a pulse applied at pin 29 of IC301 (Figure 4.10). The blanking pulse is also applied at pin 17 of IC102 to blank the OSD on the CRT screen.

The internal clocks of IC102 are synchronized to the CRT vertical and horizontal rates by the VP and HP signals applied at pins 18 and 20 (through Q101). The VP signal is also applied to IC103 at pin 4 so that IC103 updates the OSD signals during the vertical-blanking period.

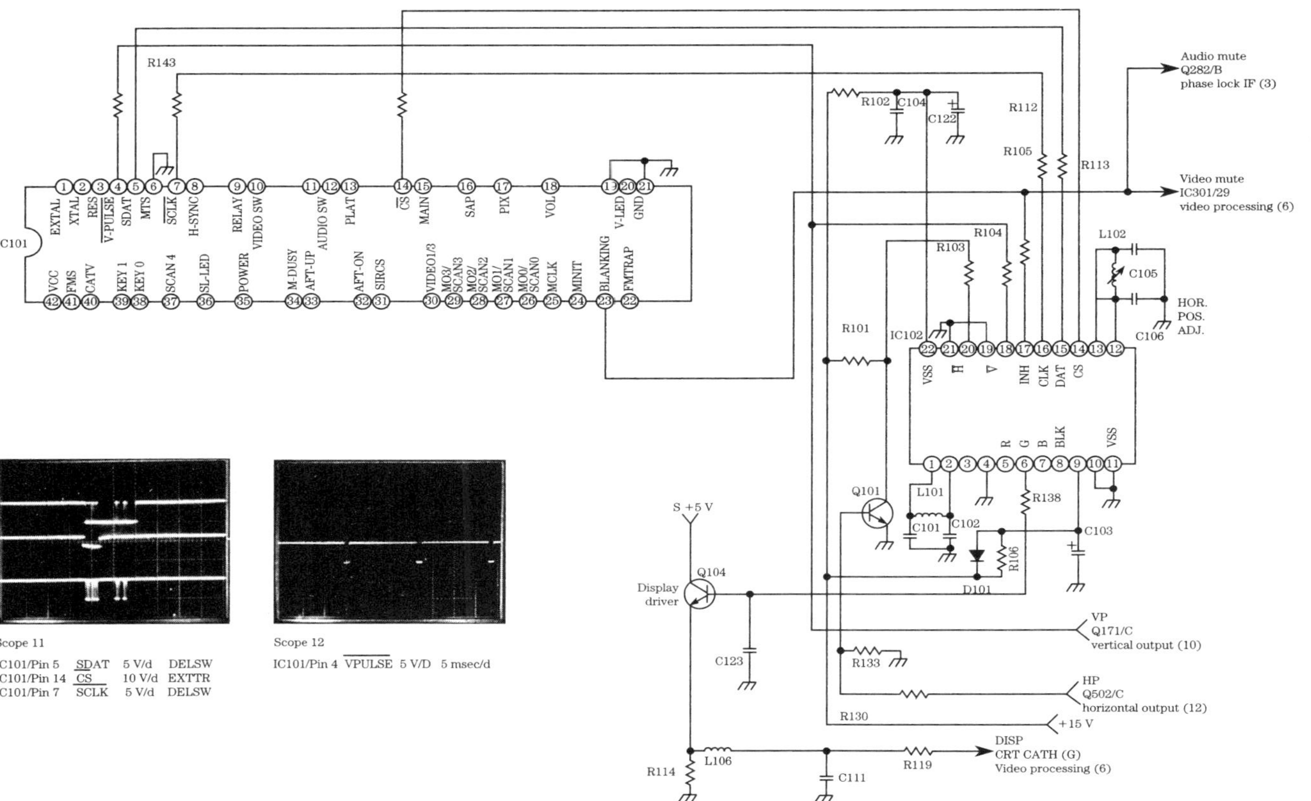

Figure 9.5 OSD circuits (2C).

Chapter

10

Special TV Circuits

This chapter is devoted to special circuits found in some TV sets. These include stereo TV, surround sound, external audio inputs, electronic volume controls, digital TV, and picture-in-picture circuits. Although there are many so-called stereo-TV sets, not all can decode and play stereo-TV broadcasts with multichannel television sound (MTS or MCS, whichever term you prefer). Likewise, there are many "surround sound" or "sound enhancer" circuits, but not all are capable of decoding TV broadcasts in true Dolby Surround. We discuss both systems here. The chapter concludes with descriptions of full digital TV, as well as descriptions of the three most commonly used picture-in-picture circuits.

10.1 Stereo-TV Basics

Before we get into stereo-TV circuits, let us review the basic stereo-TV system. Figure 10.1 shows the TV transmitter and receiver for the MTS/MCS system that delivers audio for a stereophonic TV program and separate audio program (SAP). The MTS/MCS stereo-TV broadcasts are made in accordance with the dbx Noise Reduction (NR) System and Zenith Transmission System. Figure 10.2 shows the multichannel signal spectrum used in the MTS/MCS system.

The main-channel signal is composed of the sum (L+R) signal of L (left) and R (right) signals and is the same as in the conventional-TV sound specification. As a result, the same TV sound as in conventional broadcasting can be received by an ordinary TV set when stereo TV broadcasting is in effect.

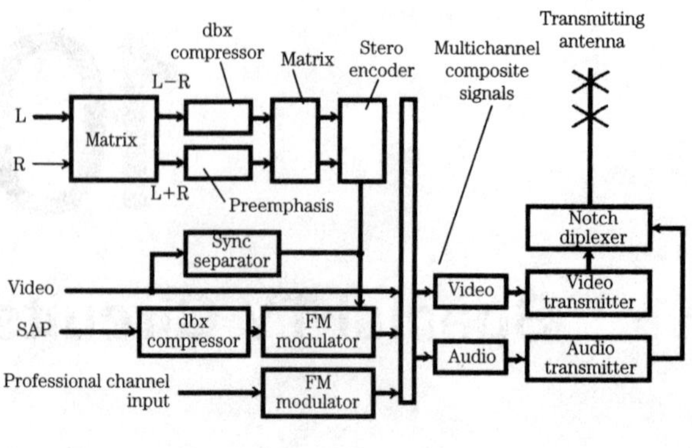

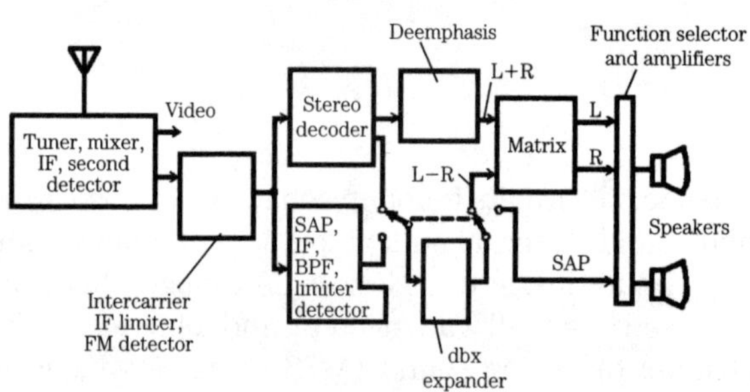

Figure 10.1 TV transmitter and receiver for the MTS/MCS system.

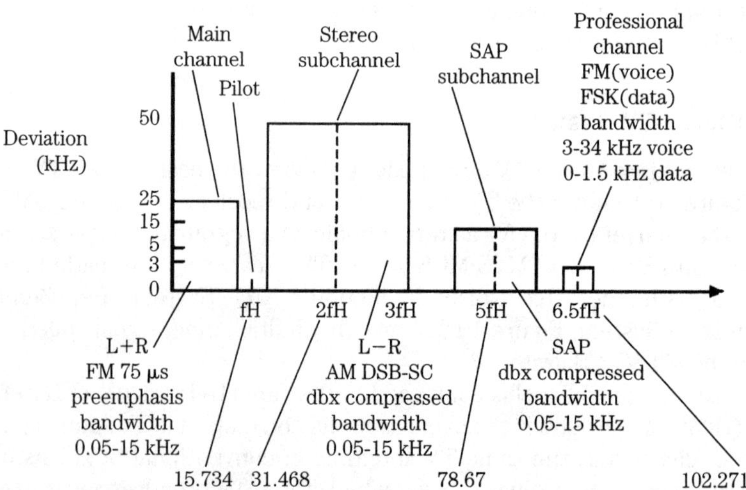

Figure 10.2 MTS/MCS multichannel signal spectrum.

When the difference between the L and R signals (L–R) is transmitted, the L and R signals are reproduced from the main-channel signal (L+R) and the stereo signal (L–R) by the TV set. The L–R signal is produced by amplitude modulating a subcarrier with a frequency double the horizontal scanning frequency (fH) using the double-sideband suppressed carrier (DSB-SC) system. As a result, the frequency division of the main-carrier is double (50 kHz) that of the main-channel signal (L+R).

The carrier of the L–R signal is suppressed, so it is necessary to transmit a reference signal to demodulate the L–R signal correctly in the TV set. To do this, a signal with a frequency equivalent to the horizontal scanning frequency (fH), called the pilot signal, is inserted between the main-channel signal (L+R) and the stereo signal (L–R). The pilot signal is also used to indicate the presence or absence of the stereo signal (no pilot, no stereo).

The SAP-channel signal frequency-modulates a subcarrier with a frequency of 5 fH and is locked to 5 fH during nonmodulation.

The nonpublic channel signal (also called the professional channel) frequency-modulates the subcarrier with a frequency of 6.5 fH. (This channel is scheduled to be used for business, not for regular TV programming.)

The sum (L+R), difference (L–R), pilot SAP-channel, and nonpublic-channel signals are added to produce the multichannel composite signal, which in turn frequency-modulates the transmitted TV broadcast signal. A noise-reduction system (NR system), which specifically encodes and transmits the L–R and SAP signals, is used to reduce noise in the TV set.

10.1.1 dbx NR system

The dbx noise reduction is an essential part of the stereo-TV system. In the simplest of terms, portions of the transmitted signal are expanded (or decoded) at the TV transmitting station. These same signals are expanded (or decoded) at the TV set. The process is known as companding (or compandoring in some literature). The following is a brief discussion of the companding process, covering both the why and how. Figure 10.3 shows simplified block diagrams of both compressor and expander circuits.

Stereo TV transmission system. As discussed, the stereo-TV system transmits the sum of the left and right stereo audio signals (L+R) in the spectrum space usually occupied by the conventional or mono TV audio signal (Figure 10.2). The stereo information is encoded by subtracting the right audio signal from the left (L–) and transmitting the difference via an AM subcarrier located at 31.468 kHz (twice the video horizontal-scanning frequency of 15,734 kHz), superimposed on the conventional FM audio carrier.

Noise reduction. Because the L–R or stereo subcarrier is at a higher frequency than the L+R signal, the stereo reception is about 15-dB noisier than mono reception, even under ideal conditions. Typically, the stereo S/N

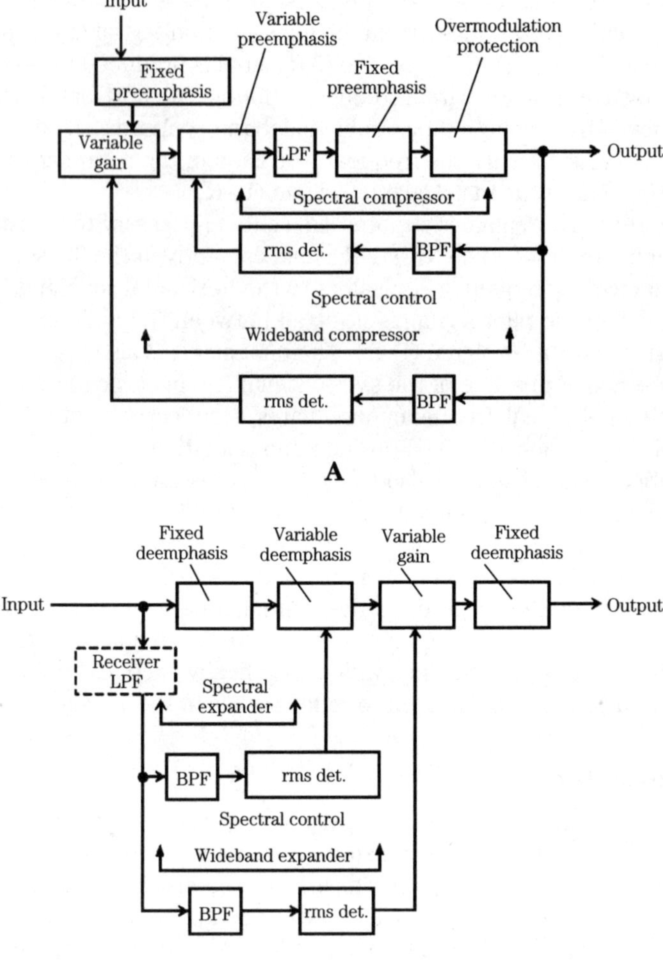

Figure 10.3 MTS/MCS compressor and expander circuit.

ratio is about 50 dB. The SAP subcarrier (at an even higher frequency of 78.67 kHz) has a typical S/N ratio of 33 dB. The practical effect of noise is to reduce the coverage area for both stereo and SAP as compared to the mono coverage area (under identical conditions).

The dbx noise-reduction system is designed to improve the S/N ratios. (Theoretically, the goal is to eliminate any noise increase when going from mono to stereo and to make SAP listenable. Notice that noise reduction is added only to the L–R and SAP channels leaving the mono L+R signal unchanged. This ensures compatibility with conventional TV broadcasts. (If you have a conventional TV set you will not know that stereo TV is being broadcast.)

Identical companding is used in both the L–R and SAP channels. This allows a single noise-reduction circuit to be switched between the L–R and SAP channels in the TV set. Operation of the noise-reduction system involves preemphasis, deemphasis, spectral companding, and wideband-amplitude companding.

Preemphasis and deemphasis. The dbx noise-reduction system uses preemphasis (at the TV broadcast transmitter) and deemphasis (at the TV set). The preemphasis circuits alter the frequency-response curve of the transmitted audio to overcome noise. The deemphasis circuit restores tonal balance to the program material and reduces hiss picked up in transmission.

Spectral companding. The dbx noise-reduction system includes a stage where preemphasis is varied to suit the signal. The function is called spectral companding. When very little high-frequency information is present in the audio, the spectral compander (Figure 10.3A) provides large high-frequency preemphasis. When strong high frequencies are present, the spectral compressor provides deemphasis, thus reducing the potential for high-frequency overload. The transmitted signal is dynamically adjusted to have high-frequency content to provide good masking.

In the TV set (Figure 10.3B), the spectral compander restores high-frequency signals to the proper amplitude. The expander also attenuates high-frequency noise when little high-frequency is present. When strong high-frequency signals are present, the signal masks the noise.

Wideband-amplitude companding. The third stage or function of the dbx noise-reduction system is the wideband-amplitude companding, which keeps the signal level in the transmission channel high at all times.

The compressor (Figure 10.3A) reduces the dynamic range of input signals by a factor of 2:1 (in dB). The transmitted signal level tends toward about 14 percent modulation, which allows transient peaks to overshoot without causing overmodulation.

During reception (Figure 10.3B), the wideband-amplitude expander restores signal levels to the proper amplitude. When the signal is low in amplitude, the decoder attenuates channel noise toward the point of inaudibility. During high-amplitude passages, the signal masks the noise.

10.1.2 Overall stereo TV decoder

Figure 10.4 shows overall stereo and SAP functions for a typical TV set in block form. Individual components of the composite stereo signal might take any of three different paths. Which path is determined by low-pass and band-pass filters (LPFs and BPFs) in the input circuit.

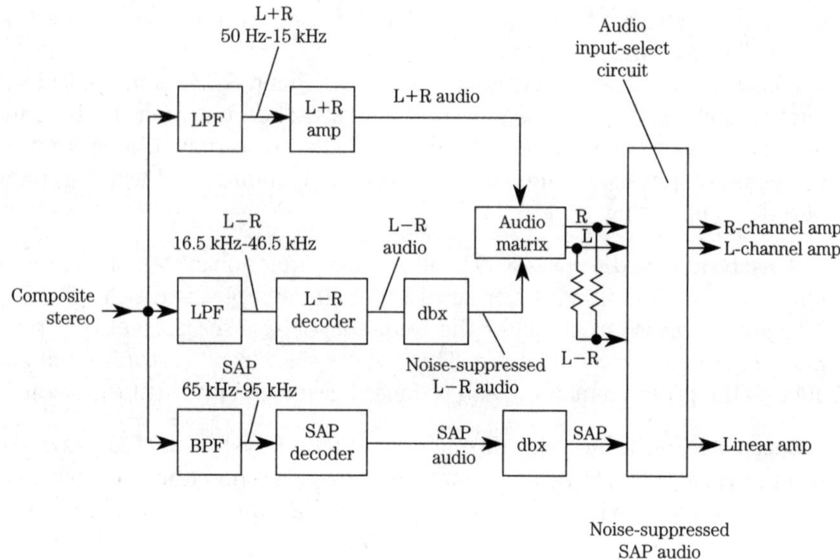

Figure 10.4 Overall stereo and SAP functions.

The L+R circuit is selected by an LPF designed to pass only those L+R signals below 15 kHz. The L+R (mono) signal is amplified and applied to the audio matrix. The second LPF passes all frequencies below 45 kHz, which includes the L+R, L–R sidebands, and the pilot signal. These signals are all applied to the L–R decoder. The L+R signal (because of the low frequency) is rejected by the L–R decoder. (Only the L–R sidebands and pilot signal are used in the L–R decoder.) The pilot signal is used to synchronize an oscillator within the L–R decoder. The oscillator produces a suppressed 31.468-kHz L–R subcarrier, thus permitting subsequent decoding of the L–R sidebands.

The output of the L–R decoder is applied to the audio matrix through a dbx noise-reduction circuit. The L–R and L+R signals are combined in the matrix to form individual left and right audio channels. The left- and right-channel audio signals are applied to the audio input-select circuits and are then selected for recording on tape by the rotating hi-fi and audio heads.

If SAP is present, the SAP audio is passed through the BPF. This BPF passes signals from 65 to 95 kHz, and thus rejects the L+R and L–R signals. The SAP signal is decoded by the SAP decoder, and applied to the audio input-select circuit through a separate dbx noise-reduction circuit.

10.1.3 L+R circuits

As shown in Figure 10.5, the composite stereo signal is amplified by Q1 and applied through filters to the stereo (L–R) and SAP signal paths. The L+R and L–R signals are rejected by BPF1 but are passed to Q20, where both

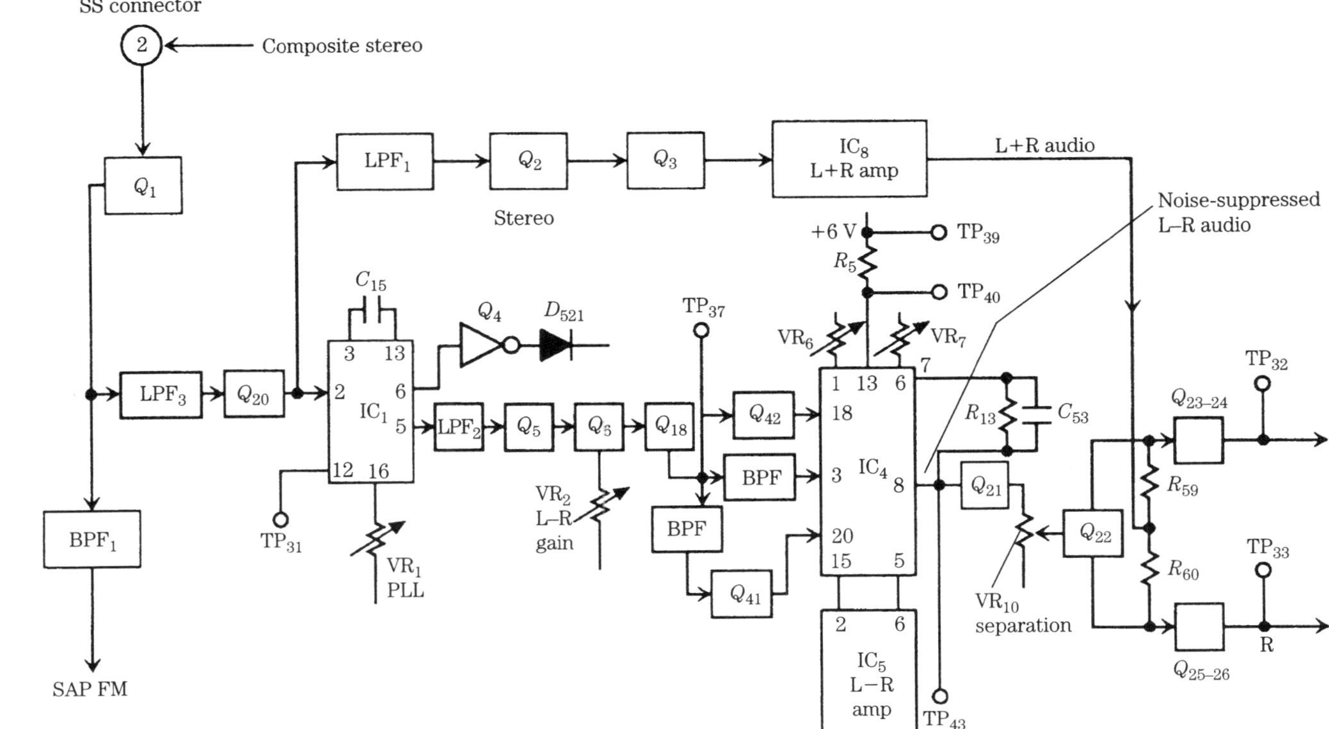

Figure 10.5 MTS/MCS combined L+R and L–R signal paths.

signals are amplified. The L+R signal is applied to amplifier IC8 through LPF1, Q2, and Q5. The L+R audio from IC8 is applied to the junction of resistors R59/R60. Both resistors are part of a matrix circuit used to combine the L+R and L–R signals.

10.1.4 L–R decoder

Most of the L–R functions are performed within IC1. These functions include detecting the presence of a stereo signal (a signal with pilot) and decoding the L–R sidebands to produce L–R audio. If the broadcast is in stereo (pilot present), IC1 produces a command at pin 6 to turn on the stereo indicator D521. The L–R sidebands and the pilot are applied to pin 2 of IC1. The L–R sidebands pass a preamp. The amplified pilot signal is applied to two comparators through C15.

A VCO in IC1 operates at 4 times the pilot frequency, or 62.936 kHz. The VCO output is divided twice by 2, resulting in a 31.468-kHz output (applied to the L–R decoder as a substitute carrier) and a 15.734-kHz output (applied to both comparators as a second input). When a stereo signal is transmitted, the pilot and the divided-comparison signals from the VCO are compared in both comparators.

The PCL comparator (phase comparator, lamp) generates a signal that is filtered by an LPF and amplified by a lamp driver to produce a low at pin 6 of IC1. If there is no pilot, pin 6 of IC1 remains high. A low at pin 6 of IC1 turns on Q4. In turn, Q4 turns on when a pilot signal is present (indicating a stereo broadcast to the listener or viewer). If pin 6 is high (no pilot, no stereo), Q4 and D521 remain off.

The PCV comparator (phase comparator, VCO) develops a correction voltage if a frequency or phase error exists between the pilot signal and the VCO comparison signal. The correction voltage is filtered, amplified, and applied to the VCO to correct any frequency or phase error. VR1, at pin 16 of IC1, is used to set the free-running frequency of the VCO to 4 times that of the pilot frequency.

The output of the VCO is applied to a divide-by-2 circuit, reproducing a phase- and frequency-corrected duplicate of the original L–R subcarrier. The resulting 31.468-kHz signal is applied to the L–R decoder, along with the L–R sidebands from pin 2 of IC1. The L–R sidebands are decoded, and the resulting L–R audio is available at pin 5 of IC1. The L–R audio is passed through BPF2, where all unwanted signals above 15 kHz are removed. The L–R audio is then amplified by Q5, Q6, and Q18 and applied to the dbx circuit.

10.1.5 dbx circuits

Although the dbx noise-reduction format used in stereo TV is the most complex part of the system, virtually all of the dbx circuits are contained within IC4. Figure 10.6 shows these circuits in block form.

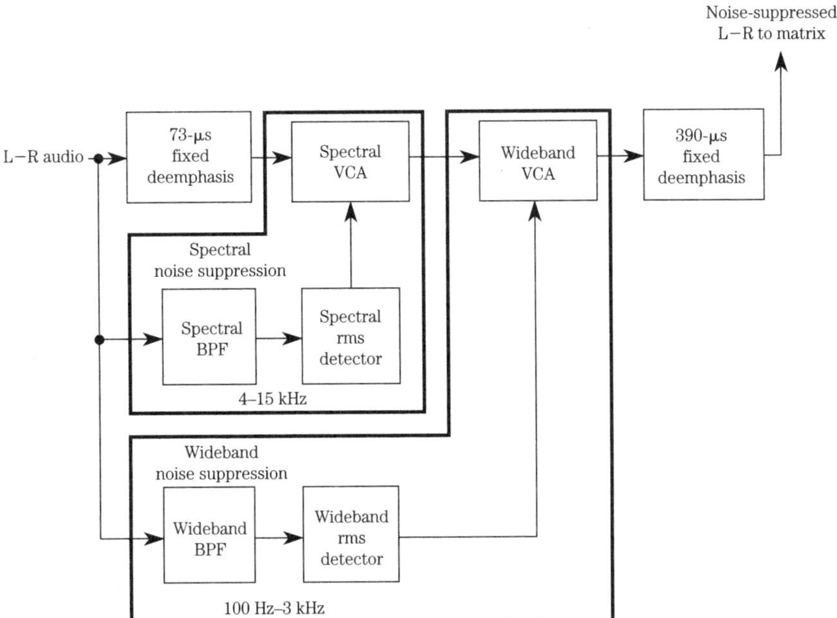

Figure 10.6 The dbx noise-reduction circuits.

A discrete-component BPF passes only those L–R signals within the spectral band (4 to 15 kHz) and applies signals to the spectral rms detector (which detects the signal amplitude). Simultaneously, a second discrete-component BPF passes those L–R signals within the wideband (100 kHz to 4 kHz) and applies the signals within this band to the wideband rms detector (which also detects the signal amplitude).

The complete L–R audio signal is deemphasized by a fixed 73-μs deemphasis network. The deemphasized L–R signal is then applied to the spectral voltage-controlled amplifier (VCA). Note that in some dbx ICs the VCAs are called current-controlled amplifiers (CCAs), just to confuse you.

The gain of the spectral VCA is controlled by the detected level of the spectral rms detector. The signals from 4 to 15 kHz are either compressed or expanded (depending on the respective amplitude) by varying the gain of the spectral VCA. The L–R signal with spectral compression is then directed to the wideband VCA.

The gain of the wideband VCA is controlled by the wideband rms detector. The signals from 100 Hz to 3 kHz are compressed by varying the gain to the wideband VCA. The L–R signal from the wideband VCA is deemphasized by a fixed 390-μs deemphasis network and is applied to the matrix circuit for mixing with the L+R audio signal.

dbx deemphasis and BPF. As shown in Figure 10.5, the L–R audio from Q19 is applied to the spectral VCA at pin 18 of IC4. Q42 and the associated parts form the fixed 73-µs deemphasis network. Q41 and the associated parts form the spectral BPF. L–R signals in the range from 4 to 15 kHz are applied to the spectral rms detector at pin 20 of IC4. A discrete-component BPF forms the wideband BPF. L–R signals in the range from 100 Hz to 3 kHz are applied to the wideband rms detector at pin 3 of IC4.

dbx processing. Two rms detectors within IC4 receive power from a constant-current generator (also in IC4). The output of the generator is adjustable by VR6 (the so-called L–R timing control), connected at pin 1 of IC4. The setting of VR6 determines the amount of control output from the rms detectors for a given signal amplitude.

Spectral processing is performed by the spectral VCA at pin 18 of IC4, controlled by the output of the spectral rms detector. The L–R signal is then applied to an L–R amplifier IC5 and an op-amp within IC4. VR7 sets the gain of the op-amp and is sometimes called the variable deemphasis (VD) control. The amplified L–R signal is then applied to another op-amp within IC4. This op-amp output at pin 8 of IC4 is applied to Q21 in the matrix circuit. C53 and R13, between 7 and 8 of IC4, produce the required 390-µs fixed deemphasis.

10.1.6 L+R and L–R troubleshooting

Here are some thoughts on troubleshooting the L+R and L–R circuits shown in Figure 10.5.

The L–R gain is set by adjustment of VR2 in the emitter of Q6, ahead of the dbx circuits. This is not to be confused with the separation adjustment VR10 after the dbx circuits, although both controls set the L–R signal level.

The dbx functions are set by adjustment of VR6 and VR7. L–R timing control VR6 sets the amount of control output from the rms detectors within IC4 for a given signal amplitude. Variable deemphasis control VR7 sets the gain of the spectral op-amp in IC4.

If there is no stereo TV but mono operation is good, the problem is in the L–R path, with IC1 and IC4 being the likely suspects. The problem could also be improper adjustment of VR1, VR2, VR6, VR7, or VR10. Signal tracing (or signal injection) through the entire L–R path should lead to the problem.

If there is excessive background noise, suspect the dbx noise-reduction circuits. IC4 is the most likely suspect, although the BPFs at the input to IC4 are also suspect because the BPFs separate the frequencies to be processed by IC4.

If there is no sound during a mono broadcast and distorted sound during a stereo broadcast, suspect the L–R audio path. During a stereo broadcast, the L–R signal falls to zero at that instant. As a result, the reproduced sound is erratic and/or distorted because there is no L–R signal at the matrix.

10.1.7 SAP circuits

Figure 10.7A shows the SAP demodulation signal path that decodes the SAP audio. This audio is applied to the audio input-select circuit through the SAP dbx processing. Figure 10.7B shows the SAP signal-detect path that turns on the front-panel SAP indicator and produces an SAP-present signal (also applied to the audio input-select circuit). Notice that the circuits of Figure 10.7 are for a VCR equipped with stereo TV capability, but are identical to stereo circuits found in a TV set (except for the output to the head amplifier from pin 4 of IC34).

AP demodulation path. The composite stereo signal is applied to BPF1, which passes only those signals with frequencies between 65 to 95 kHz. The output of BPF1 is applied to the SAP signal-detect path and to pin 6 of IC2, which acts as a limiter. The SAP FM from pin 3 of IC2 is applied to the SAP demodulator (consisting of T1, D1, and D2) through Q7. SAP audio from the SAP demodulator is applied to the SAP dbx processing circuits (at Q44) through LPF4, Q8, Q9, Q10, and Q19. Notice that VR3 in the emitter of Q10 sets the level of the SAP audio signal.

SAP demodulation troubleshooting. What might appear to be a loss of SAP audio and a defect in the SAP demodulation path can be caused by a problem in the SAP signal-detect path (discussed next). For example, if the SAP signal-detect circuit does not apply the SAP-present signal to the audio input-select circuit, the SAP signal cannot be directed to the audio output circuit.

SAP signal-detect path. As shown in Figure 10.7B, the heart of the SAP signal-detect circuit is PLL IC3. The SAP FM signal is applied at pin 6 of IC3 through BPF1, Q11, and Q12. The VCO in IC3 does not oscillate at the SAP carrier frequency of 78.67 kHz. Instead, the VCO oscillates at 44 kHz. If the input to IC3 is 44 kHz, the output of the phase detector is zero. However, because the SAP signal extends from 46 to 95 kHz, the SAP signal is always higher than the VCO frequency.

The phase detector produces a negative voltage when the incoming signal is higher than the VCO. This negative voltage, generated only when the SAP is present, is amplified and filtered. The negative or low output from IC3 at pin 7 is applied through Q13, Q14, Q15, and Q16 and appears as a high at the collector of Q16. The high drives Q33 into conduction, turning the SAP indicator D520 on and informing the user that a SAP signal is being broadcast.

The high at Q16 is also applied to Q17, and drives the output of Q17 low. This output is a status command (called SAP) applied to the audio input-select circuit. If the SAP line does not go low, the output of the SAP demodulator is not applied to the audio output circuits (loudspeakers), making it appear as though no SAP is present. This point should be remembered when troubleshooting for a loss of SAP. With Q17 low, the SAP audio signal

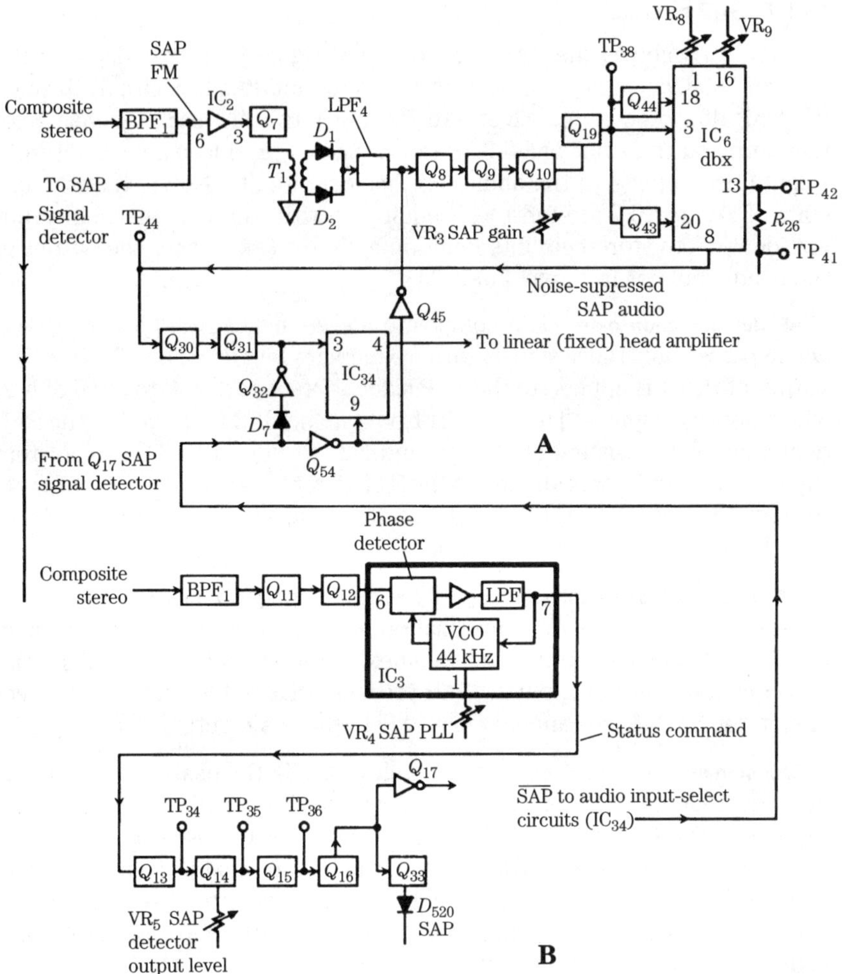

Figure 10.7 MTS/MCS SAP signal paths.

at Q31 is applied to the audio output. With Q17 high, the audio is muted (at Q31 through Q32 and at Q8 through Q45).

10.2 Stereo-TV Circuits

Figures 10.8 through 10.11 show the audio circuits of a Sony 19-inch TV set. These circuits include an MPX decoder, as well as a noise-reduction IC, for stereo TV. The circuits also include an audio controller and audio-output amplifier. The audio controller is an IC version of the audio input-select circuits discussed in section 10.1.

10.2.1 MPX decoder

Figure 10.8 shows the MPX decoder circuits used in some Sony TV sets (such as the KV-1981R). Notice that the majority of the circuits are continued within a single IC, designated as the CX20112.

The functions performed by IC810 include the SAP bandpass filter (5 fH), the SAP VCO, and the SAP FM detector. Also included are the stereo low-pass filter (3 fH), the pilot detector (1 fH), the VCO (4 fH) with the necessary dividers, and the AM stereo detector. All necessary switching for detecting SAP or stereo to the noise-reduction circuit, and for directing the appropriate signal to the built-in matrix circuit, is also included.

Operating modes are selected by the switch-control inputs at pin 16, 17, and 20 of IC810. The possible operating modes are: auto stereo or mono, main, SAP, or both. A high at pin 20 forces the matrix into mono operation. A low at pin 17 switches the matrix to stereo, if the stereo pilot is present. If there is no stereo pilot, the matrix operates in mono, passing the main audio channel (L+R).

A low at pin 16 of IC801 switches the matrix to process the SAP audio channel in mono (SAP at both loudspeakers). A low at both pins 16 and 17 causes the matrix to process SAP on the right channel, and main L+R audio on the left channel. The SAP mute circuit Q872/Q877 operates when SAP is present to enable the SAP detector by making pin 14 low.

The signal path for the L+R audio is in at pin 1, output at pin 37, in at pin 34 through an internal L+R amplifier and low-pass filter, out at pin 29 through the 73-µs deemphasis circuit, and in at pin 28 to the internal matrix. The signal path for stereo L–R is in at pin 1, out at pin 37, in at pin 34 to the stereo detector, out at pin 40 through the level set (RV316), and in at pin 41. The L–R then passes through a low-pass filter, out at pin 24 to the noise-reduction circuit, back from the noise reduction circuit, and in at pin 27 to the matrix. The matrixed L+R and L–R signals develop the left and right outputs at pins 22 and 23.

The SAP signal flow is in at pin 1 to the SAP bandpass filter, out at pin 6, in at pin 8 to the FM detector. The detected SAP is then output at pin 25, across the level set (RV814, SAP GAIN), in at pin 26, through another internal LPF, out to the noise-reduction circuit at pin 24, and back from the noise reduction at pin 27 to the matrix where SAP is processed as mono audio (both speakers). In the BOTH mode, the main (L+R) and SAP signal flows are the same, except that the matrix processes SAP as right-channel audio and L+R as left-channel audio.

Figure 10.9 shows the noise-reduction circuits used in some Sony sets (such as the KV-1981R). Notice that the majority of circuits are contained within two ICs, dbx IC830 and matrix ICs discussed thus far in this chapter. These circuits are common to those used in the TV sets of many manufacturers, so I will not describe the circuits in detail. However, the circuits are included here to show the tie-in to the decoder circuits (Figure 10.8).

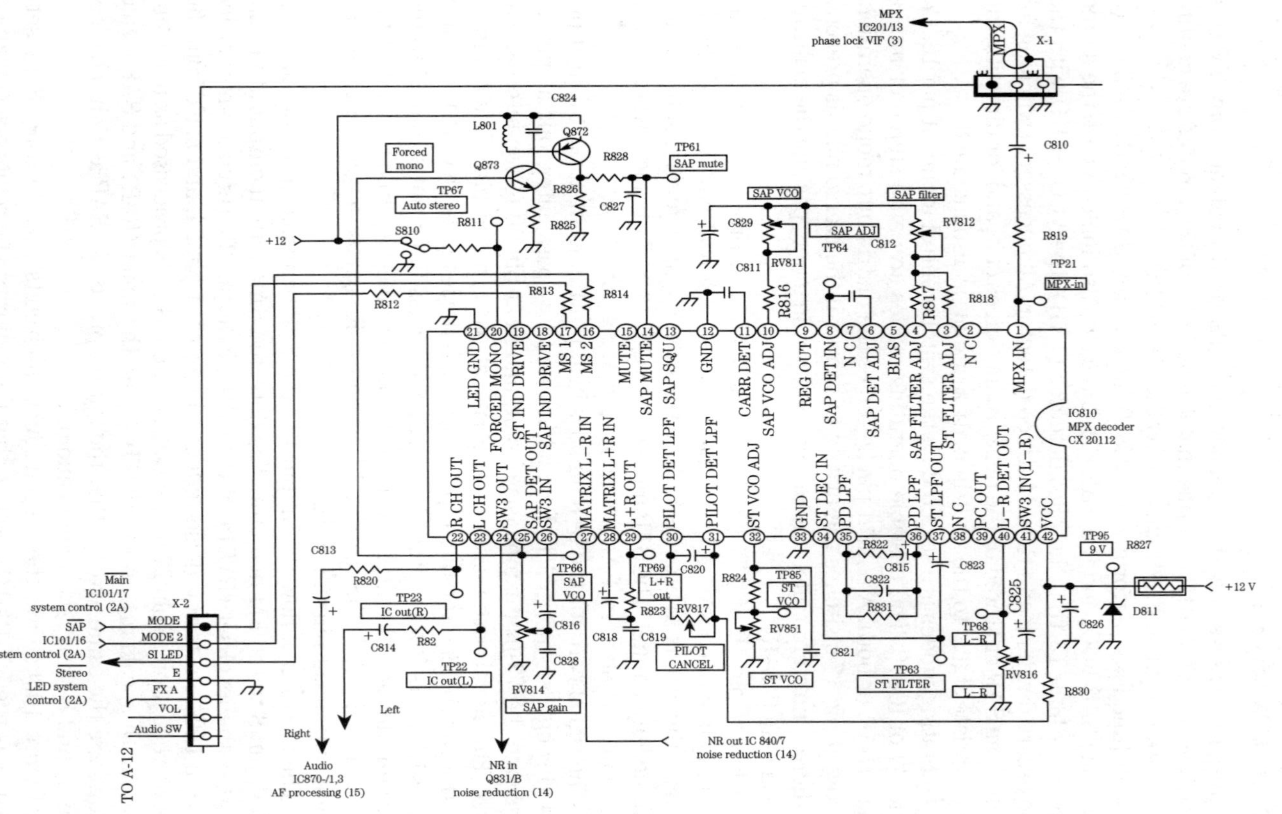

Figure 10.8 MPX decoder circuits (3).

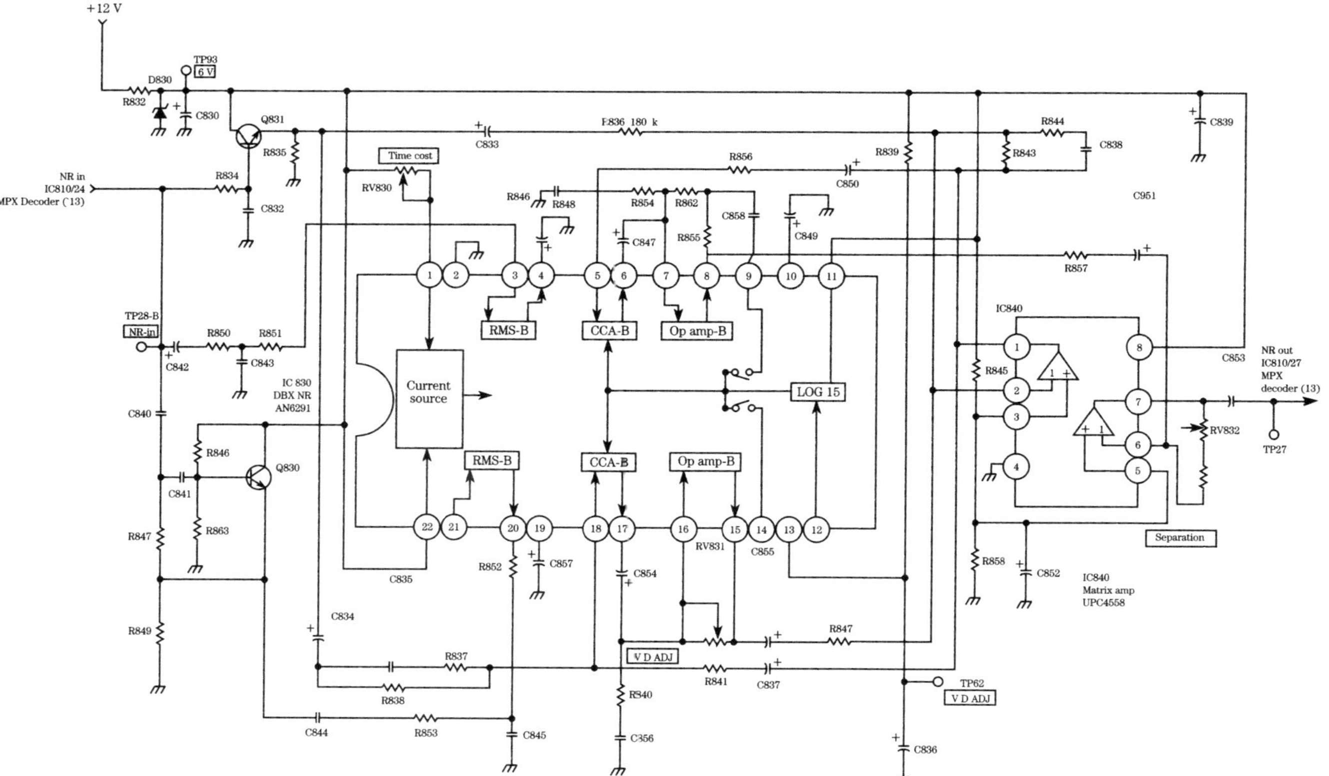

Figure 10.9 Noise reduction circuits (14).

The SAP and stereo signal (L–R) are input to the noise-reduction circuit at Q831. Transistor Q831 and the associated circuit provide the fixed deemphasis. The SAP and L–R are also applied through an LPF to pin 3 of IC830 (to one rms detector), and through Q830 and a high-pass filter to pin 20 (to the other rms detector).

Spectral deemphasis is performed on the signal by IC830. The output, after deemphasis, is applied to pin 2 of IC840 from pin 15 of IC830. The output from pin 1 of IC840 is applied to the wideband-processing circuits of IC830 at pin 5.

Wideband processing is performed on the signal by IC830. The output, after wideband processing, is applied to pin 6 of IC840 from pin 8 of IC830. The output from pin 7 of IC830 is applied to the decoder circuits (at pin 27 of IC810). The time-constant control RV830 at pin 1 of IC830 sets the level of current produced by the constant-current source or generator. The VD adjust control RV831 at pins 15/16 of IC830 sets the level of the spectral deemphasis op amp. The separation control RV832 connected across pins 6/7 of IC840 sets the gain of the noise-processed signal.

10.2.2 Audio processing

Figure 10.10 shows the audio-processing circuits for a typical TV set (Sony 19 inch). The left and right audio outputs from the MPX decoder IC810 are input to audio controller IC870 at pins 1 and 3. The left and right audio from the external audio inputs (section 10.4) is applied at pins 25/26 of IC870. The logic at pin 8 of IC870 determines which audio source is used. In turn, the logic is determined by system control IC101 (Figure 7.3). Operation of the IC101 audio-select function is determined by audio switch S870.

As an example, if S870 is set for TV audio, pin 41 of IC101 is returned to ground. Under these conditions, pin 11 of IC101 goes low, Q874 remains off, pin 8 of IC870 is high, and the TV audio signal at pins 1/3 is passed to the audio output (Figure 10.11) through pins 16/17 of IC870. The opposite occurs when S870 is set to EXT. That is, pin 11 of IC101 is high, Q874 turns on, pin 8 of IC870 is low, and the external audio signal at pins 25/26 is passed to the audio output.

No matter which audio is selected, the audio is affected by the bass, treble, and balance control signals at pins 10, 11, and 13, as well as by the volume-control signal at pin 14 (from pin 18 of IC101). Bass, treble, and balance are set manually by RV870, RV871, and RV872. Volume is set electronically by system-control IC101. (Electronic volume controls are discussed further in section 10.5.)

10.2.3 Audio output

Figure 10.11 shows the audio-output circuits for a typical TV set (Sony 19 inch). Both the left and right audio signals from the audio controller

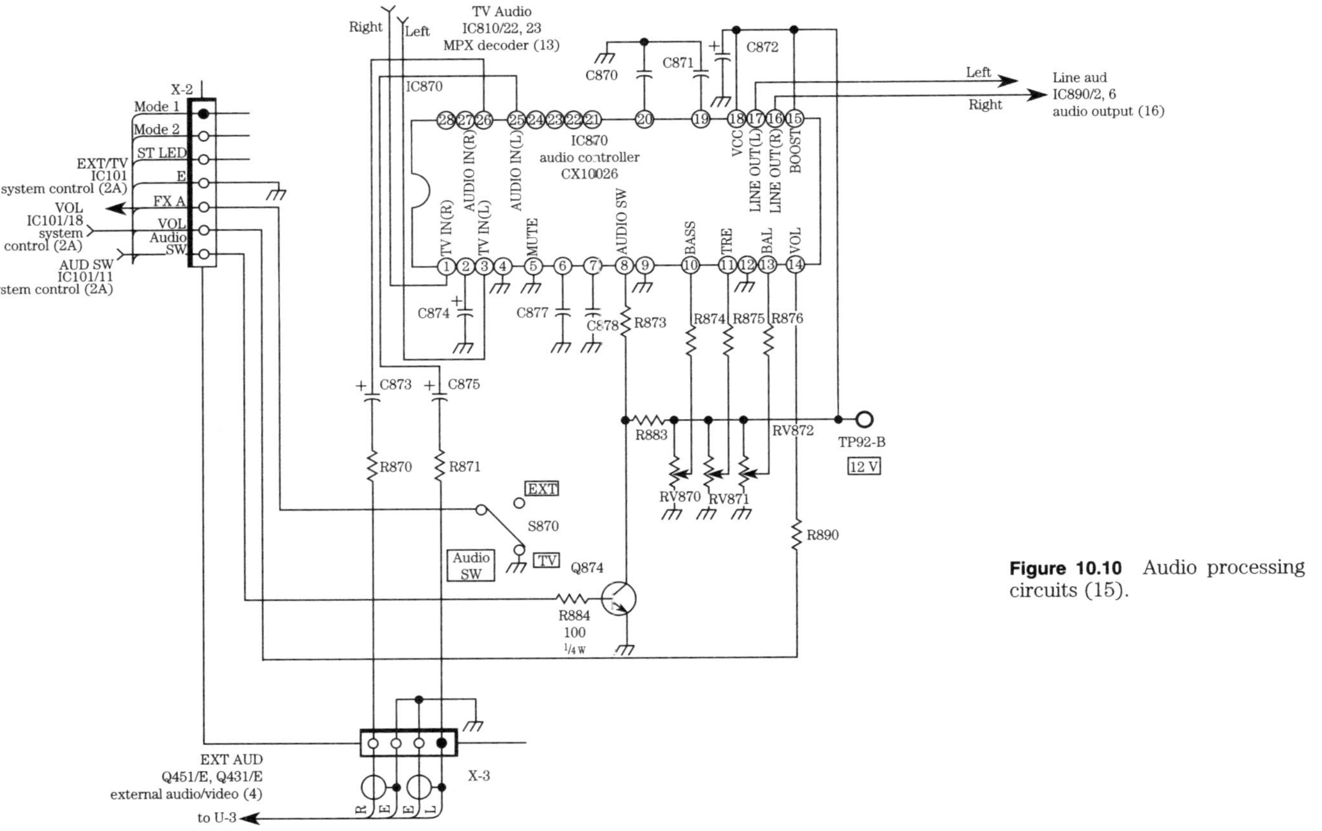

Figure 10.10 Audio processing circuits (15).

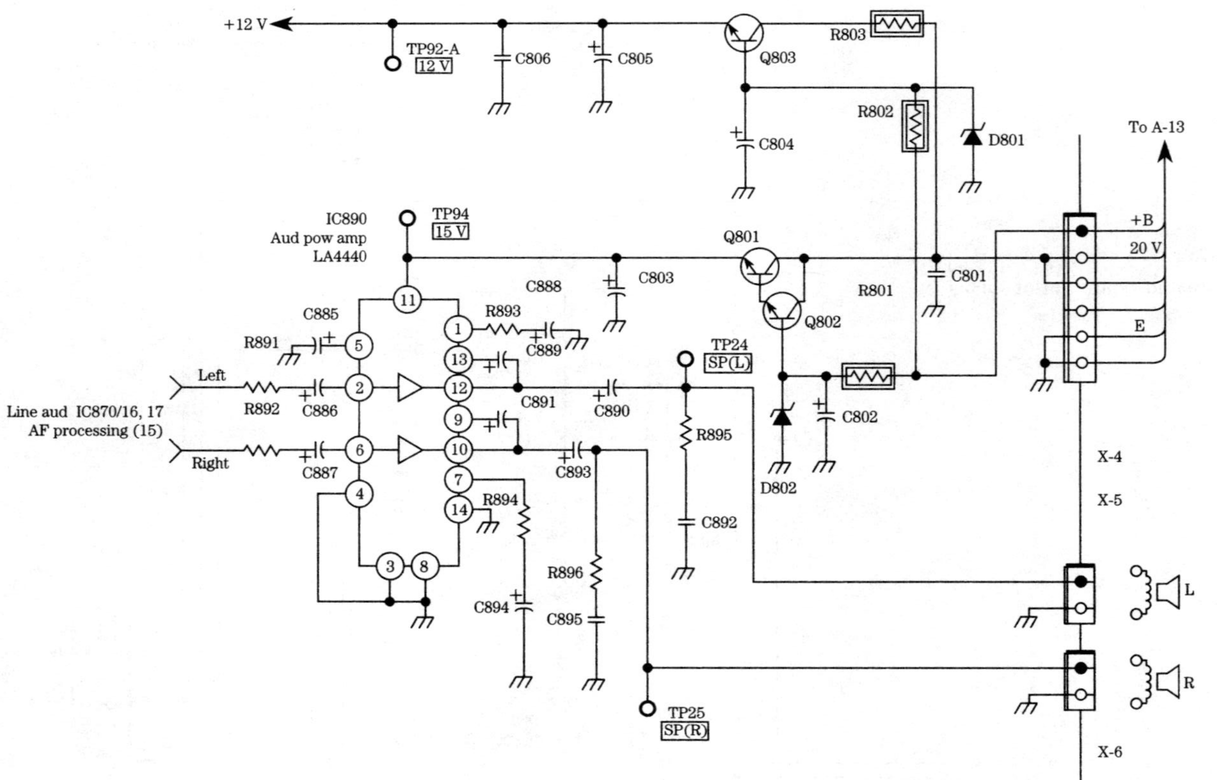

Figure 10.11 Audio output circuits (16).

(Figure 10.10) are applied to the corresponding loudspeakers through amplifiers in IC890.

Notice that 135 V from the main power supply (Figure 8.6) is applied to regulators Q802/Q803 through current-limiting resistors. The reference levels for the two regulators are set by zeners D801/D802. Also, 20 V is applied to the collectors of Q801–Q803. The 12-V output from Q803 is used throughout the audio-processing stages. The 15-V output of Q801 is used only by IC890.

10.3 Surround-Sound Circuits

Figure 10.12 shows some typical surround-sound configurations available to present-day stereo, TV, and VCR listeners and/or viewers. With surround sound (which is not directly related to stereo TV), it is possible to simulate the sound of motion-picture theaters in the home.

This section describes three of the most popular surround-sound configurations: matrix surround, Hall matrix, and Dolby Surround. These configurations might be called by other names (music surround, mono enhance, simulated stereo, time-delay surround, etc.) and can be provided by a built-in circuit (in TV or stereo receivers) or by external surround processors. Although all of the configurations produce a surround effect, only the Dolby systems (Dolby Surround or Dolby Prologic) are capable of decoding Dolby surround sound recorded on tape and film (and occasionally broadcast by TV stations).

10.3.1 Matrix surround

Figure 10.13A shows a basic matrix surround circuit. Such a circuit extracts "ambiance" information from a pair of stereo signals by finding the difference between the signals. When the speaker A switch is on, normal stereo sound is heard from both speakers connected to the A terminals. When speaker B is on, the sound is also available from the speakers connected to the B terminals. When both switches are on, the A speakers reproduce the normal stereo signal, but the B speakers produce the difference between the left- and right-channel signals (essentially the same as L–R found in stereo TV).

If the B speakers are placed behind the listener (Figure 10.12) ambiance is simulated or synthesized. Many matrix systems also include amplification. In some matrix-surround configurations (with amplification), 2L+R is produced from the front left speaker, 2R+L from the front right speaker, 2L–R from the rear left speaker, and 2R-L from the rear right speaker.

10.3.2 Hall surround or Hall matrix

Hall surround or Hall matrix (Figure 10.12B) is similar to the basic matrix, except that a time delay is added to the rear (surround) speakers. This

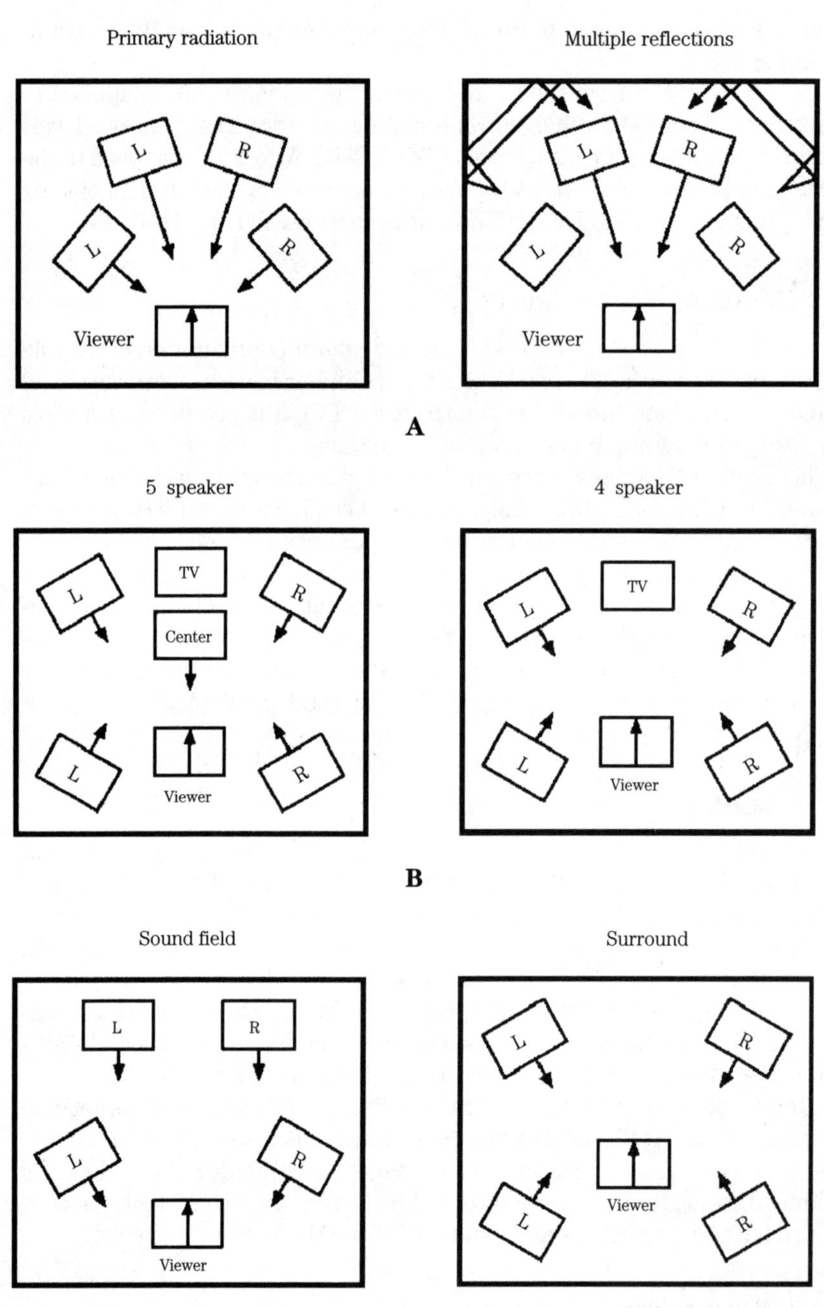

Figure 10.12 Typical surround-sound configurations. (A) Time delay or Hall; (B) Dolby; (C) matrix or simulated.

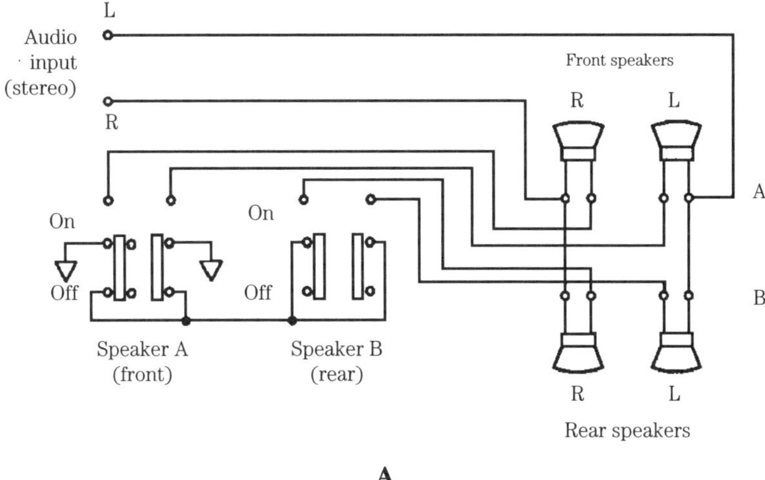

A

L
Audio
input
(stereo)
R
L
Front
speakers
R
L
Rear
speakers
R
Time delay

B

Figure 10.13 Matrix and Hall-surround circuits.

gives the illusion of spaciousness (such as being in a concert hall, thus the name Hall surround). The time delay might also produce sound reverberations (in some listening areas).

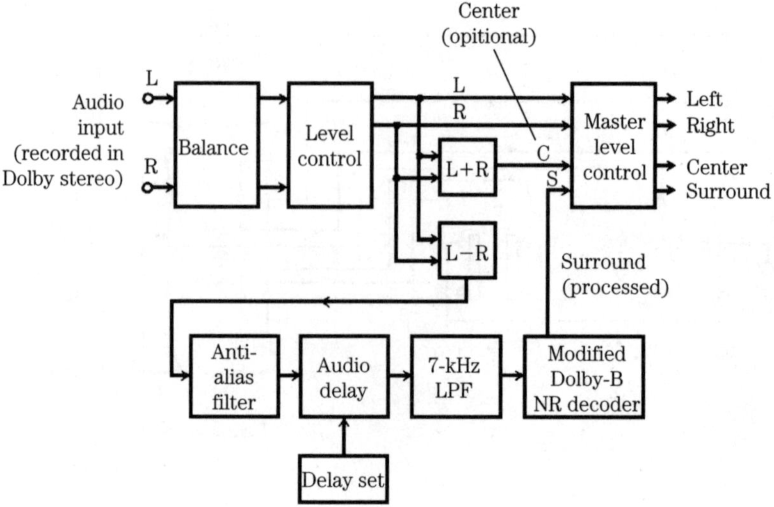

Figure 10.14 Basic Dolby Surround decoding circuit.

10.3.3 Dolby Surround processing

To understand either form of Dolby Surround (usually found in surround processors but not built into TV), it is necessary to understand the basics of the theatrical Dolby stereo encoding and decoding system. The basic Dolby Surround decoding circuit is shown in Figure 10.14. Notice that four channels of sound are encoded on only two sound tracks.

The basic Dolby decoder derives sound information by subtracting the right channel from the left (L–R). Unlike either Hall surround or matrix surround, Dolby also produces a center channel by summing the left and right channels (L+R). The mono center channel prevents a center "hole" when the L and R speakers are widely separated.

Like the Hall matrix, the basic Dolby decoder adds time delay to the surround output. The delay (of 15 and 30 ms) is to take advantage of the Haas effect. (The Haas effect causes the mind to identify the sound source as that from which the sound is first heard and to ignore the same sound arriving later at the ear.)

Unlike Hall matrix, the Dolby decoder cuts off the surround sound at 7 kHz and adds a modified form of Dolby B (5 dB of Dolby processing, instead of the usual 10 dB found in normal Dolby B). In addition to providing noise reduction, the modified Dolby prevents the rear surround signal from altering the front L and R audio.

10.3.4 Dolby Prologic

The ultimate Dolby surround system is the Dolby Prologic, which contains active circuits that provide steering logic for the recorded audio. As shown in Figure 10.15, this steering logic is part of an adaptive matrix (also called a directional-emphasis circuit and is found in IC form). The steering logic senses the direction of soundtrack dominance (from which direction the loudest sound on the track seems to originate) and generates control signals that increase the gain in the appropriate combinations of channels (left, right, center, surround).

By comparing the left, right, center, and surround signal pairs and taking the logarithms of the values, a pair of bipolar control signals is generated. (Logs are used, in part, because human hearing works in a logarithmic manner.) The bipolar control signals adjust gain of eight voltage-controlled amplifiers (VCAs), four VCAs for each channel. The output of the VCAs, together with the L and R signals, produces a total of 10 control signals.

When the control signals are applied to the four output channels, a total of 40 summed directional-audio components are available. Separation between any pair of channels, adjacent or opposite, is 30 dB. This compares to Dolby Surround's 3 dB of adjacent separation and 40 dB of opposite separation. Prologic decoders are two-speed devices. When only one sound source dominates, the circuits are in a slow mode. When there are two distinct sound sources, the circuits go into a fast time-division multiplexing mode, where the circuits control one source and then the other. The de-

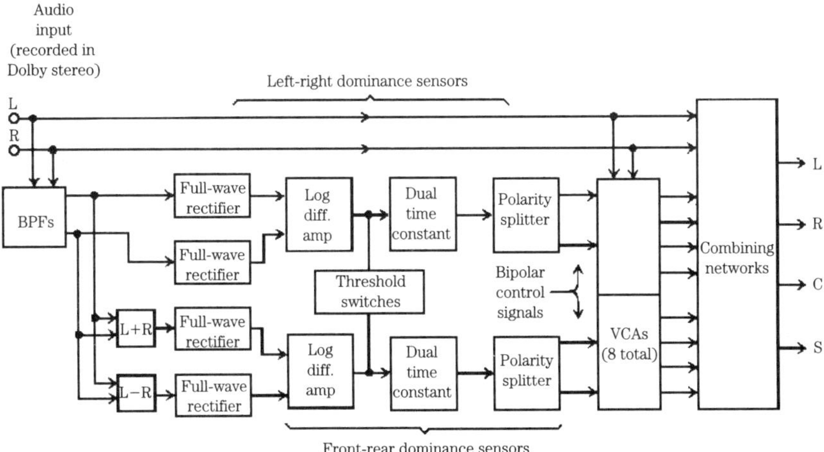

Figure 10.15 Dolby Prologic circuits.

coder switches back and forth between the two sources so quickly that the effect is unnoticed by the listener.

10.3.5 Surround-sound troubleshooting

Troubleshooting for the surround-sound circuits is essentially a matter of signal tracing, even for the more complex Dolby Prologic. A possible exception is when the circuits must be adjusted. Of course, the service-literature procedures must be used for adjustment (and there are no general or typical procedures for surround-sound adjustment). Notice that some circuits (particularly those in surround processors with Dolby Prologic) have built-in signal-generator circuits to produce simulated Dolby recordings during test.

10.3.6 Typical surround-sound circuits

Figures 10.16 and 10.17 show some typical surround-sound circuits found in TV sets (several different Hitachi models).

The circuit of Figure 10.16 applies the difference-signal components L–R and R-L to the built-in stereo speakers so that a sound with "presence" is produced on either side of the TV set. The L and R signals are passed through difference amplifier Q4501/Q4502 to produce the L–R and R–L signals. The L–R and R–L signals are then added to the L and R signals, respectively, and the resultant sum signals are output to the speakers (through the set's stereo amplifiers).

The surround-sound effect is turned on when electronic switch Q4056 is actuated (base high) by a signal from the tuner microprocessor (in this particular set). The SURROUND ADJ control R4057 equalizes the levels of the L and R signals applied to the difference amplifier Q4051/Q4052 in the monaural mode so that no signal is output from the amplifier and surround-sound circuits.

In the circuit of Figure 10.17, the R and L audio signals taken from the signal-switching circuit are applied directly to the audio control (and then to the speakers) when the surround switch S4051 is set to OFF (connecting the bases of Q4051/Q4052 to ground through R4075). When S4051 is set to ON, difference amplifier Q4051/Q4052 is turned on, and the difference of the two inputs applied to the bases develops at the corresponding collectors. L–R develops at the collector of Q4051 and R–L develops at Q4052. These R–L and L–R signals are combined with the R and L signals to produce the surround-sound effect.

10.4 External-Audio Circuits

Figure 10.18 shows the external audio circuits of a Sony 19-inch TV set. Notice that the circuits also provide for external video, and all three circuits

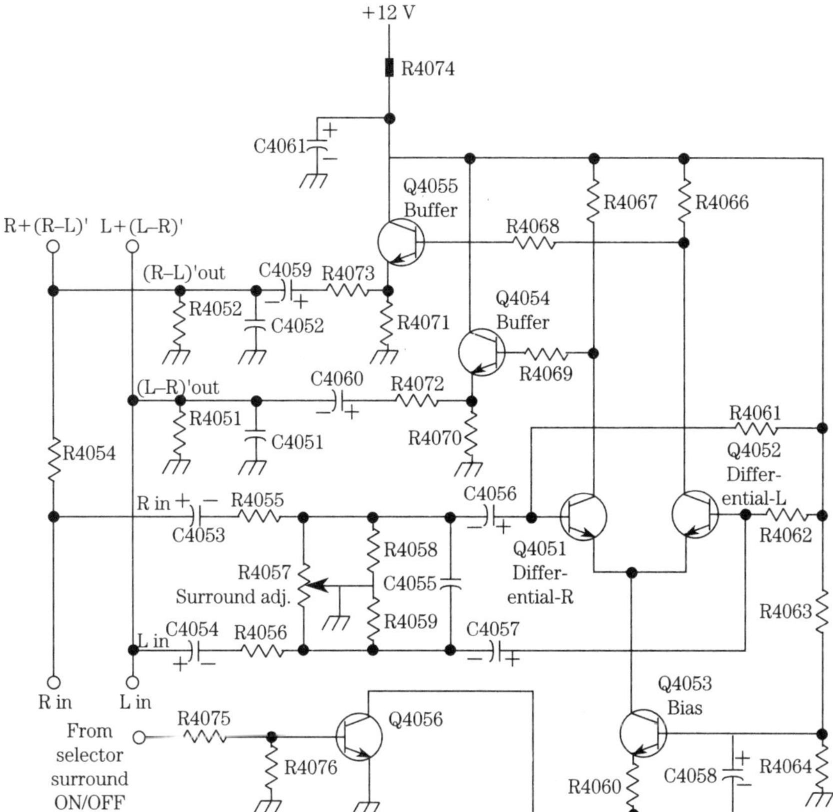

Figure 10.16 Surround-sound circuit.

(video, left channel, right channel) use optocouplers to isolate the external signal sources from the TV set circuits. External video at the VIDEO IN jack is applied to the video-processing circuits (Figure 10.10), and is adjusted by RV430 (right channel) and RV450 (left channel).

Because of their simplicity, troubleshooting for the external-audio circuits should present no problem. Basic trouble isolation and signal tracing (or signal injection) should be sufficient. For example, assume that video and left-channel audio are good, but there is a problem in the right-channel audio.

First try to correct the problem by adjusting RV430. Then try signal tracing through the corresponding circuits. With an identical audio signal at both VIDEO IN jacks, compare the signals at the emitters of Q420/Q440, pin 2 of IC421/IC441, collectors of Q430/Q450, and the emitters of Q431/Q451. The signals should be substantially the same at each test point. If not, any difference in signal should point to the problem. For example, if the signals at the emitters of Q420/Q440 are identical, but there is no signal at pin 2 of

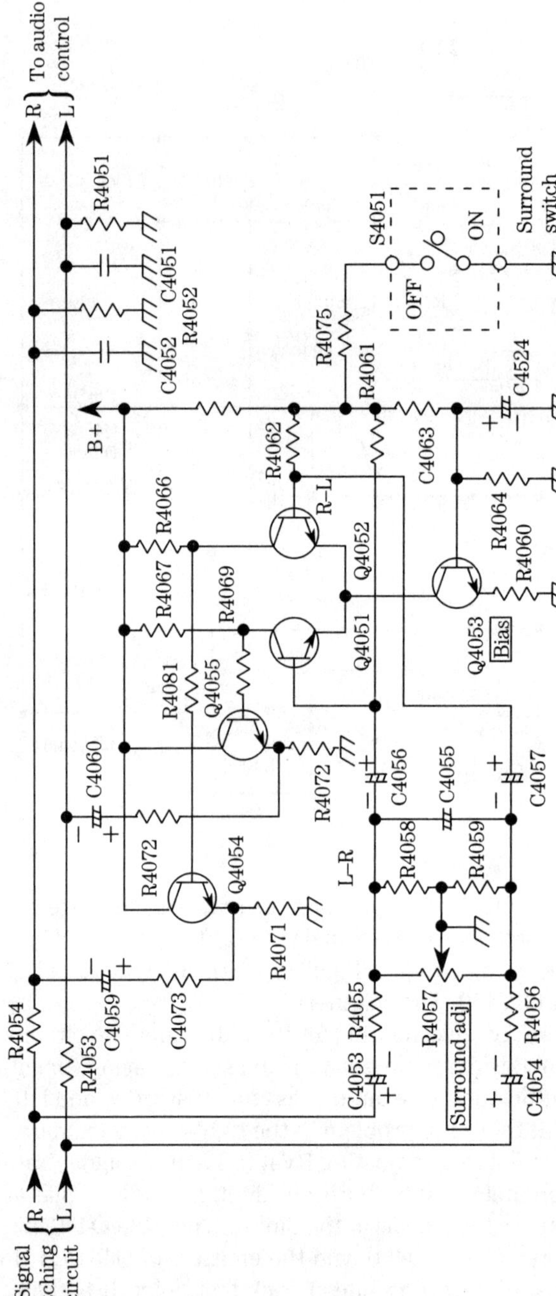

Figure 10.17 Surround-sound circuit with manual switch.

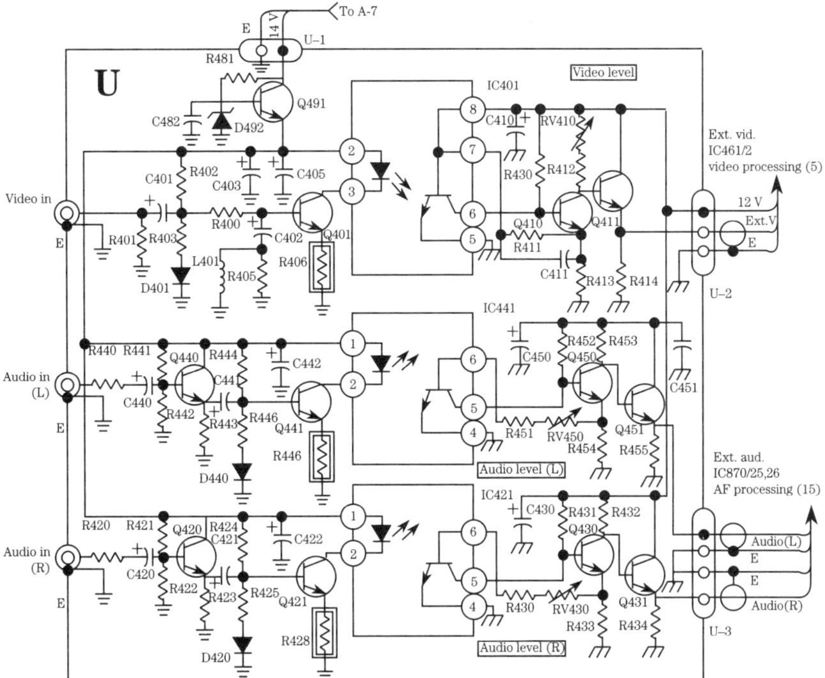

Figure 10.18 External audio/video circuits (4).

IC421 (with a signal at pin 2 of IC441), suspect Q421. Notice that if there is a total lack of signals beyond the IC optocouplers in all three channels, suspect regulator Q491 or D492, because these components apply power to the LEDs in all three optocouplers.

10.5 Electronic Volume-Control Circuits

The audio circuits of most present-day TV sets are provided with an electronic volume control. In some cases, the audio volume is set by an audio processor IC (such as in Figure 10.10), as directed by system control. In turn, system control receives audio volume commands from front-panel buttons and/or the remote transmitter. In other cases, there are separate ICs to control audio. I have described such a system here.

Figure 10.19 shows the electronic volume-control circuit for our set (which has stereo TV and is part of a modular home-entertainment system). Notice that only the left stereo channel is shown.

The volume up, volume down, muting, and loudness functions are controlled by IC604 (sometimes called a volume control, other times called an attenuator). Volume is adjusted in 40 steps, including full muting. (Notice

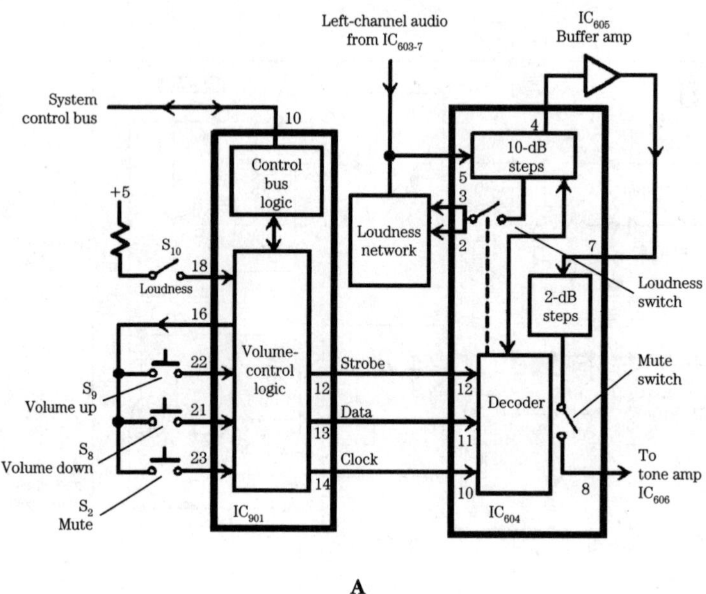

A

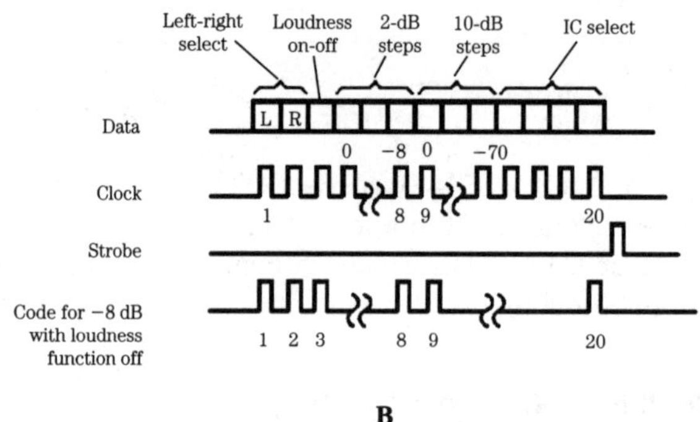

B

Figure 10.19 Electronic volume-control circuits.

that most electronic volume controls provide for adjustment of the volume in steps, rather than continuous adjustment.) Volume control IC604 is under control of microprocessor IC901.

Commands from front-panel switches (or system control) are applied to IC901, which generates a volume-control code that is applied to IC604. The code includes clock, data, and strobe information at pins 10, 11, and 12 of IC604 respectively. The code is applied to decoder circuits within IC604 and causes IC604 switches and attenuators to be selected.

Figure 10.19B shows a typical volume-control code. IC604 receives 20 bits of serial data from IC901. The 20 bits are transmitted at the clock rate (4 MHz), or 1 bit per clock pulse. Once the 20 bits are transmitted, the strobe signal is transmitted and instructs IC604 to produce the correct amount of attenuation.

As an example, assume that the volume is already at –10 dB, without the loudness function, and it is desired to turn the loudness function on, with a volume of –8 dB. The front-panel loudness button is pressed (once), and the volume up button is held until the desired –8 dB is obtained. Under these conditions, bits 1, 2, 3, 8, 9, and 20 are selected (data line high).

Audio from circuits ahead of IC604 is applied to 10-dB attenuators in IC604 through pin 5. The output of the 10-dB attenuators is applied to 2-dB attenuators through buffer amplifier IC605. The output of the 2-dB attenuators is applied to the audio output circuits through a mute switch in IC604.

The 10-dB attenuator in IC604 has eight steps, 0 to –70 dB, while the 2-dB attenuator has six steps, 0 to –8 dB. With the volume-control setting at minimum volume (–78 dB attenuation), continuously pressing (or holding) the volume up button causes the 2-dB attenuator to step up from –8 to 0 dB, in 2-dB steps. Once the 2-dB attenuator reaches 0 dB, the attenuator resets to –8 dB. At the same time, the 10-dB attenuator switches to –68 dB. This results in a 2-dB step, from –70 to –68 dB.

At low-volume settings, pressing the loudness button causes the loudness switch in IC604 to close, connecting a loudness network between pins 2, 3, and 5 of IC604. The loudness circuit attenuates the midrange audio frequencies and passes the bass and treble frequencies. This has the effect of supplying (or reinforcing) positive feedback to the audio at pin 5 of IC604 (at high and low frequencies).

Pressing the mute button opens the mute switch in IC604 to interrupt the audio. However, this does not affect the attenuators. When mute is pressed again, the audio is restored at the same level of attenuation (unless the attenuation is changed during mute).

10.5.1 Electronic volume-control troubleshooting

If the volume control does not operate, check for audio at pin 5 of IC604. If there is none, check the audio source (in this case, left-channel audio at pin 7 of IC603). Remember that the electronic volume control is in the audio path, typically just ahead of the final audio-amplifier IC and loudspeakers. So before you condemn the volume-control circuits, check the audio source ahead of the control.

If there is audio at pin 5 of IC604, check for audio at pin 4 while operating the volume up and volume down buttons. The audio volume should increase in 10-dB steps at pin 4 each time you press the corresponding button. (If you hold the volume buttons down, the audio should increase or decrease steadily in 10-dB steps.)

If there is no audio at pin 4 of IC604, with audio at pin 5, suspect IC604. If there is audio at pin 4 but there is no change when the volume buttons are pressed, monitor strobe (pin 12), data (pin 13), and clock (pin 14) outputs from IC901 while holding the volume buttons.

Although you probably cannot decode the information on the control lines, the presence of pulse activity on the lines usually indicates that IC901 is good. If any one of the lines shows no activity with the volume buttons operated, suspect IC901. On the other hand, if there is pulse activity on all three lines but there is no change in the volume at pin 4 of IC604 (with the volume buttons held down), suspect IC604. If there is audio at pin 4, and the volume changes, check for audio at pin 7 of IC604. If the audio is absent, suspect IC605.

If there is audio at pin 4 of IC604, check for audio at pin 8, and make sure that audio changes in 2-dB steps when the volume buttons are operated. If there is no audio, suspect IC604. If there is audio but there is no change with the volume buttons operated, suspect either IC901 or IC604. Notice that IC604 is the most likely suspect if the audio is good at pin 7, but it is possible that IC901 is not generating the correct code to produce 2-dB changes in volume.

If there is audio at pin 8 and the audio changes in 2-dB steps when the volume buttons are operated, press mute and check that audio is cut off at pin 8. If not, suspect the mute switch circuit, IC901, or IC604. You can also check for pulse activity on the data-code outputs from IC901 (but you probably cannot decipher the code). Press mute again, and check that audio is restored at pin 8 and is at the same level (if you have not pressed the volume buttons during the mute condition). If audio is not restored, suspect the mute circuit, IC901, or IC604.

Note that the loudness network is part of the electronic volume control (although there are external components). Also note that a failure symptom for the loudness functions is usually difficult to define. This is because the loudness function attenuates the midrange signals so that the ear hears what appears to be the same level across the audio range. (In some circuits, the loudness function boosts the treble and bass, but this is rare.)

Unfortunately, all ears are not the same, and not all loudness networks define midrange at the same frequencies. Usually, the customer complains that "there is no difference when the loudness function is on or off." To troubleshoot such a symptom, apply an audio signal (say between 7 and 10 kHz) with loudness off. Then press loudness and check for a drop of about 20 dB in level at pin 4 of IC604. Repeat the test at 50 Hz and 20 kHz. There should be substantially no change in level (at pin 4 of IC604) at the low and high ends of the audio range (unless a boost circuit is used), even though there is a change at the midrange.

No matter what type of loudness circuit is used, if there is no change in audio level at any frequency when the loudness function is switched in and out, there is a problem in the loudness circuit. Start by checking the loudness switch circuit, IC901, IC604, and the network connected at pins 2 and 3 of IC604.

10.6 Digital-TV Circuits

Some present-day TV sets include digital circuits. Such circuits receive conventional TV signals and produce corresponding pictures and sound (although both video and sound are generally superior to those of conventional TV).

The key to digital TV is in analog-to-digital (A/D) and digital-to-analog (D/A) conversion. In simple terms, the analog signals (composite video and sound) at the output of a TV tuner or front end are converted to the digital equivalent by an A/D process somewhat similar to that found in compact discs. The resulting digital signals are then processed to produce the corresponding video and audio in digital form. When the processing is complete, the digital signals are restored to analog form by D/A converters and applied to the CRT and audio circuits. The A/D conversion, digital processing, and D/A conversion all take place in digital ICs.

10.6.1 Digital TV basics

The obvious advantage of digital processing is in the quality of the signal reproduced. As in the case of compact discs, the tuner signals to be processed are sampled (at high frequency), the sampled signals are stored digitally, and after processing, the restored signals are retrieved.

Another advantage is that a digital TV set can easily be adapted to the three basic TV systems because the sampling clock for the A/D converter is phase-locked to the broadcast color-burst frequency. Simply by changing the frequency of the clock, the system can accommodate NTSC (3.58-MHz), or PAL (4.42-MHz) color-burst systems. The same digital TV can also be used for SECAM in black and white. However, the system must be modified for color SECAM (because the PAL and SECAM color techniques are different).

For those not familiar with the three TV broadcast systems, the NTSC system (525 line) is used exclusively in North America and widely in Latin America and Japan. PAL (Phase Alternate by Line, 625 line) is used in the United Kingdom and most of Western Europe, Australia, New Zealand, and South Africa. SECAM (sequential color with Memory, or in French, Sequential Coleur avec Memorie, 625 line) is used in France and Eastern Europe.

Because the TV picture can be stored (in digital form), a digital TV can reduce flicker caused by interlaced scanning and can increase the apparent resolution of the picture. Digital TV eliminates interlace problems by storing all 525 lines and displaying the complete picture on the CRT screen all at once instead of having only half the scan lines on the screen for each field of video, as is the case with conventional TV.

In addition to these obvious advantages, digital TV is generally superior because sync is checked on each horizontal line with PLL circuits, and there are a minimum of capacitors, inductors, and RC circuits to break down or to distort video signals.

10.6.2 Special effects

In addition to superior performance, the big selling feature of digital TV is the ability to display special effects. These effects are programmed into the digital ICs. The following is a brief summary. Notice that all of the effects described here are not necessarily available on all digital TVs.

The mosaic and paint effects are available on some digital TV systems. With mosaic, the screen is a picture composed of small blocks. The paint effect is similar to an oil painting and is also called posterization.

The freeze mode freezes the current picture being viewed. Either field or frame reproduction can be selected to minimize jitter. Simultaneously, a real-time picture can be set into one corner of the CRT screen if desired.

The preview mode displays the still picture of nine channels in sequence, arranged in three rows and three columns on the screen. At predetermined times, the display is changed so that all channels programmed into the tuning-system memory are scanned.

The picture-in-picture mode inserts a ⅙-normal-size real-time picture from an extra video input (possibly a second tuner) in the corner of the screen. The main picture and insert picture can be exchanged at any time by pressing a single button. Notice that it is possible to have the picture-in-picture function without full digital TV operation. Picture-in-picture circuits are discussed further in section 10.7.

The strobe mode displays eight time-sequenced still pictures at once, while showing the real-time picture in the lower corner of the screen. The editing mode allows the user to change the still pictures displayed in the strobe mode.

10.6.3 Five-chip digital TV

Figure 10.20 shows the block diagram of a five-chip digital TV system (the Zenith Digital System 3). Compare this block to those of conventional TVs shown throughout this book. If you look closely, you will find seven chips or ICs. However, the clock chip is generally not counted, and there are typically two identical audio-processor chips (for stereo TV operation).

There are two points to consider when studying the five-chip system. First, not all digital TVs have all five chips. Second, although analog signals

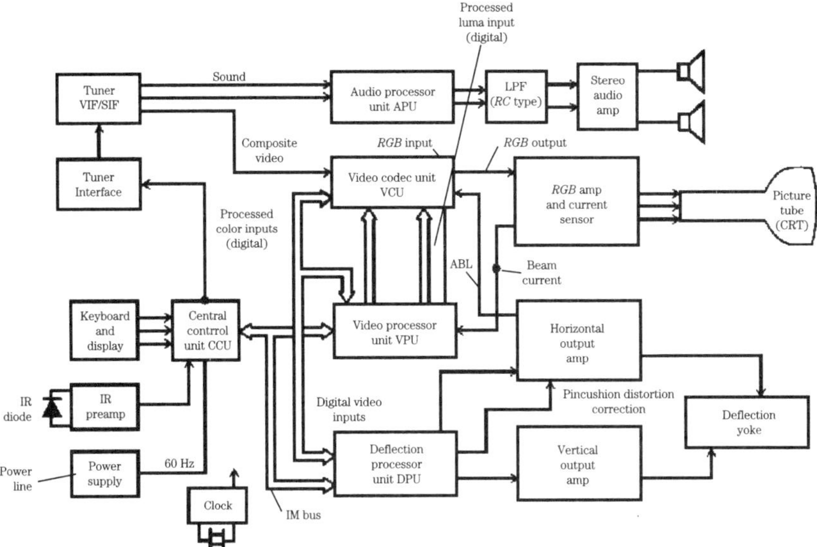

Figure 10.20 Basic digital TV system.

are fed in and analog signals come out, all signal processing is done digitally in the five-chip system. This is not necessarily true for some of the digital TVs that do not use the five-chip system.

I will not describe full operation of all chips here (only what is necessary to understand the troubleshooting approach for digital TV). If you want an exhaustive discussion of digital-TV ICs, read *Lenk's Video Handbook* (McGraw-Hill, 1991).

The circuits for the Digital System 3 are contained on six modules. Five of these modules include conventional circuits and/or perform TV set operations that are used on other (nondigital) Zenith sets. I will not concentrate on the five conventional modules here, except for the input/output relationship to the digital circuits. All circuits that are unique to the digital functions are contained on one module, designated as the digital main module 9-535, and generally referred to as the "9-535" or "main" module.

Figure 10.21 shows the IC layout of the 9-535 module. From a troubleshooting standpoint, there are several important features to remember concerning the digital main module.

First, with one exception, all of the ICs on the 9-535 (both digital and nondigital) are plug-in type. The exception is the ADC IC1404, which is permanently installed on the module. As with any plug-in device, if all other troubleshooting steps fail, you can replace the ICs one at a time until the problem is cleared (or you can replace the entire module at unbelievable expense to the customer).

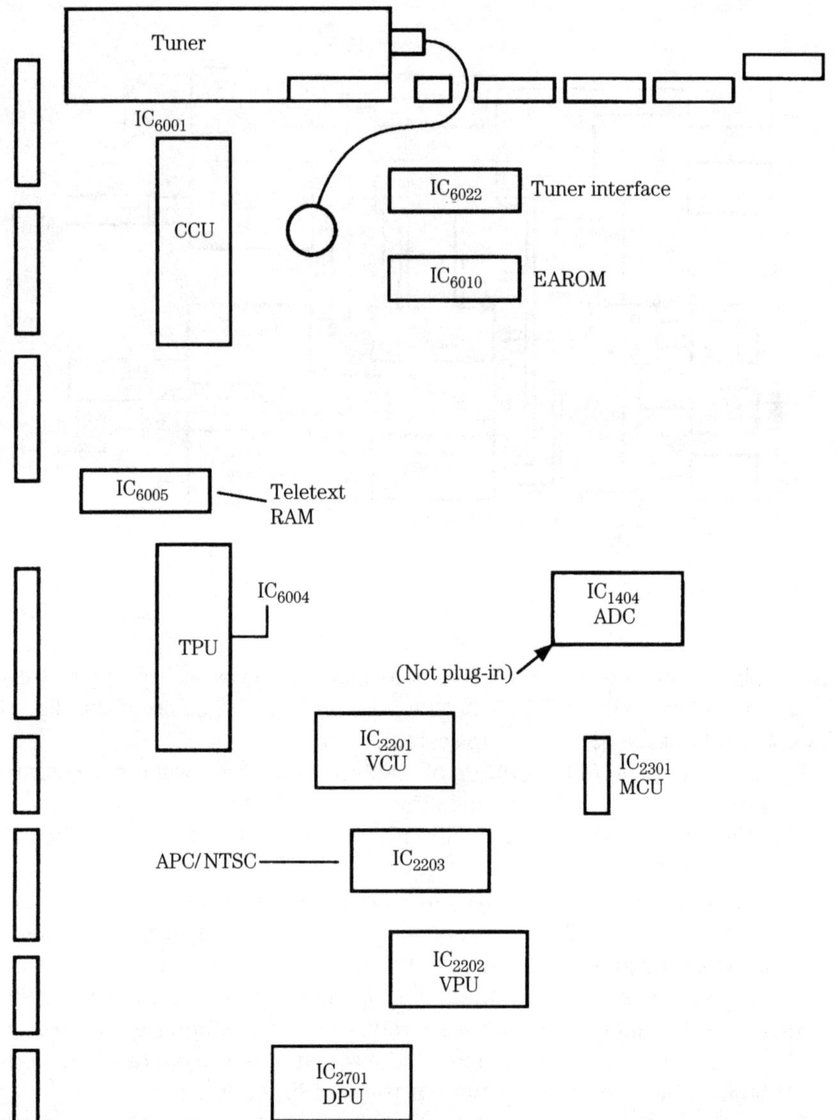

Figure 10.21 Digital main-module layout.

All the voltages and signals to and from the digital ICs can be measured at this module. If the voltages and/or signals are incorrect, you can trace back to the source from the module. That is the approach used here.

10.6.4 Basic digital TV troubleshooting

Before getting into troubleshooting details for the Zenith Digital System 3, let us review the basic troubleshooting approach for any digital TV (or for any TV with digital circuits). Remember that (in this section) we are concerned with locating troubles caused by the digital video ICs, not with troubleshooting the remaining circuits (which are the same as those found in other nondigital TVs).

Plug-in IC replacement. Because all the Zenith digital ICs plug in (except for the ADC), it is practical to try correcting the problem by replacement. (This is not practical where ICs are wired in.) By replacing each IC in turn with a known-good IC and checking to see if the trouble symptoms are eliminated, you can cure about 90 percent of the problems in a digital TV (except for those problems caused by adjustment, which are covered in the service literature).

Of course, it is helpful if you start by replacing the ICs most likely to cause the problem. For example, in the Zenith set, if the problem is bad video, with good audio, start by replacing the VCU, VPU, APC, and NTSC chips. If the problem appears to be in deflection, either horizontal or vertical, start with the DPU. If the problem cannot be easily localized, start with the CCU.

Obviously, to make a logical choice for replacement of ICs, you must be able to group those ICs that perform specific functions. I've done just that for the Zenith Digital System 3 in the following troubleshooting discussions. Now let's consider the terrible possibility that all of the digital ICs have been replaced with known-good ICs, all adjustments have been made (in accordance with the service literature), but the problem remains.

Modular replacement. At this point, if a known-good digital main module is available, you can try substitution. If this cures the problem, you can then trace the problem on the defective module. During replacement, you might find that the problem is one of poor electrical connection (such as loose connectors, dirty contacts, etc.).

Unfortunately, modular replacement might not always be practical. You might not have a replacement module readily available, or you might have only one shop-standard module (that you will not surrender at any price). A shop-standard module is recommended if you plan to service a particular model of digital TV. Of course, you might not want to invest in dozens of known-good modules for a variety of digital TVs.

If you choose modular replacement, take special care when reinstalling the shielding systems. Digital video signals are quite high in amplitude and frequency. This can cause interference in nearby electronic equipment or in the picture being received. Such problems are most evident in areas where signal levels are weak, and where antennas are used (instead of cable). To eliminate problems, make certain that all shield covers are reinstalled and that locking clips are tight. All ground connections should be carefully resoldered.

Checking the ICs. When the problem is not corrected by substitution or adjustment, check all power and ground connections to the ICs. Then check all reset connections and clock signals. As shown in Figure 10.22, there is one reset line, but two clock lines, for the ICs in the Zenith set. As discussed in other chapters, if any one IC does not have proper ground/power connections, does not have a reset command, or does not receive the required clock signals, the entire digital operation is impaired.

Input-output signals. Once you are certain that all ICs are good, and have proper power and ground connections, and that all reset and clock signals are available, the next step is to monitor all input and output signals at each IC. This can be done with a scope, although a logic or digital probe might prove convenient. There are a number of probes on the market (including a Zenith Digi-Probe).

10.6.5 Digital horizontal-drive circuits

As shown in Figure 10.23A, the horizontal-drive circuits originate with the DPU. When IC2701 is turned on, drive pulses appear at pin 31. These pulses are applied to the sweep-module circuits through Q2700, Q2701, pin 5 of connector 3A4, and pin 1 of connector 4A3.

The drive pulses are applied to flyback circuits on the sweep module through horizontal predrive XQ3202 and drive XQ3201. The circuits produce flyback pulses for the CRT and develop secondary supply voltage in the usual manner (chapter 5). I do not discuss flyback circuits here, because such circuits are essentially the same as for other nondigital Zenith sets.

Digital horizontal drive troubleshooting. Start by checking for proper power, clock, and reset signals to IC2701, as discussed in section 10.6.4. Next check for horizontal-drive pulses at pin 31 of IC2701. Then check for pulses at pin 1 of connector 4A3 on the sweep module.

If the pulses are absent at pin 31 of IC2701, with IC2701 on, suspect IC2701. If pulses are present at pin 31, but not at pin 5 of connector 3A4, suspect Q2700 or Q2701.

If drive pulses are available at pin 1 of connector 4A3, but there is no horizontal sweep (or any other indication that the set is off, such as lack of secondary supply voltages for the CRT), check the sweep-module circuits, starting with XQ3202.

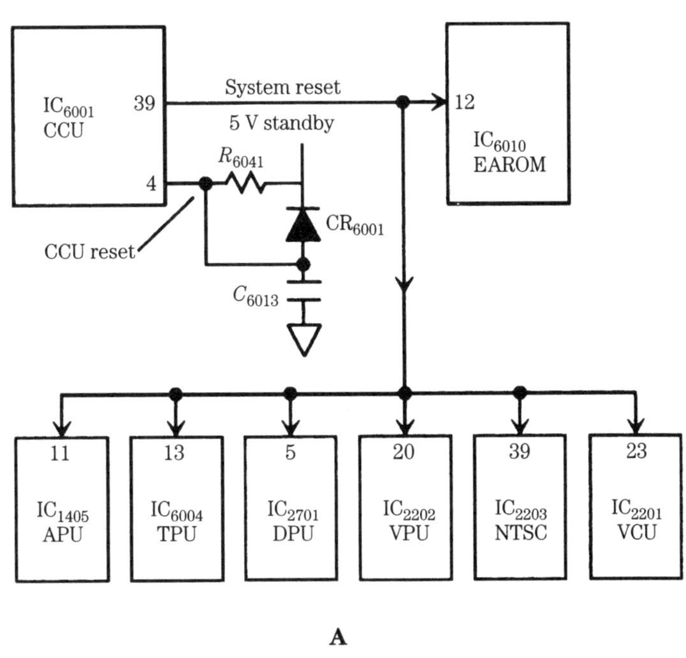

A

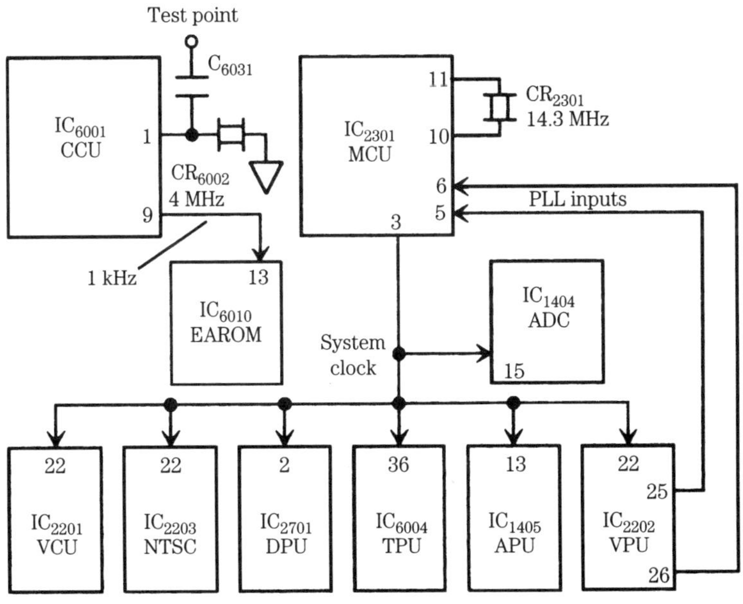

B

Figure 10.22 Reset and clock signals for digital ICs.

10.6.6 Digital vertical sweep circuits

Figure 10.23B shows the vertical-sweep circuits. Generally, there is no point checking the vertical circuits until the horizontal circuits have been checked. The vertical sweep is produced by circuits on both the 9-535 and 9-370 modules. Notice that the 9-370 module provides both power-supply and vertical-signal functions (in addition to other functions not related to the video ICs).

The vertical-sweep signal is generated in the DPU and appears at pins 26 and 27. These outputs are combined in RC circuits to produce a typical vertical sawtooth (chapter 5) at the emitter at Q2100. The sawtooth signal is applied to vertical amplifier IC2100 on the sweep module. The output of IC2100 is applied to the vertical yoke in the usual manner.

Digital vertical-sweep troubleshooting. Troubleshooting for the vertical circuits is straightforward. Check for pulses at pins 26/27 of IC2701, for a sawtooth at Q2100, and for a sawtooth at the input of IC2100. Notice that IC2100 is the IC version of the vertical-yoke drive found on many sets (as discussed in chapter 5).

If the input to Q2100 is not good, suspect IC2701 or the RC components. If the input to IC2100 is not good (no sawtooth), suspect Q2100 or the related circuits. The sawtooth should be about 4 V peak to peak. If the input to IC2100 is good but there is no vertical sweep, suspect IC2100 or the vertical yoke.

10.6.7 Digital video circuits

Figure 10.23C shows the digital video circuits. As discussed, most of the video processing takes place in the VCU, VPU, APC, and NTSC. Likewise, most of the processing is in the form of digital signals. In basic terms, the composite-video signal is converted to digital form in the VCU IC2201. The digital composite video is separated into luma and chroma signals in VPU IC2202. The separated luma and chroma signals are processed in APC and NTSC IC2203 and returned to IC2201. (Notice that IC2203 is called the APC in some sets, and the NTSC in others.)

All of the signals in the video-processing loop are difficult to monitor. Also, both IC2202 and IC2203 are under control of IM BUS signals (from CCU IC6001) that are equally difficult to monitor. You can check that the signals are present on each line, but not that the signals are correct.

Digital video troubleshooting. Although troubleshooting for the video circuits appears to be difficult (in a digital TV), remember that the inputs to the video loop (pins 35/37 of IC2201) are conventional video/audio signals that can be monitored and traced back to the source (VIF/SIF, chapter 3, or external video jacks). Similarly, the output from the loop (pins 26, 27, and 28 of IC2201) are conventional RGB signals (chapter 6) that can be monitored and traced to the video-output circuits or module.

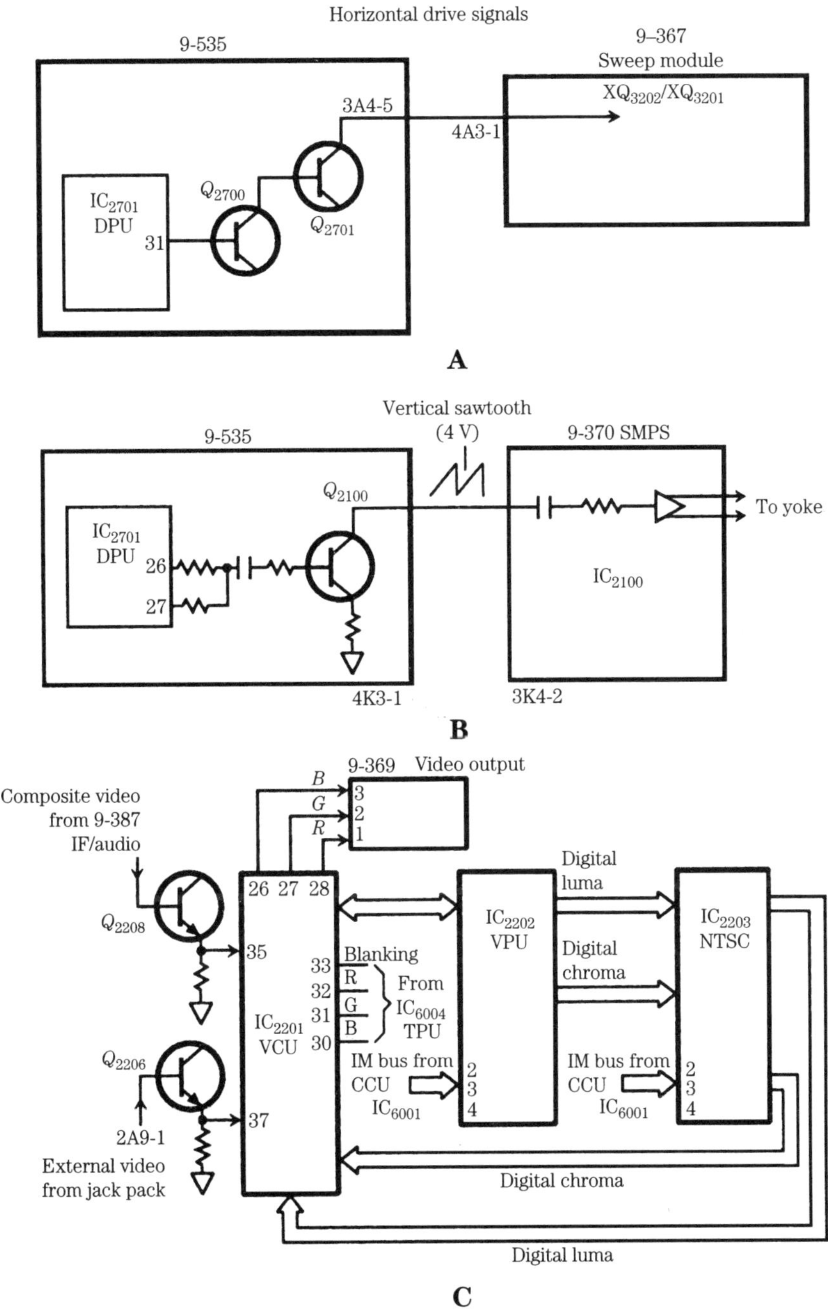

Figure 10.23 Horizontal drive, vertical sweep, and video processing for digital TV.

In simple terms, if the signals applied to pins 35/27 of IC2201 are good, but the signals at pins 26, 27, and 28 are bad, you have isolated video problems at IC2201, IC2202, or IC2203 or to the loop circuits (typically PC wiring and connectors). Of course, it is possible that CCU IC6001 is producing false information on the IM BUS, input to IC2202 or IC2203, the bus is probably at fault (shorted or stuck line). If you want a thorough discussion of digital and bus troubleshooting, read *Lenk's Digital Handbook* (McGraw-Hill, 1993).

Notice that on-screen teletext information developed in TPU IC6004 is added into the video-processing loop at pins 30 through 33 of IC2001. If TPU IC6004 is suspected of causing a video problem, remove IC6004. The set should operate normally, as far as video is concerned, with IC6004 removed. Of course, if all video functions except teletext are normal, you have isolated the problem to IC6004 and the related circuits.

In addition to processing the composite video signal, the circuits of Figure 10.23C are also the point at which viewer selections are processed. For example, if the tint button is pressed, commands are applied to the CCU through the remote-control system. These commands are converted to digital signals on the IM BUS (by the CCU) and are applied to IC2202 and IC2203 (at pins 2, 3, and 4).

10.7 Picture-in-Picture Circuits

As discussed in section 10.6, it is not necessary for a TV set to have all digital features (special effects) to produce the picture-in-picture (PinP) function. It is possible to provide a conventional TV set with PinP using a few added ICs. One such circuit (found in some Hitachi models) is discussed in this section.

10.7.1 Basic PinP processing

Figures 10.24 and 10.25 show the basics of PinP processing. In this basic circuit, PinP is a two-picture display (for example, the subpicture from tuner 2), located in one of the four corners of the main picture (the picture from tuner 1). The subpicture area is reduced to $\frac{1}{9}$ or $\frac{1}{12}$ of the main picture. The choice of main and subpicture is controlled by A/V switching (under direction of the remote control, in the usual manner).

The original picture (before PinP processing) is formed from the selected subpicture input with 49 μs (H) and 460 lines (V), as shown in Figure 10.24. The selected original picture is written into a DRAM (in 4-bit data units or groups) using the A/D-converter section of the PinP processor circuits. The data bits are read out from the DRAM and inserted into the main picture after conversion by the D/A portion of the PinP unit. Timing for insertion into the main picture and for digital processing is controlled by fsc (the chroma subcarrier frequency) and the sync signal.

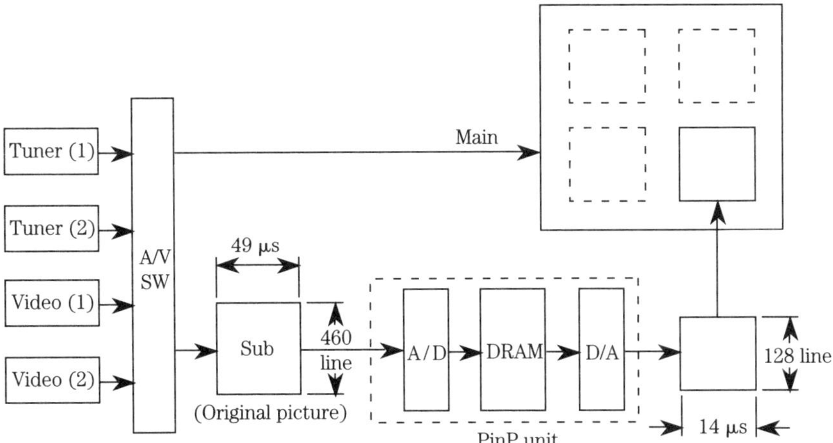

Figure 10.24 Principle of picture-in-picture.

The diagram on the upper left of Figure 10.25 shows the original picture for one field. The vertical axis is shown as the number of horizontal-scanning lines, while the horizontal axis is shown as number C (columns) by which the horizontal period (1H) is counted down with ⅔ fsc (⅔ times the chroma subcarrier frequency).

The sampling area of the original picture used to form the subpicture is shown by the solid line. Vertical area is from 36H to 228H, while the horizontal area is from 7C to 152C. The upper and lower sections of the effective picture (14.5H to 36H, and 228H to 259H) are not sampled because of memory limitations.

The composite video is separated into Y and C components, and is processed with a sampling frequency synchronized to fsc of the main picture. (The Y signal is approximately 2.4 MHz, while the C signal is about 0.6 MHz.) The video is quantized to 6-bit units by the A/D converter.

As shown in the lower left of Figure 10.25, the quantized data bits are stored in memory (DRAM) as the PinP signal. The diagram on the lower right of Figure 10.25 is composed from 64H (V) and 48C(H) held in memory. The DRAM memory is an IC with four memory-cell arrays. Each cell has a memory capacity of 65,536 bits (256 columns × 256 lines). Each memory-cell array has 256 vertical addresses, which are divided by 4 for a 4-field subpicture writing area. As a result, the writing area is (64 vertical × 256 horizontal) × 4 fields.

The digital information of the picture (for the field that matches the scanning of the main picture in memory) is selected and is synchronized to the scanning of the main picture. The digital information is read out at the speed of 4.5 fsc. The data bits read out from memory are D/A converted, and converted to the video signal is one of the four corners of the main pic-

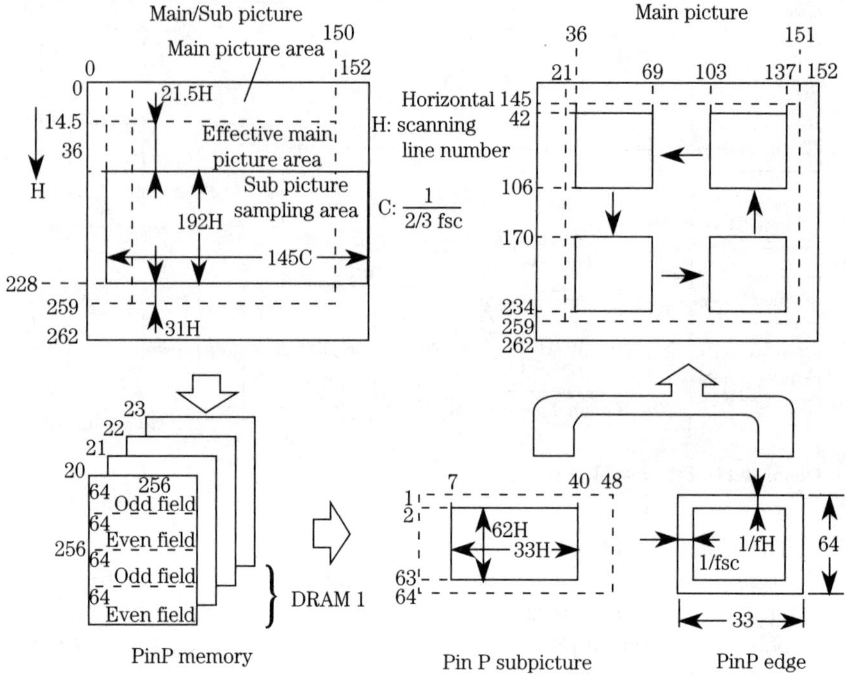

Figure 10.25 Relationship of fields and display in PinP.

ture. The section 64H × 33C from the 64H × 48C of the subpicture data is displayed in the main picture.

The remaining section of data is used for two purposes. First, the information is superimposed in the main picture so as not to lose the video signal at the changeover part when scanning is changed from the main picture, and vice versa. Second, the information matches the pedestal level of the signal that is demodulated into the color-difference signal. (This matches the tint of the main picture and subpicture.)

The areas immediately surrounding the subpicture are edged in white. This is done to emphasize the border of the main and subpictures. The edging line is set as 1/fH vertically and 1/fsc horizontally. When the circumference is edged, the dc-voltage equivalent to 100% white level is superimposed on the video signal. The position where the subpicture is inserted into the main picture can be selected (with the remote control) to one of four corners of the screen, as shown on the upper right of Figure 10.25.

10.7.2 PinP signal-processing sequence

Figure 10.26 shows the PinP signal-processing sequence. The composite video is first applied to an NTSC decoder, where the video is separated in Y,

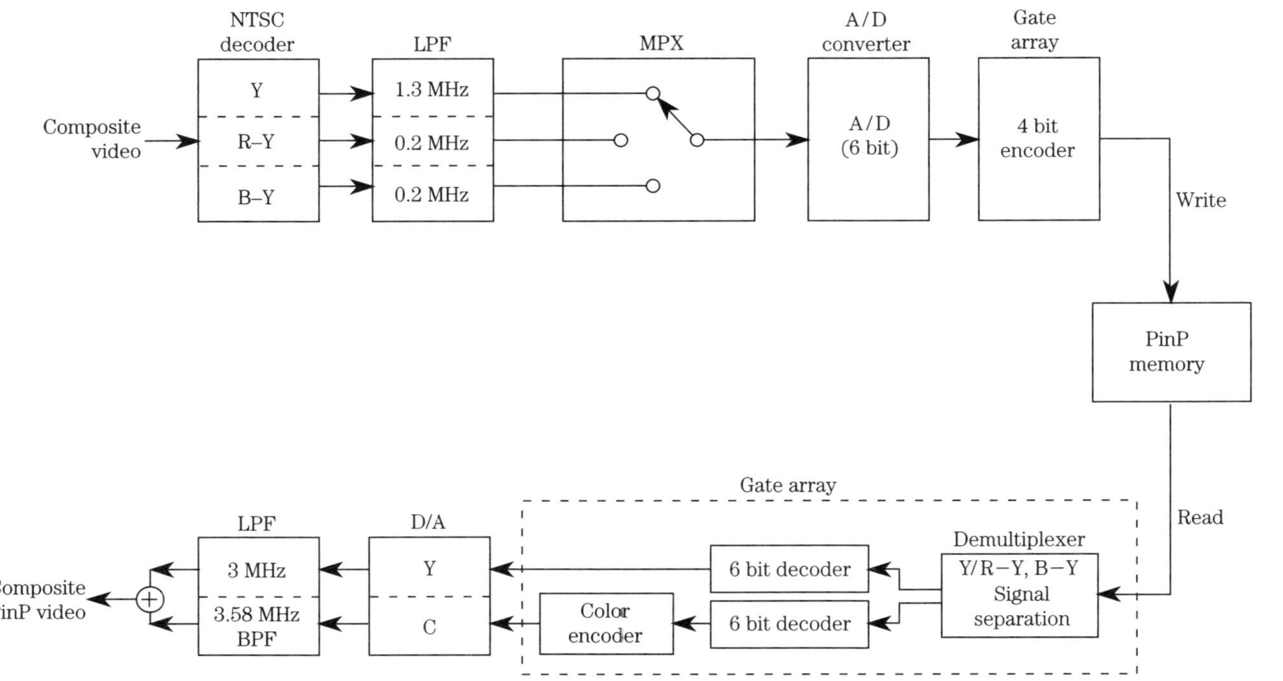

Figure 10.26 PinP signal-processing sequence.

R–Y, and B–Y components. This is done so that brightness (Y) and color difference (R–Y, B–Y) components can be sampled individually. If the composite video was sampled as is, the correct phase could not be obtained because the sampling rates are changed to reduce the subpicture to ⅑ size. (To reduce the original subpicture to ⅑, the sampling rate of the horizontal axis to the read-out time is set at 1:3, while the number of vertical axis samplings is set to ⅓ of the total scanning lines.)

After decoding, the bands of the Y, R–Y, and B–Y signals are limited by an LPF and applied to the multiplexer. The purpose of the LPF is to make the maximum frequency of each signal ½ or less of the sampling frequency. (The sampling frequency for Y is about 2.4 MHz, or fsc × ⅔, while the sampling frequency for R–Y/B–Y is about 0.6 MHz, or fsc/6.).

Because the sampling frequency of the color-difference signals can be ¼ that of the Y component, and so save memory capacity, a multiplexer is used following the LPF. The multiplexer samples the signals on a time-share basis. The Y signal is sampled once per three horizontal-scanning lines. Then, to sample the chroma signals from the following scanning line, the B–Y and R–Y signals are alternately sampled by time sharing at ⅙ fsc. This processing method is known as Line Sequence Method, and is shown in Figure 10.27.

The A/D converter that follows the multiplexer converts the multiplex output to a 6-bit digital signal, and supplies the signal to a gate array before being written into the PinP memory. This process is called Quantization. The gate arrays arrange the 6-bit signal to a 4-bit signal (which can be written more easily).

When the desired picture-insertion position is reached, the digital Y, R–Y, and B–Y components read from the PinP memory are separated from each other by the demultiplexer and are converted to 6-bit signals (from 4-bit

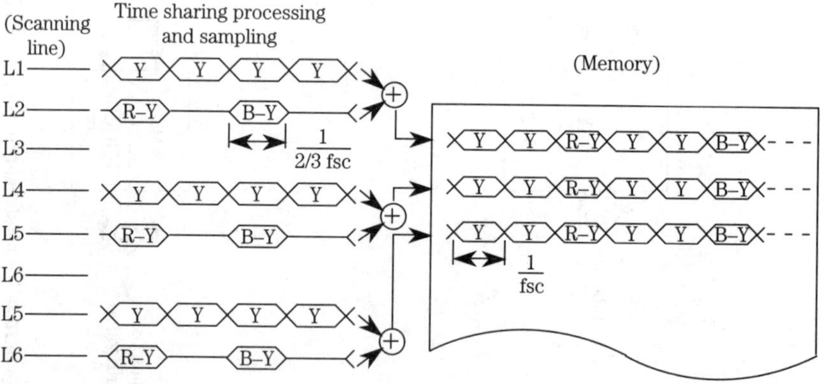

Figure 10.27 Principle of line-sequence method.

signals) by the 6-bit decoder. Then, the components are converted back to the original Y and C signals by the D/A converter. Because the 6-bit decoder outputs B–Y and R–Y color-difference signals, a color encoder is necessary to restore the original C or chroma signal.

To frame the subpicture correctly in the specified position of the main picture, and to get the same tint, the sync and burst signals (separated from the video used for the main picture) are used for timing the PinP memory and for producing the chroma subcarrier signal required for NTSC modulation. All of these processes are executed according to a procedure programmed in the PinP gate array.

10.7.3 Typical PinP circuits

Figure 10.28 shows the circuit configuration of the PinP unit. Notice that there are 10 ICs involved, including the PinP memory IC10. The following is a brief description of the IC functions.

This PinP unit uses a digital-video process, with a 16-MHz (4.5 fsc) high-amplitude pulse signal. Because such a pulse signal can cause interference with nearby circuits, the PinP unit is housed in a shielded case. As discussed in section 10.6, always make sure that all shields are in place, and properly grounded, to prevent interference.

The video signal is input and output to and from the PinP unit through the PB connector. The control signals for the PinP functions are input through the PA connector. The PinP function is on when pin 2 of PA is high, and off when pin 2 is low. Pin 1 of PA is the input terminal for moving the position of the subpicture (to each of the four corners of the main picture, in turn, as shown in Figure 10.25). Each time pin 1 of PA is made high, the subpicture position shifts in the counterclockwise direction.

Main/subvideo input/output switch IC1. The input switch of IC1 swaps the video signals of the main and subpictures. However, in this configuration, the function is performed at a stage prior to the PinP unit, so the switch is not used. The output switch of IC1 determines whether the subpicture is to be inserted into the main picture. If PinP is selected, the output switch also adds the edging signal (equivalent to 100% white) to emphasize the border of the main and subpictures. The timing control for these functions is performed by the PinP controller IC6.

NTSC demodulator, sync separator, Y/C multiplexer IC2. This IC contains the 3.58-MHz (fsc) generator or VCO, sync separator, H/V pulse generator, and Y/C multiplexer.

The Y/C multiplexer multiplexes the video signal in the subpicture. The H/V sync pulses generated after the main sync separator are fed to controller IC6, and are used to determine the subpicture display position. The pulses also control the timing of write-read data to and from the PinP memory.

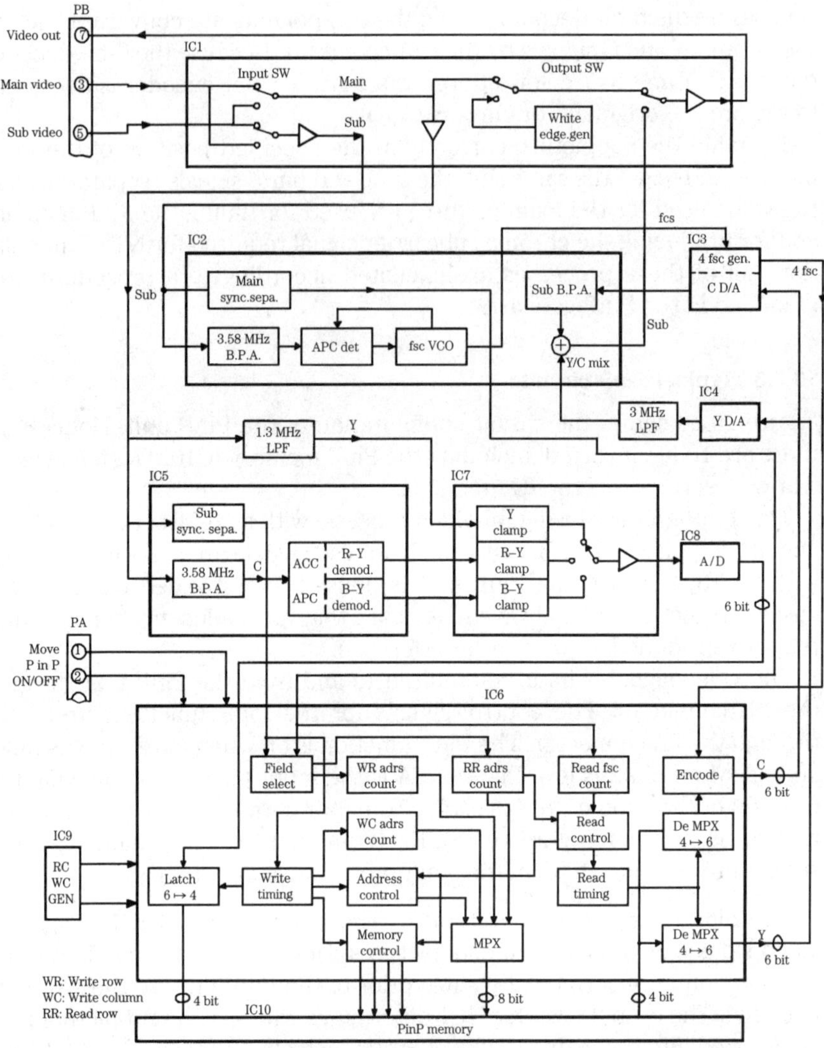

Figure 10.28 Typical picture-in-picture circuit configuration.

NTSC demodulator, sync separator IC5. This IC performs NTSC demodulation of the subpicture video signal and feeds the signal to multiplexer IC7. The IC also performs sync separation, and delivers H/V sync pulses to controller IC6.

Clamper, multiplexer IC7. This IC sets the amplitudes of the Y, B–Y, and R–Y signals before multiplexing the three signals into a serial data stream (on a time-share basis). The (3N + 1) H scanning line of the Y signal is time-shared at ⅔ fsc. The color-difference signals of R–Y and B–Y are alternately time-shared by ⅓ fsc. This means that the sampling frequency of the Y signal is ⅔ fsc and the sampling frequency of the color-difference signal is ⅓ fsc. Both the clamping time of the clamper, and the time-sharing of the multiplexer, are controlled by IC6.

A/D converter IC8. This IC converts the subpicture video signal (separated into R–Y and B–Y signals with a maximum amplitude of 1 V p-p) to a 6-bit quantized digital signal. The clock signal needed for A/D conversion is supplied from IC6.

PinP controller IC6. Control operation of IC6 is performed by two control signals (PinP and PinP position) from the tuner or channel-selector microprocessor (chapter 2). The functions of IC6 are as follows:

- Selects the video signals of the main and subpictures when the PinP mode is set.
- Determines the position where the subpicture is to be inserted.
- Generates the control signal for the multiplexer.
- Determines the clamp timing of the color-difference signals.
- Generates the clock signal for A/D conversion.
- Receives the sync signals separated from the video of the main and subpictures, and generates the control signals for writing and reading to/from the PinP memory IC10.
- Arranges the 6-bit data to 4-bit data units and performs writing/reading to and from IC10.
- Converts the digital subpicture data (read from IC10) to a Y signal, and color-difference signals (also read from IC10) to a C signal. This includes encoding the color-difference signals to restore the original chroma or C signal.

PinP memory IC10. This memory has a capacity of 262,144 bits and can store data for four fields. Writing and reading is performed in real time (or apparent real time).

Y-signal D/A converter IC4. This IC converts a 6-bit quantized digital Y signal to an analog signal. The analog output is shaped by an LPF, and is limited within the band specified by NTSC.

Chroma signal D/A converter IC3. This IC converts the 6-bit quantized digital chroma (C) signal to an analog chroma signal. The output becomes a chroma signal (within the NTSC band) through processing by a BPF in IC2. The 4 fsc VCO (generator) and phase detector are part of IC3. Phase control of the 4 fsc

VCO is locked to the fsc of the main-picture video from IC2. The 4 fsc signal is supplied to IC6, as the reference clock for digital encoding.

Clock signal generator IC9. This IC generates the clock signals required for writing and reading to and from PinP memory IC10. The writing frequency is 14.3 MHz (4 fsc), while the reading frequency is 16.1 MHz (4.5 fsc).

10.7.4 PinP circuit troubleshooting

Although troubleshooting for the PinP circuits appears difficult, remember that the inputs to the PinP loop (pins 3 and 5 of the PB connector, Figure 10.28) are conventional video signals that can be monitored and traced back to the source (typically VIF or external-video circuits). Similarly, the output from the loop (at pin 7 of the PB connector) are conventional video signals that can be monitored and traced to the video-output circuits.

In simple terms, if the signals applied to pins 3/5 are good, but the signal at pin 7 is bad, you have isolated the trouble to the PinP unit (which includes the 10 ICs shown in Figure 10.28). In some cases, the PinP units can be replaced as a package. In other cases, individual ICs can be replaced. Either way, it is a case of digital troubleshooting, such as described in *Lenk's Digital Handbook* (McGraw-Hill, 1993).

Before you condemn the PinP circuits, make certain that they are turned on. Pin 2 of the PA connector must be high for the PinP circuits to operate. If pin 2 is not high, trace back to the source (which is probably a signal from system control in most sets, but could be from a user switch).

If the PinP circuits are operating, but the subpicture does not move from corner to corner, check that pin 1 of the PA connector goes high each time the appropriate button is pushed. If not, trace back to the source (probably from the remote transmitter through system control).

Chapter

11

TV Test Equipment and Procedures

This chapter is devoted to test equipment used in TV service and to general procedures for use of the equipment during service. (The procedures for use of test equipment in troubleshooting are described in chapter 12.)

The discussions here are based on the assumption that most TV service can be performed with meters (digital or analog), scopes, and generators (RF, sweep, NTSC, stereo TV), as well as assorted probes, clips, patch cords, etc. It is further assumed that you already have test equipment of your own, and that you know how to operate all controls and read all indicators. If not, study the user manuals. With that in mind, this chapter discusses the why and how of test equipment in TV service.

11.1 Safety Precautions

In addition to a routine operating procedure, certain precautions must be observed during the operation of any test equipment used in TV service. Many of these precautions are the same for all types of test equipment; others are unique to special test instruments such as meters, scopes, and signal generators. Some of the precautions are designed to prevent damage to the test equipment or to the TV circuit. Other precautions are to prevent injury to you. Where applicable, special safety precautions are included throughout the various chapters of the book.

Study the following general safety precautions and then compare them to any specific precautions called for in the test-equipment manuals, and in related chapters of this book. Refer to chapter 12 for safety precautions related to troubleshooting.

Many service instruments are housed in metal cases. These cases are connected to the ground of the internal circuit. For proper operation, the ground terminal of the instrument should always be connected to the ground of the TV set. Make certain that the TV set ground is not connected to either side of the ac line, or to any potential above ground. If there is any doubt, connect the set to the power line through an isolation transformer.

Remember that there is always danger when servicing circuits that operate at hazardous voltages (such as the high-voltages used by the CRT). Keep this firmly in mind as you pull off a TV set back panel or case and apply power through a "cheater" cord. Always make some effort to familiarize yourself with the set before servicing it, bearing in mind that high voltages might appear at unexpected points in a defective set.

It is good practice to remove power before connecting test leads to high-voltage points. It is preferable to make all service connections with the power removed. If this is impractical (as is usually the case), be careful to avoid accidental contact with circuits that are grounded. Working with one hand away from the equipment and standing on a properly insulated floor lessens the danger of electrical shock.

Capacitors can store a charge large enough to be hazardous. Discharge filter capacitors before attaching test leads (but after you have removed the power).

Leads with broken insulation offer the additional hazard of high voltages appearing at exposed points along the leads. Check test leads for frayed or broken insulation before using the leads.

To lessen the danger of accidental shock, disconnect test leads immediately after the test is completed.

Remember that the risk of severe shock is only one of the possible hazards. Even a minor shock can place you in danger of more serious risks, such as a bad fall or contact with a higher voltage.

The experienced service technician guards continuously against injury and does not work on hazardous circuits unless another person is available to assist in case of accident.

Even if you have considerable experience with test equipment, always study the service literature of any instrument with which you are not thoroughly familiar.

Use only shielded leads and probes. Never allow your fingers to slip down to the metal probe tip when the probe is in contact with a "hot" circuit.

Avoid vibration and mechanical shock. Most test equipment is delicate.

Study the TV circuit being serviced before making any test connections. Try to match the capabilities of the instrument to the circuit being serviced.

11.2 Signal Generators for TV Service

The signal generator is an indispensable tool for TV service. Without a generator, you depend entirely on TV broadcast signals and are limited to signal tracing only. This means that you have no control over frequency, amplitude, or modulation of such signals, and have no means for signal injection.

With a generator of the appropriate type, you can duplicate transmitted signals or produce special signals required for alignment and test of all circuits within the TV set. Also, the frequency, amplitude, and modulation characteristics of the signals can be controlled so that you can check the operation of a TV circuit under various signal conditions (weak, strong, normal, abnormal, etc.).

At one time, the sweep/marker generator was used widely in TV service (both to check overall performance of the video circuits and to align the RF/IF stages). For example, in sets where the IF and video detector are as shown in Figure 3.1, a sweep/marker generator is essential for alignment. In a present-day set where the IF/AGC/video-detector/video-amplifier circuits are combined in an IC such as shown in Figure 3.2, a sweep/marker generator is of little value. As a result, the sweep/marker generator has generally been replaced by the analyst/pattern/color generators such as described in section 11.3. However, because the service literature for some TV sets (even those as shown in Figure 3.2) still recommends sweep alignment, the basic procedures are summarized here.

11.2.1 Sweep/marker generator

The main purpose of the sweep/marker generator in TV service is sweep-frequency alignment as shown in Figure 11.1. A sweep/marker generator capable of producing signals of the appropriate frequency is used with a scope to display the bandpass characteristics of a TV circuit (tuner, IF, video amplifier, etc.).

The sweep portion of the generator is essentially an FM generator. When the sweep generator is set to a given frequency, this is the center frequency. The output varies back and forth through the center frequency. The rate at which the frequency modulation is typically 60 Hz. The sweep width, or the amount of variation from the center frequency, is determined by a control, as in the center frequency.

The marker portion is essentially an RF generator with highly accurate dial markings that can be calibrated precisely against internal or external

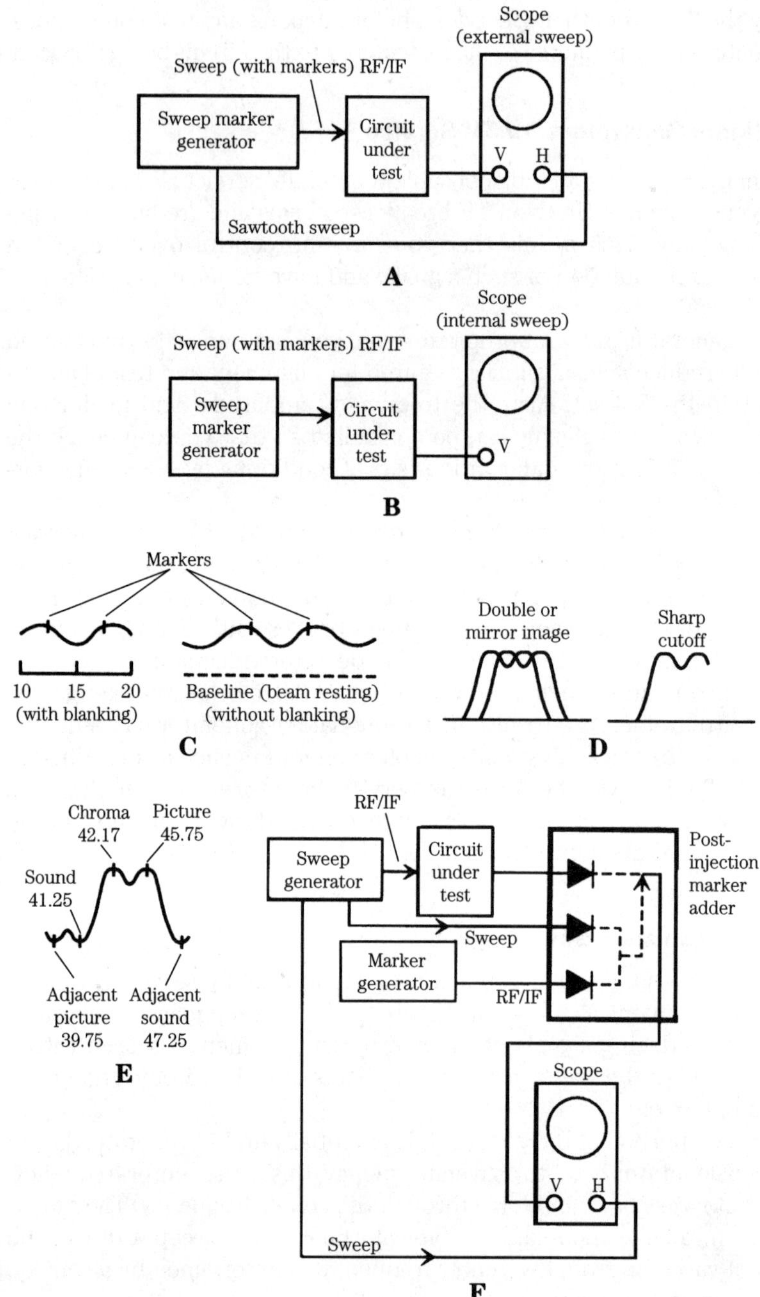

Figure 11.1 Basic sweep-frequency alignment procedures.

signals. Usually, the internal signals are crystal controlled. Marker signals are necessary to pinpoint frequencies when making sweep-frequency alignments and are usually produced by a built-in marker added.

In addition to the basic sweep and marker outputs, the generator might have other special features. For example, the generator might provide a variable bias to disable the AGC circuits, and a blanking circuit that permits a zero reference line to be observed on the scope during the retrace period.

11.2.2 Basic sweep-frequency alignment procedure

Figure 11.1 shows the relationship between the sweep/marker generator and scope during sweep-frequency alignment. If the equipment is connected as shown in Figure 11.1A, the scope sweep is triggered by a sawtooth output from the generator. The scope's internal sweep is switched off, and the scope sweep selector and sync selector are set to external.

Under these conditions, the scope sweep represents the total sweep spectrum as shown in Figure 11.1C, with any point along the horizontal trace representing a corresponding frequency. For example, the midpoint on the trace represents 15 kHz. If you want a rough approximation of frequency, adjust the horizontal gain control until the trace occupies an exact number of scale divisions on the scope screen (such as 10 cm for the 10- to 20-kHz sweep signal). Each centimeter division then represents 1 kHz.

If the equipment is connected as shown in Figure 11.1B, the scope sweep is triggered by the scope's internal circuits (both the sweep and sync selectors are set to internal). Certain conditions must be met to use the test connections shown in Figure 11.1B.

If the scope has a triggered sweep (section 11.3), there must be sufficient delay in the vertical input, or part of the response curve might be lost. If the scope is not a triggered sweep, the generator must be swept at the same frequency as the scope (usually the line frequency of 60 Hz). Also, the scope or generator must have a phasing control so that the two sweeps can be synchronized. If the phase adjustment is not properly set, the sweep curve might be prematurely cut off, or the curve might appear as a double or mirror image, as shown in Figure 11.1D.

As shown in Figure 11.1E, the markers provide accurate (crystal-controlled) frequency measurement, with a number of markers at significant frequencies. Figure 11-1E shows the bandpass response curve of a typical video circuit (a TV tuner and VIF package). The markers can be selected (one or several at a time) as needed.

The response curve e (scope trace) depends on the circuit under test. If the circuit has a wide bandpass characteristic, the generator is set so that the sweep is wider than that of the circuit. (The bandpass of the circuit shown in Figure 11.1E is about 6 MHz.) Under these conditions, the trace starts low at the left, rises toward the middle, and then drops off at the right.

The sweep/marker method of alignment tells at a glance the overall bandpass characteristics of the circuit (sharp response, irregular response at certain frequencies, and so on). The exact frequency limits of the bandpass can be measured with the markers (often called pips).

As shown in Figure 11.1F, the sweep-generator output is applied to the circuit. A portion of the output is also mixed with the marker output in a mixer/detector known as a post-injection marker adder. The post-injection (sometimes known as bypass-injection) method for adding markers minimizes the chance of overloading the circuit, and permits use of a narrowband scope.

11.2.3 Analyst and pattern generators

An analyst generator is used for signal substitution (a form of signal injection). A bench-type analyst generator provides outputs that duplicate all essential signals in a TV set. Such generators have RF, IF, composite video (including sync), sound IF, audio separate sync, possibly flyback and yoke test signals, etc.

The RF, IF, and video signals are usually in the form of a pattern (or patterns), typically lines or bars (horizontal and vertical), dots, crosshatch lines, square pulses, blank rasters, or color bars (section 11.4). Some typical patterns are shown in Figure 11.2. Such patterns can be used in TV troubleshooting, as described in chapter 12.

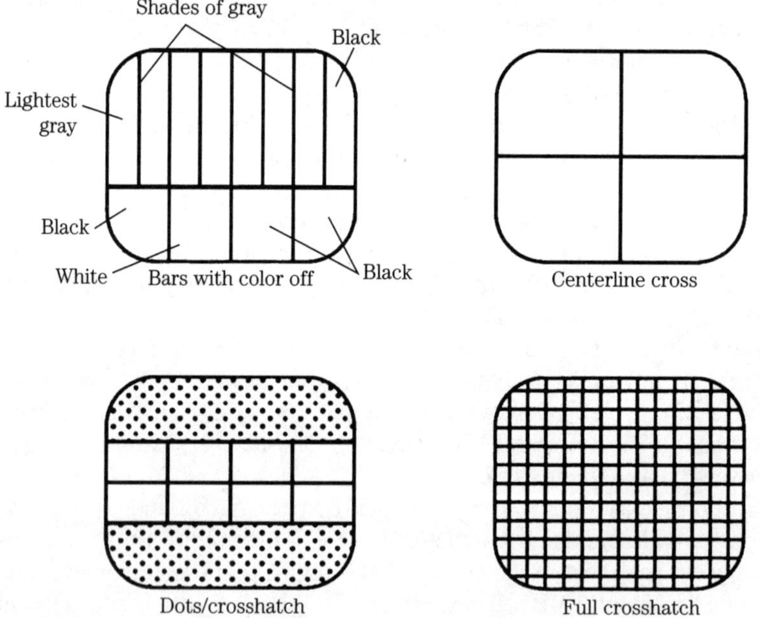

Figure 11.2 Typical analyst and NTSC generator patterns.

At one time, the test-pattern generator was popular for black-and-white TV service. Such generators differ from present-day analyst/pattern generators in that the test-pattern generator reproduces positive transparencies of various pictures that might be inserted into the generator. The analyst/pattern generator described here are generally replaced test-pattern generators.

11.2.4 Typical analyst/pattern generator

The typical analyst/pattern generator produces several patterns, including color at RF, IF, and video frequencies. In effect, the instrument generates many signals normally transmitted by a TV station, and those produced within a TV set, for point-to-point troubleshooting. For example, a typical instrument generates:

VHF and UHF signals on various channels for testing the RF tuner (and overall performance of the set).

IF signals from 20 to 48 MHz for testing the IF portion of TV sets.

A 4.5-MHz sound-channel test signal that is frequency-modulated by a 1-kHz tone.

A separate 1—1-Hz tone.

Positive and negative composite video signals (including sync pulses) for injection into video stages.

NTSC color bars such as described in section 11.4.

11.3 Oscilloscopes for TV Service

Ideally, a scope used for TV service should have a bandwidth of 25 MHz or more, although you can probably get by with a 10-MHz scope (or even a 6-MHz scope). Minimum sweep time should be 0.1-μs per division, although a 0.2-μs sweep will probably do the job.

The scope should have a triggered sweep in addition to conventional internal and external sweep synchronization. With a triggered sweep, the scope is synchronized by the signals applied to the vertical input. The sweep remains at rest until triggered by the signal. This ensures that the signals are always synchronized, even if the waveform is of varying frequency. The triggered sweep threshold should be fully adjustable so that the desired portion of the waveform can be used for triggering.

A dual-tracer scope facilitates some measurements but is not absolutely required. Such refinements as calibrated voltage scales, calibrated sweep rates, Z-axis input (for intensity modulation), and sweep magnification and illuminated scales are, of course, always helpful in TV service (as they are in any type of electronic work).

A few scopes have special provisions for TV service. Such features include the TV sync or sweep (usually identified as TV-horizontal and TV-vertical, or TVH and TVV, or some similar term) and the vectorscope provision, which is discussed next.

11.3.1 Vectorscopes

A vectorscope is sometimes used with a color generator to check the response of a color TV set to color signals. The true vectorscope is used more extensively in TV broadcast work.

A vectorscope permits the phase relationship (and amplitude) of color signals to be displayed as a single pattern on the scope screen. When used for TV service, the vectorscope monitors the color signals directly at the color-gun inputs of the CRT. This makes the vectorscope adaptable to any type of color circuit (and any type of color demodulation). By comparing the vectorscope display with that of an ideal display (for phase relationship, amplitude, and general appearance), the condition of the TV color circuits can be analyzed. A conventional scope can be used as a vectorscope. However, this requires special connections to the horizontal and vertical deflection system, as described in section 11.16.

11.4 Color Generators

Color generators are essential when servicing any type of color TV, because such generators produce signals that are equivalent to the color-test signals used in TV broadcast work. At one time, the keyed-rainbow generator was popular for color TV service. Today, the NTSC color-bar generator has replaced the rainbow generator. Color generators are often combined with the analyst/pattern generator discussed in section 11.2.3.

11.4.1 NTSC color-bar generator

An NTSC generator produces standard EIA colors at the NTSC-prescribed luma level (brightness), chroma phase angle (hue), and chroma amplitude (saturation). Generally, the most useful signal produced by an NTSC instrument is the standard NTSC bar pattern with a –IWQ signal occupying the lower quarter of the pattern, as shown in Figure 11.3.

Notice that Figure 11.3 shows both the pattern (as produced on a TV screen when the NRSC signal is applied to the antenna input) and the waveforms (as produced on a scope connected to various points in the TV circuits) for one horizontal line of the pattern. Also notice that the –IWQ waveform is generated for the bottom 25 percent of the display, while the color-bar waveform is on for the top 75 percent of the pattern. When using a scope, the two waveforms can be superimposed, as discussed in section 11.4.3.

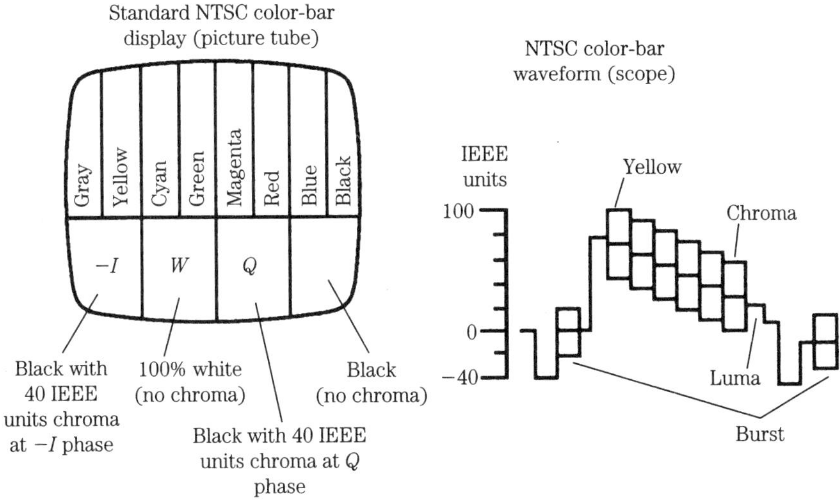

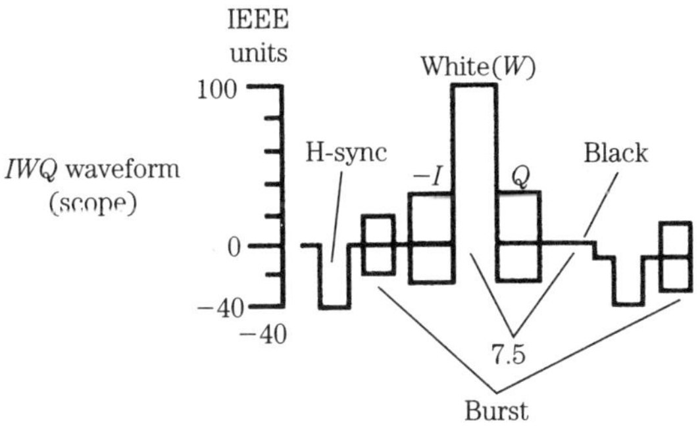

	Burst	H-sync	−I	White(W)	Q	Black
Luma amplitude	0	−40	7.5	100	7.5	7.5
Chroma amplitude	40	- - -	40	- - -	40	- - -
Vector (from B–Y)	180%	- - -	303°	- - -	33°	- - -

Figure 11.3 NTSC color-bar pattern with –IWQ.

A typical color generator also produces a number of other patterns, including various convergence, linear staircase, and full-color rasters. Use of these patterns is discussed throughout the remainder of this chapter. The following sections also describe how the patterns and waveforms can be used in troubleshooting (chapter 12).

11.4.2 NTSC color-bar pattern

The basic pattern in Figure 11.3 is the one most often used for testing, troubleshooting, and adjusting TV circuits. Analysis of the pattern shown on the screen of a TV set under test can often localize a problem to a few specific circuits. Here are a few examples.

Overall performance. An overall performance test of a TV set can be conducted using the NTSC color-bar pattern. If all of the colors are good, as well as the black-and-white levels, the set is performing properly. If there is color distortion, particularly in the vividness of colors (color saturation), it is possible that the circuits following the comb filter (Figures 4.5 and 4.6) are not properly combining the video and chroma signals.

Video and chroma delay. Another problem is a difference in delay between the video and chroma signals because of some circuit defect following the comb filter. Such delay difference can cause fuzziness along the edges of the color bars. The delay problem might be more pronounced along the edges of the white bar in the –IWQ portion of the pattern.

RF and IF performance. The color-bar RF/IF outputs of the generator can be used effectively to troubleshoot RF and IF sections by comparing the patterns. For example, performance of a TV should be nearly as good when using the RF signal as when applying composite video directly into the IF. For example, the CRT display should be substantially the same when an RF signal is applied at the antenna input of TU101 (Figure 2.5) and when an IF signal is applied to the input of the VIF circuits at W221 (Figure 3.4). If the display is substantially reduced when the NTSC generator output is applied through the tuner, as compared to when a signal is applied directly to the IF, the tuner is suspect, and you have a good starting point for troubleshooting.

Color adjustments. The NTSC color-bar pattern provides a standard reference for color adjustments. The pattern contains bars of the three primary colors: red, blue, and green. These are a good reference for checking 3.58-MHz phase problems. White, yellow, cyan, and magenta help define problems when the mix of colors is not in the correct proportion.

In some troubleshooting applications, it is helpful to know if the problem is chroma or luma related. Many generators are provided with a chroma-off function, which produces the bars and waveforms shown in Figure 11.4A (normal bar pattern but without color).

A TV set can be checked by comparing the displays of Figures 11.3 and 11.4A. If the display of Figure 11.4A is good, but the one in Figure 11.3 is not, the problem is in the chroma circuits. Notice that in Figure 11.4A the

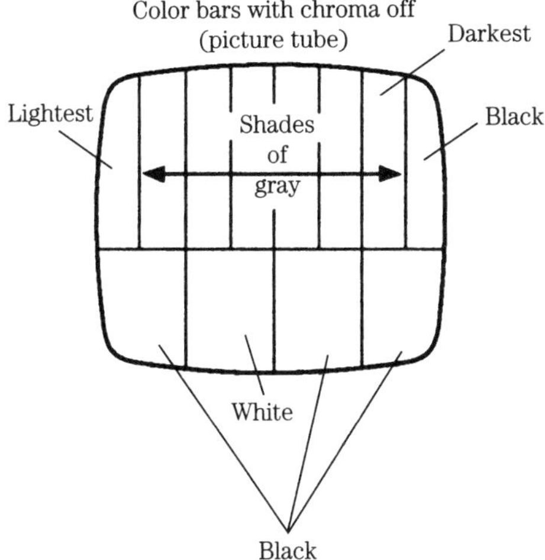

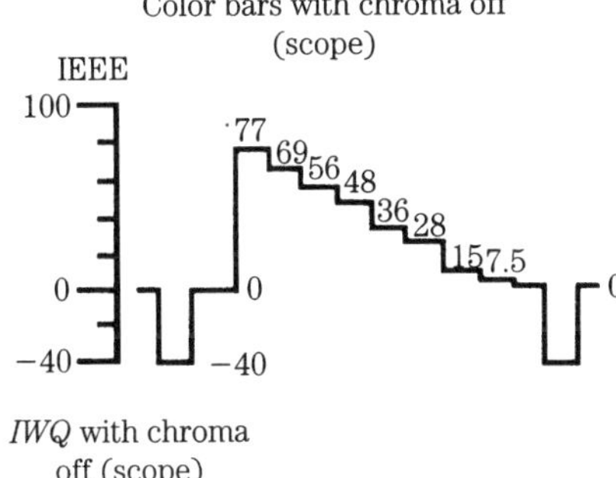

IWQ with chroma off (scope)

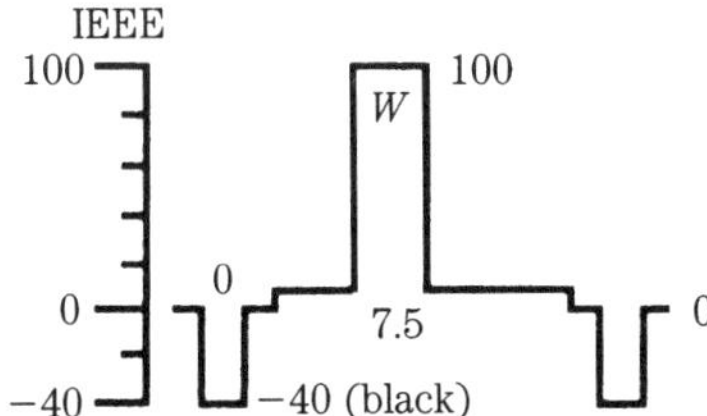

Figure 11.4 Normal color-bar pattern (but without color).

chroma-off waveform is generated for the top 75 percent of the pattern, and the –IWQ waveform is on for the bottom 25 percent of the display. The two waveforms are superimposed on the scope display.

Phase lock of the 3.58-MHz oscillator. The 3.58-MHz oscillator should be locked in phase whenever the color burst is applied. This can be checked using the top-burst-off function available on some generators. When the top-burst-off function is selected, the color is unsynchronized for the top quarter of the pattern, but there should be no delay or color distortion where the color bars start (middle of the pattern) as shown in Figure 11.4B.

Notice that in Figure 11.4B the bars with the burst-off waveform are generated for the top 25 percent of the pattern, the bar waveform is generated for the middle 50 percent, and the –IWQ waveform is generated for the bottom 25 percent. The three waveforms are superimposed on the scope display.

Color-killer function. As discussed in chapter 4, the color-killer circuit disables the color functions when there is no color burst present. In present-day sets, the color killer is part of an IC (such as IC301 in Figures 4.6 and 4.11). The color-killer function can be checked using the full-burst-off function available on some generators. When this function is selected, the bars should appear (in various shades of gray) but there is no color, as shown in Figure 11.4C. Notice that the bars with full-burst-off are generated for the top 75 percent of the pattern, and the –IWQ is generated for the bottom 25 percent. The two waveforms are superimposed on the scope display.

Audio or sound functions. A 1- or 3-kHz audio signal can be added to the RF output of many generators. In a properly operating set, this audio modulation should not affect the normal display. For example, if you select the pattern in Figure 11.3, there should be no change in the display when the audio modulation is applied or removed. If you see some interference (typically diagonal lines across the display) when audio modulation is applied, sound is leaking into the chroma circuits. Possibly one or more of the sound traps are not properly adjusted, or the sound filters are not operating properly. The 1- and 3-kHz signals can also be used to test the sound or audio circuits.

Vectorscope applications. The standard NTSC color-bar output (Figure 11.3) can be used with a vectorscope (or a conventional scope) to analyze color circuits. The procedures are discussed in section 11.16.

11.4.3 –IWQ patterns

The basic –IWQ pattern appears on a split field with the NTSC color-bar pattern, with the –IWQ pattern appearing on the lower quarter of each field, as shown in Figures 11.3 and 11.4. When viewing horizontal lines of video (waveforms) on a scope, both the bars and the –IWQ signals are superimposed, which is very handy in many applications. For example, the 7.5 percent black level (right-hand side) and the 100 percent white level (second from left) of the –IWQ pattern are the key luma-amplitude references.

These references are used for black-and-white level adjustments in some sets. The black-and-white levels are also used whenever luma and chroma ratios are being adjusted or checked during troubleshooting. (The –I and Q signals are used primarily in video cameras and studio equipment for setting up the phase and amplitude of the –I and Q signals and maintaining the proper relationship between the two.)

11.4.4 Staircase patterns

Figures 11.5 and 11.6 show the various staircase patterns and waveforms available with many generators. The following describe some typical uses of the staircase format.

Comb filter adjustments. The staircase pattern can be used to check and adjust the circuits that follow the comb filter (to make certain that there is no shift in phase when the video and chroma are recombined after processing). Notice that not all sets have such adjustments.

White and black levels. The staircase pattern is desirable for setting the white and black levels (on sets where such adjustment is possible). Because the top step is 100 percent white level, the step provides the correct reference for white-clip adjustment.

Amplifier linearity checks. Typically, the staircase pattern contains five equal steps of increasing luma, with a constant chroma amplitude and phase. With the chroma off (Figure 11.5C and D), only the luma steps are generated. This luma-only pattern is valuable for checking linear in amplifier circuits. The amplitude of each step should be equal at the output of an amplifier or other circuit because the amplitude is equal at the input. Nonequal steps at the output of a circuit represent nonlinear distortion.

Differential gain and differential phase lock. The staircase patterns are most effective when they are used to check both differential gain and differential phase. Excessive differential gain or phase can be the cause of color distortion. (Both conditions are checked often in studio equipment that processes video signals.)

Theoretically, the chroma amplitude should not change as luma is varied from 0 to 100 percent. Any interaction is called differential gain. To check for differential gain, the output of a chroma circuit is displayed on a precision waveform monitor while the staircase-pattern input is applied from the generator. As the luma signal steps from 0 to 100 percent in 20 percent increments, any differential gain causes changes in chroma amplitude, in sync with the luma steps.

Sometimes, the degree of differential gain is affected by the peak-to-peak amplitude of the chroma signal. This characteristic can be checked by switching between the low-staircase (20 IEEE units of chroma amplitude), and high-staircase (40 IEEE units) pattern.

11.4.5 Convergence patterns

Figures 11.7 and 11.8 show the various convergence patterns and waveforms available with many generators. The convergence patterns are used

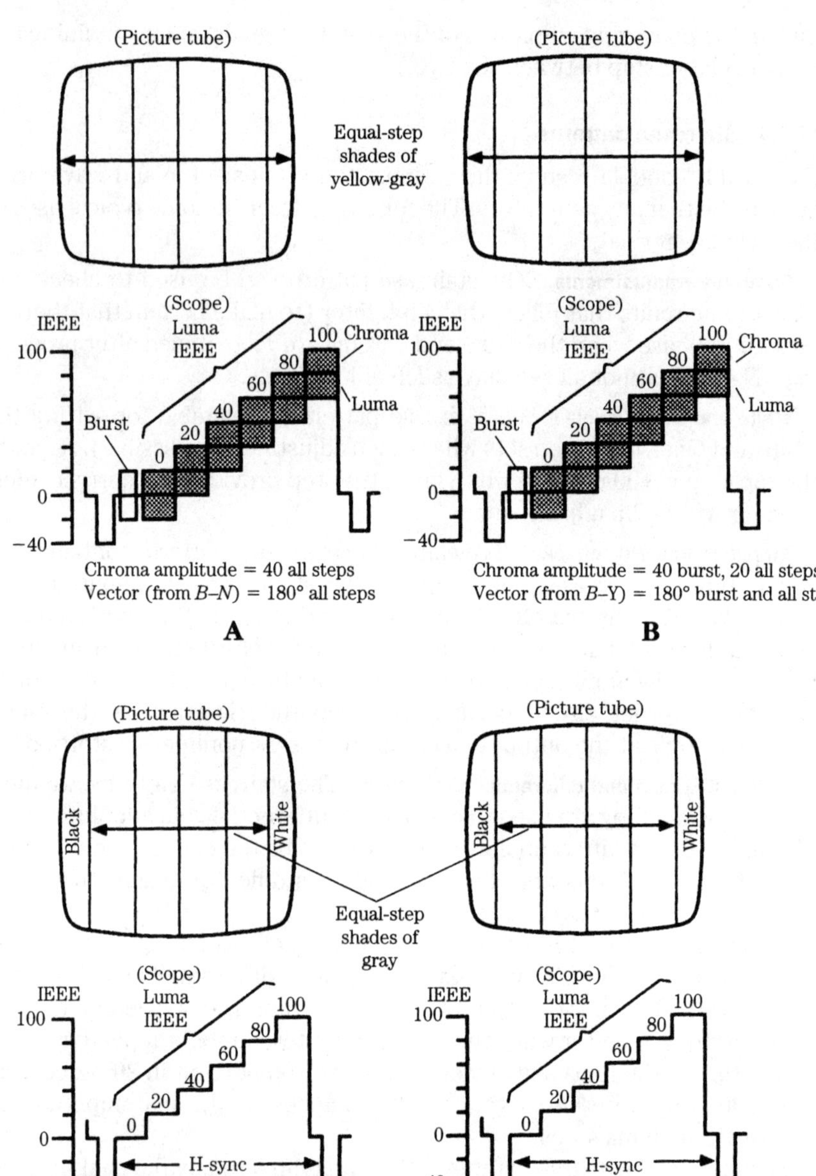

Figure 11.5 Staircase patterns.

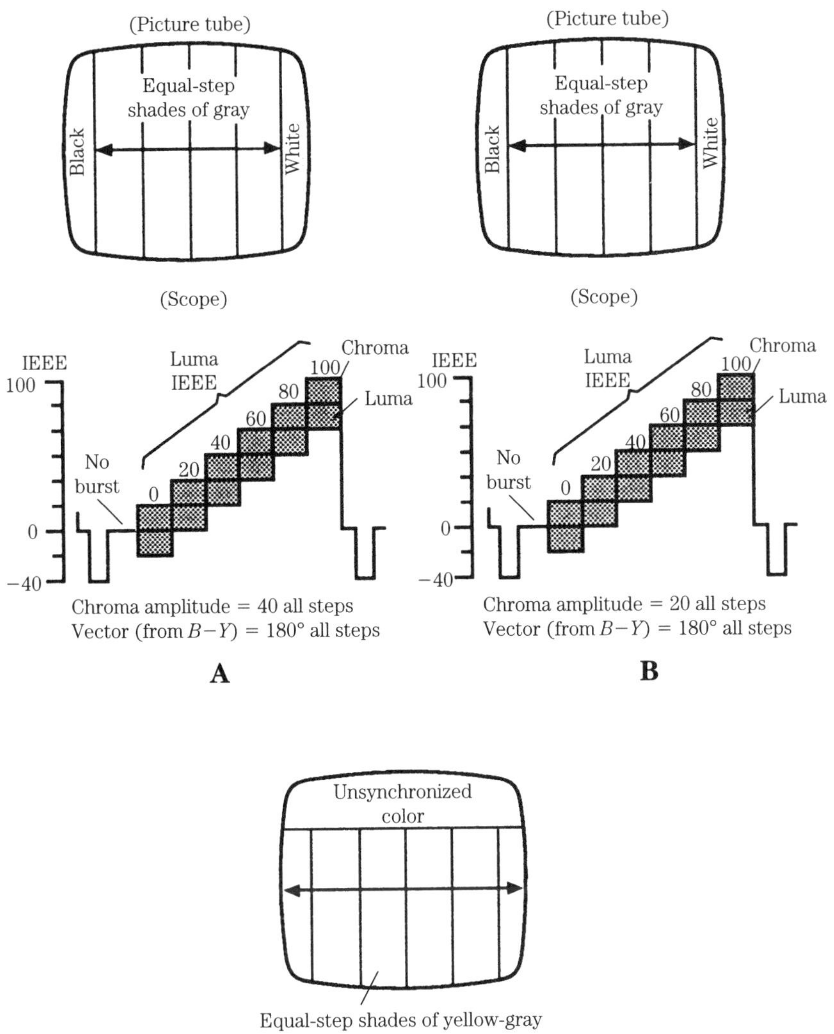

Figure 11.6 More staircase patterns.

primarily for static and dynamic convergence of color TV, as discussed in section 11.10.

Center cross. The center-cross pattern (Figure 11.10A) should intersect at the center of the screen, and there should be no tilt of the horizontal line. Improper centering indicates the need for centering adjustment or a possible deflection-circuit fault. Tilt might require repositioning of the deflection

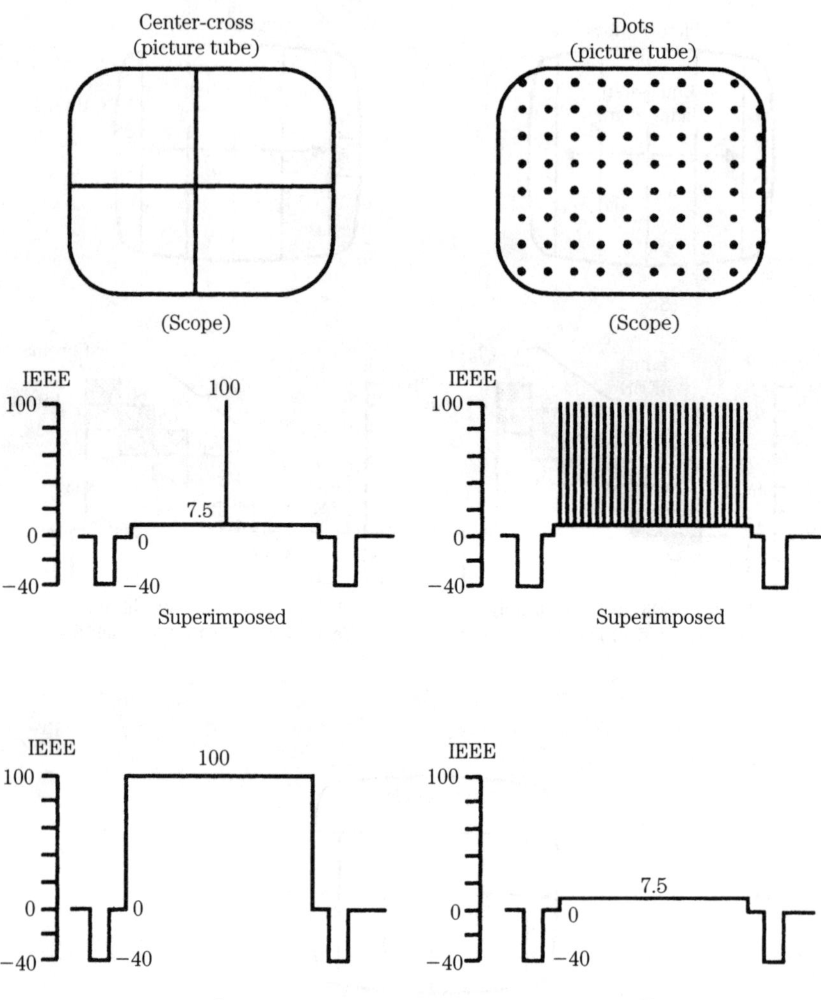

Figure 11.7 Center-cross and dots convergence patterns.

yoke for correction. The center-cross pattern also provides a good general check of vertical and horizontal sync.

Dots pattern. The dots pattern (Figure 11.10B) is used for static convergence, usually by converging the center dot of the pattern. Most TV sets have overscan, so that all dots are not visible except possibly under low-voltage conditions. Some TV sets have a tendency toward a greater amount of overscan than other sets. Typically, it is desirable to display at least a 17- by 13-dot pattern.

Crosshatch pattern. The crosshatch pattern (Figure 11.8A) is usually preferred for dynamic convergence, although some technicians prefer the dots pattern for both static and dynamic convergence. The crosshatch pattern checks both vertical and horizontal linearity. Each square of the crosshatch pattern should be the same size, which is a convenient reference for making linearity adjustments. The crosshatch pattern is also used to check so-

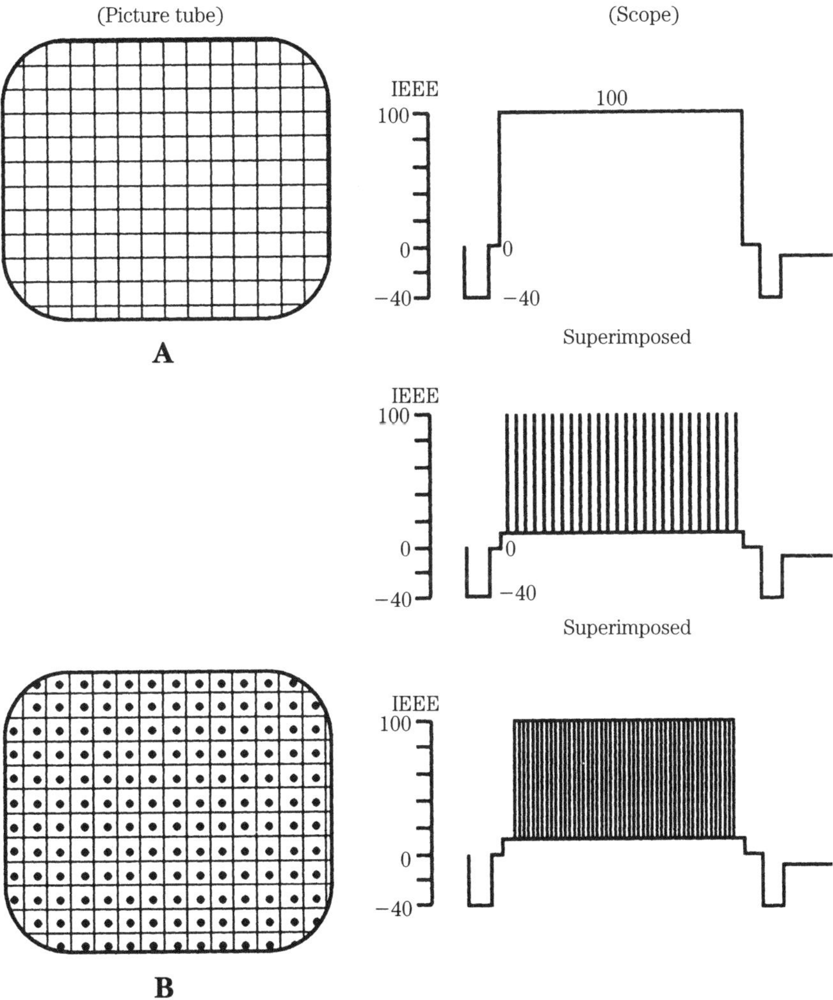

Figure 11.8 Crosshatch (A) and dothatch (B) patterns.

called pincushion distortion, which sometimes appears at the outside edges of large-screen picture tubes as bends in the lines. Refer to chapter 5 for a discussion of pincushion-correction circuits.

Dothatch pattern. The dothatch pattern (Figure 11.8B) combines the dots and crosshatch patterns for a quick overall check of static and dynamic convergence, linearity, overscan, and pincushion distortion from a single pattern. There might be some slight amount of vertical jitter when monitoring the convergence patterns with some generators. This jitter is the result of interlaced scan and is normal for an NTSC generator (because the NTSC signal is interlaced). The jitter does not degrade the accuracy of the convergence adjustments and does not indicate a malfunction associated with vertical sync.

11.4.6 Raster patterns

Figure 11.9 shows the raster patterns and waveforms available with many generators. As shown, one raster fills the entire screen, while the other raster pattern (with top burst off) fills the bottom 75 percent of the screen, with the top 25 percent occupied by unsynchronized color.

The raster patterns are valuable for checking and adjusting purity (section 11.10). Not only can the white raster be used, but the three separate guns of the CRT can be individually adjusted (in some circuits) using a continuous chroma signal of red, blue, or green.

Analysis of the hue and saturation problems can be simplified by analyzing each primary color or the yellow, cyan, and magenta hues, which are the equal mixing of two primary colors without the third one. On some generators, a "black burst" test signal is generated when none of the three primary colors is selected. The luma and chroma components for each raster color are identical to the corresponding bar from the color-bar pattern, but each color can be selected individually for analysis.

11.4.7 Sync signals

Every pattern produced by an NRSC generator contains NTSC sync pulses. These pulses are the same as the sync signals produced by a TV broadcast station. The sync amplitude of 40 IEEE units is often the reference against which the remainder of the luma signal is compared. Circuits can be checked for sync clipping by monitoring the staircase pattern on a scope and checking to see if the sync-pulse amplitude remains at 0.4 compared to the 100 percent white step, which has a reference of 1.0.

AGC circuits, which respond to sync-pulse amplitude, can be checked with the composite-video output available on many generators. Although

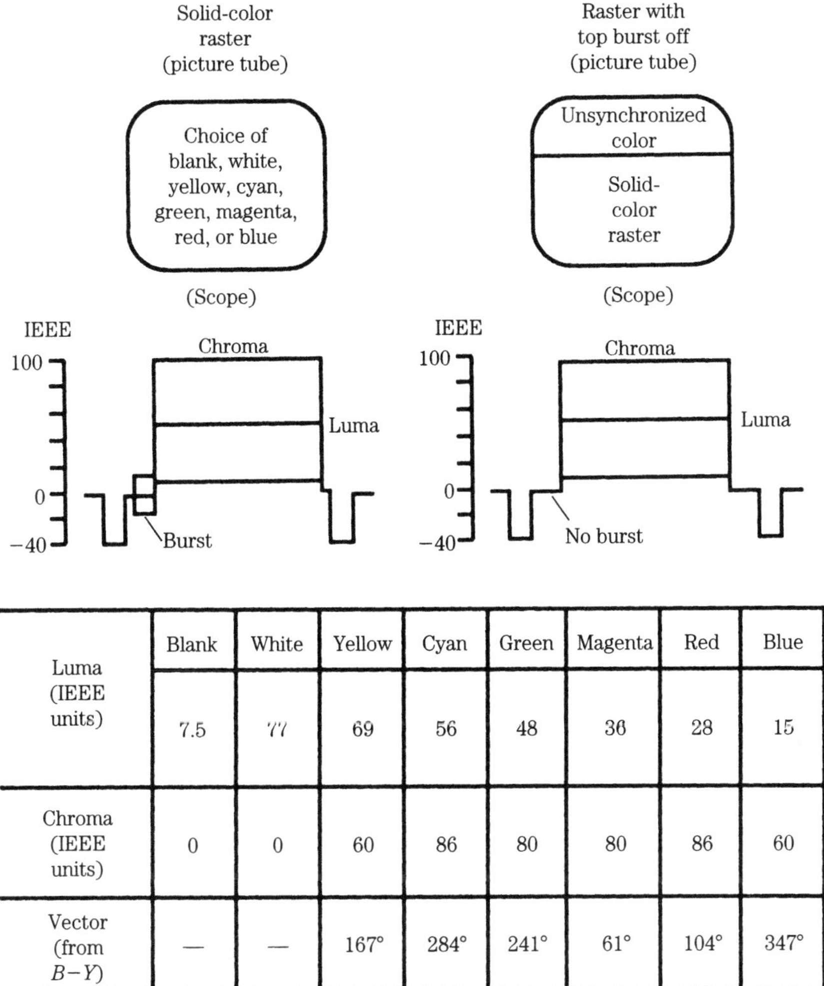

	Blank	White	Yellow	Cyan	Green	Magenta	Red	Blue
Luma (IEEE units)	7.5	77	69	56	48	36	28	15
Chroma (IEEE units)	0	0	60	86	80	80	86	60
Vector (from $B-Y$)	—	—	167°	284°	241°	61°	104°	347°

Figure 11.9 Raster patterns.

the overall amplitude of the video is varied, the sync-pulse amplitude is a constant percentage of the total video amplitude.

11.5 Miscellaneous Test Equipment

In addition to a good scope, with a color generator (preferably with pattern and analyst features), and possibly a sweep/marker generator, most TV service functions can be performed with conventional test equipment found in other electronic service fields. These include meters, possibly transistor and diode testers, RC-substitution boxes, and assorted adapters, clips, and probes. The following sections concentrate on those items of test equipment that apply to most TV service problems.

11.5.1 Probes (for meters and scopes)

Most meters and scopes used in TV service operate with some type of probe. In addition to providing for electrical contact to the circuit under test, probes modify the voltage being measured to a condition suitable for display on a scope or meter. For example, assume that the CRT high voltage must be measured and that this voltage is beyond the maximum input limits of the meter or scope. A voltage-divider probe can be used to reduce the voltage to a safe level for measurement. The voltage is reduced by a fixed amount (known as the attenuation factor, typically 10:1, 100:1, or 1000:1).

Basic probe. The basic probe is a test prod (possibly with a removable alligator clip). Basic probes work well on circuits carrying dc or audio. If the circuit contains high-frequency signals, or if the scope gain is high, it might be necessary to use a special low-capacitance probe.

The hand capacitance in a simple probe can cause hum to be picked up. This condition is offset by shielding in low-capacitance probes. Of more importance, the input impedance of a meter or scope is connected directly to the circuit under test by a simple probe. Such input impedance can disturb circuit conditions.

Low-capacitance probes. The low-capacitance probe contains a series of capacitors and resistors that increase the meter or scope impedance. In most low-capacitance probes, the resistors form a divider (typically 10:1) between the circuit under test and the meter or scope input. Thus, low-capacitance probes serve the dual purpose of capacitance and voltage reduction. When low-capacitance probes are connected at the inputs of meters or scopes, remember that the voltage indications are one-tenth (or whatever value of attenuation is used) of the actual voltage.

High-voltage probes. Most high-voltage probes are resistance-type voltage-divider probes. Such probes are similar to the low-capacitance probe except that the frequency-compensating capacitor is omitted. Usually, the conventional resistance-type probe is used when a voltage reduction of 100:1, or greater, is required and when a flat frequency response is of no particular concern.

High-voltage probes for TV service must be capable of measuring potentials at or near 30 kV (typical), usually with a 1000:1 voltage reduction. In certain isolated cases, the resistance-type probe is not suitable because of stray conduction paths set up by the resistors. A capacitance-type probe can be used in those cases. Such probes contain two (or more) capacitors, with values selected to provide the desired voltage reduction and to match the input capacitance of the meter or scope.

Special high-voltage probe. There are high-voltage probes that need not be connected to a scope or meter. Such probes have a built-in meter that permits CRT voltages to be measured directly.

Radio-frequency probes. An RF probe is required when the signals to be measured are beyond the frequency capabilities of the scope or meter. RF probes convert (rectify) the RF signals into a dc voltage that is equal to the peak RF voltage (or possibly equal to the RMS of the RF voltage). The dc output of the probe is applied to the meter or scope input and is displayed as a voltage readout in the normal manner.

Demodulator probes. The circuit of a demodulator probe is essentially like that of the RF probe. However, the demodulator probe produces both ac and dc outputs. The RF carrier is converted to a dc output voltage equal to the RF carrier. If the carrier is modulated, the modulating voltage appears as ac (or pulsating dc) voltage at the probe output.

The meter or scope is set to measure direct current, and the RF carrier is measured. The meter or scope is then set to measure alternating current, and the modulating voltage (if any) is measured. Generally, demodulator probes are used for signal tracing, and the output is not calibrated to any particular value. However, some demodulator probes produce calibrated outputs equal to both carrier and modulation (and might be described as RF and demodulator probes).

11.5.2 Frequency counters

The most common use for a frequency counter in TV service is to check or adjust the various 3.58-MHz oscillators in the chroma circuits. Counters can also be useful when checking the clock oscillators of microprocessors.

In reality, the so-called 3.58-MHz oscillator is locked (in frequency) to a color TV broadcast at a frequency of 3.579545 MHz. The oscillator remains locked at this frequency, even though the phase and color hue might shift. A seven-digit counter is required to get the full-frequency resolution. The counter should also be capable of reading low frequencies, such as 60 and 30 Hz. Notice that the color reference oscillator is sometimes operated at a harmonic or multiple of 3.58 MHz (such as the 14.3-MHz oscillator in Figure 4.6).

11.5.3 Picture tube (CRT) testers and rejuvenators

These instruments provide for test and possible rejuvenation of picture tubes. The instruments are also known as tube or CRT renewers. With these instruments it is possible to check each element of the CRT for such factors as shorts, opens, leakage, and proper emission. It is also possible to rejuvenate some CRTs by application of high voltages, heavy currents, etc. to the proper elements.

Because of the highly specialized nature of tests and rejuvenators, and because you receive a detailed set of instructions with the instrument, the units are not described here.

11.6 Analyzing the Composite-Video Waveform

Probably the most important waveform in TV service is the composite waveform consisting of the video signal, the blanking pedestals, and the sync pulses. Figure 11.10 shows typical scope traces when you are observing the composite-video signals synchronized with horizontal-sync and vertical-blanking pulses. The trace shown is one horizontal line of an NTSC color-bar signal.

You can observe the composite video signals at various stages in the TV set to determine if the circuits are operating properly. (Typical monitoring points are at TP12 in Figure 3.4, and at the emitter of Q203 in Figure 3.6.) A knowledge of the waveform makeup, the appearance of a normal waveform, and the causes of various abnormal waveforms helps you to locate and correct many problems. You should study such waveforms in a set known to be in good operating condition, noting the waveform at various points in the video circuits following the video detector.

Notice that the pattern of Figure 11.10 might appear inverted. The output of some detectors is positive, whereas other video detectors produce negative outputs. Also, it is possible to invert the pattern for a positive or

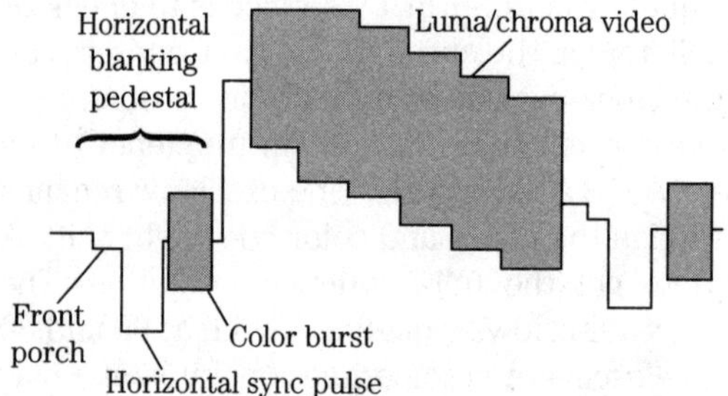

Figure 11.10 Analyzing the composite video waveform.

negative display (whichever is convenient) by adjusting the scope controls (on most scopes).

The pattern of Figure 11.10 assumes that a color signal from an NTSC generator is applied to the antenna terminals. The video portion of the pattern contains black-and-white luma levels as well as the chroma and luma levels for the standard NTSC colors.

Figure 11.10 also assumes that the composite-video waveform is being monitored at the video-detector output. This is the usual starting point. However, the composite video can also be checked at other points in the video circuits. For example, the composite-video signal can be monitored anywhere between the video-detector output (pin 20 of IC201 in Figure 3.4) to the comb filter (base of Q351 in Figure 4.5) through various buffers and amplifiers (Q469, Q467, Q466, IC402, and Q481, Figure 6.1).

Notice that the waveforms shown in Figure 11.10 are theoretical and will almost never appear exactly as shown in any circuit, even directly at the video-detector output. Figure 11.11 is a reproduction of the composite-video signal at TP12 (the emitter of Q201 in Figure 3.4), and at the input to the comb filter (base of Q351, Figure 4.5).

11.7 Testing Video Transformers, Yokes, and Coils

Transformers, yokes, and coils are often difficult components to test. Of course, if one of these components has a winding that is completely open or shorted, or that has a very high resistance, this shows up as an abnormal voltage or resistance during troubleshooting. However, a winding

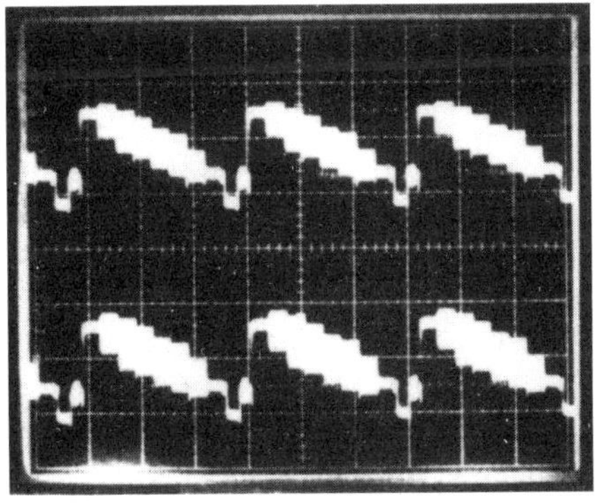

(S6) Q201/E COMPVID 1 V/d 20 μsec/d
Q351/B COMPVID 1 V/d 20 μsec/d
Color bars

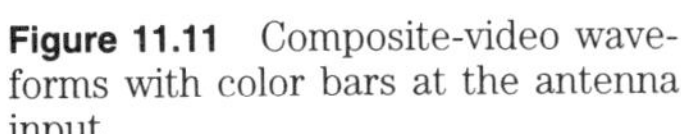

Figure 11.11 Composite-video waveforms with color bars at the antenna input.

with a few turns shorted, or leakage to another winding, can produce voltage or resistance indications that appear close to normal and pass undetected.

A scope can be used to perform a ringing test on such components. In this procedure, a pulse is applied to the component under test. The condition of the component can then be evaluated by the amount of damping observed in the waveform.

The best way to get reliable results from a ringing test is to compare the ringing waveform of the part being tested with the waveform of a known-good duplicate part. However, because duplicate parts are rarely available for comparison, you must judge the part being tested by a study of the waveform. To gain experience in evaluating the ringing waveform, it is helpful to try the procedure several times, both with good parts and with parts that you have purposely shorted with various resistance.

11.7.1 Basic ringing-test procedure

Figure 11.12 shows the various ringing-test connections as well as typical waveforms. The basic procedure is as follows:

1. Remove power from the part or circuit to be tested. Do not apply power to the circuit at any time during the test procedure.
2. Disconnect the part to be tested from the related circuit. Although many parts can be tested in-circuit, it is usually necessary to remove circuit connections, especially where diodes and transistors have a loading effect on the part.
3. Connect the scope to the part terminals as shown. Use a low-capacitance probe.
4. Connect a pulse source to the part through a capacitor as shown in Figure 11.12A or through a few turns of wire as shown in Figure 11.12B. All scopes have circuits that produce pulses suitable for use in ringing tests (if you do not have a pulse generator). These pulses are the same ones used to trigger the scope horizontal sweep. Unfortunately, the pulses are not always accessible without modification of the scope. However, scopes designed specifically for TV service have an external terminal that provides for connection to the pulse source.
5. Operate the scope controls to produce a ringing waveform as shown in Figure 11.12C or 11.12D. The waveform of Figure 11.12C is a typical ringing pattern for a normal coil or transformer winding. The waveform of Figure 11.12D is typical for patterns taken from defective coils and transformers (with shorted turns, or leaking between windings). Remember that these waveforms are typical. Experience is necessary to judge the waveforms when no duplicate parts are available for comparison.

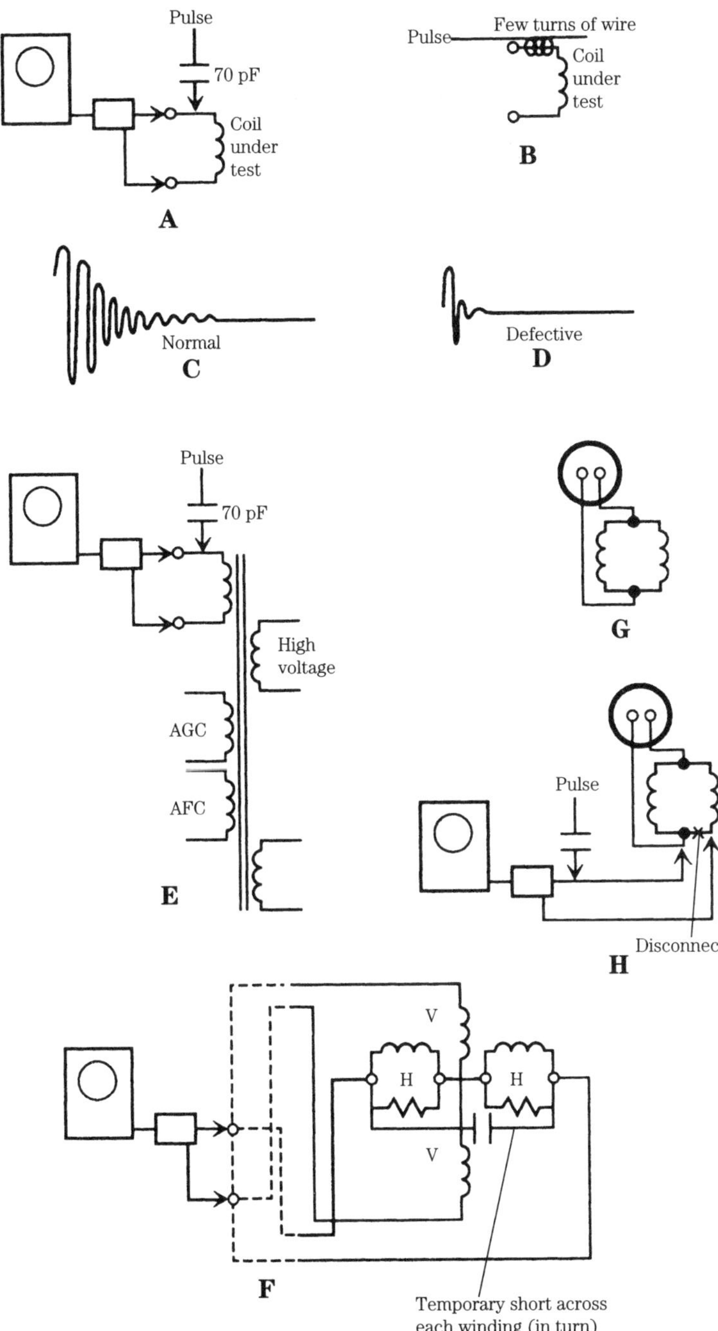

Figure 11.12 Basic ringing-test connections.

6. To help you make the judgment, connect a resistance across the part terminals (or make a loop consisting of a few turns of older wire and pass the loop around the part). Note any change in the pattern. There should be a drastic change in the pattern when the resistor (or solder coil) is added. If not, the part is probably defective.

11.7.2 Ringing test for high-voltage (flyback) transformers

To test the high-voltage transformer in the horizontal system (chapter 5), disconnect the transformer leads. Using the procedure described in section 11.7.1, get a ringing waveform across the transformer primary. Typical test connections are shown in Figure 11.12E.

With the ringing pattern established, connect a short across each of the transformer secondary windings in turn, and check the ringing pattern. If a significant change in the waveform is noted for all windings, the transformer is probably good. Little or no waveform change indicates that the transformer is probably defective.

11.7.3 Ringing test for picture-tube yoke

To test the horizontal and vertical windings of the deflection yoke, disconnect the yoke from the circuit. Using the procedure described in section 11.7.1, get a ringing pattern across each winding in turn. Typical test connections are shown in Figure 11.12F.

Notice that the horizontal winding of the yoke in Figure 11.12F consists of two sections. Alternately short each section. If the winding is good, the effect on the waveform should be the same as each section is shorted.

The vertical winding of the yoke also consists of two sections, with a damping resistor connected across each section. Disconnect these damping resistors, as well as the capacitor in parallel across both vertical sections. Then alternately short each section. If the winding is good, the effect on the waveform should be the same as each section is shorted.

On some yokes, the two sections of each winding are connected in parallel as shown in Figure 11.12G and H. In such cases, disconnect the sections at one end. Connect the scope across both sections, as shown in Figure 11.12H, and then short each section in turn. Again, the effect on the waveform should be the same as each section is shorted.

11.7.4 Ringing test for picture-tube coils

Use the procedure described in section 11.7.1 to test the width, linearity, focus, etc. coils of a picture tube if you suspect shorted turns or leakage. The solder loop usually produces the best results. Again, if the older loop causes a significant change in the waveform, the coil under test is probably good. However, experience is the best judge when making any ringing test.

11.8 Purity, Convergence, and Linearity Adjustments

All color sets require some form of purity, convergence, and linearity adjustments. These adjustments must be made in addition to the adjustments for black and white (height, width, centering, linearity, etc.). Generally, the adjustments are made when the set is placed in operation and should be checked when the set has been subjected to extensive repair, such as replacement of the CRT and/or deflection coils. Typically, the setup procedure for color circuits consists of degaussing the color CRT, followed by purity, linearity, and convergence adjustments.

It is recommended that the manufacturer's service instructions be followed exactly when color setup procedures are involved (when all else fails, follow instructions). However, in the absence of manufacturer's instructions, and to show what typical color setup procedures require, a description of a complete setup for color TV circuits follows, as recommended for a particular set. With the examples of the controls shown, you should be able to relate the procedures to a similar set of controls on most color TV sets.

11.8.1 Black-and-white adjustments

Use the following procedure for all black-and-white picture tubes and as a starting point for color CRTs.

1. Measure from opposite diagonal corners of the CRT screen, and mark the physical center of the screen with a grease pencil or similar marker, as shown in Figure 11.13A.
2. Connect the generator to the antenna terminals and select a center-cross pattern (Figure 11.7A).
3. Adjust the ion trap or centering adjustment of the CRT so that the center cross of the test pattern coincides with the physical center of the screen. Notice that if the pattern is badly distorted, the linear adjustments (or troubleshooting and repair if necessary) should be performed before the center adjustment is complete.
4. Switch to a crosshatch or dot-batch pattern (Figure 11.8). If the circuit has a width adjustment, adjust it so that the raster just fills the screen. (Some manufacturer's literature specifies the number of horizontal and vertical lines.)

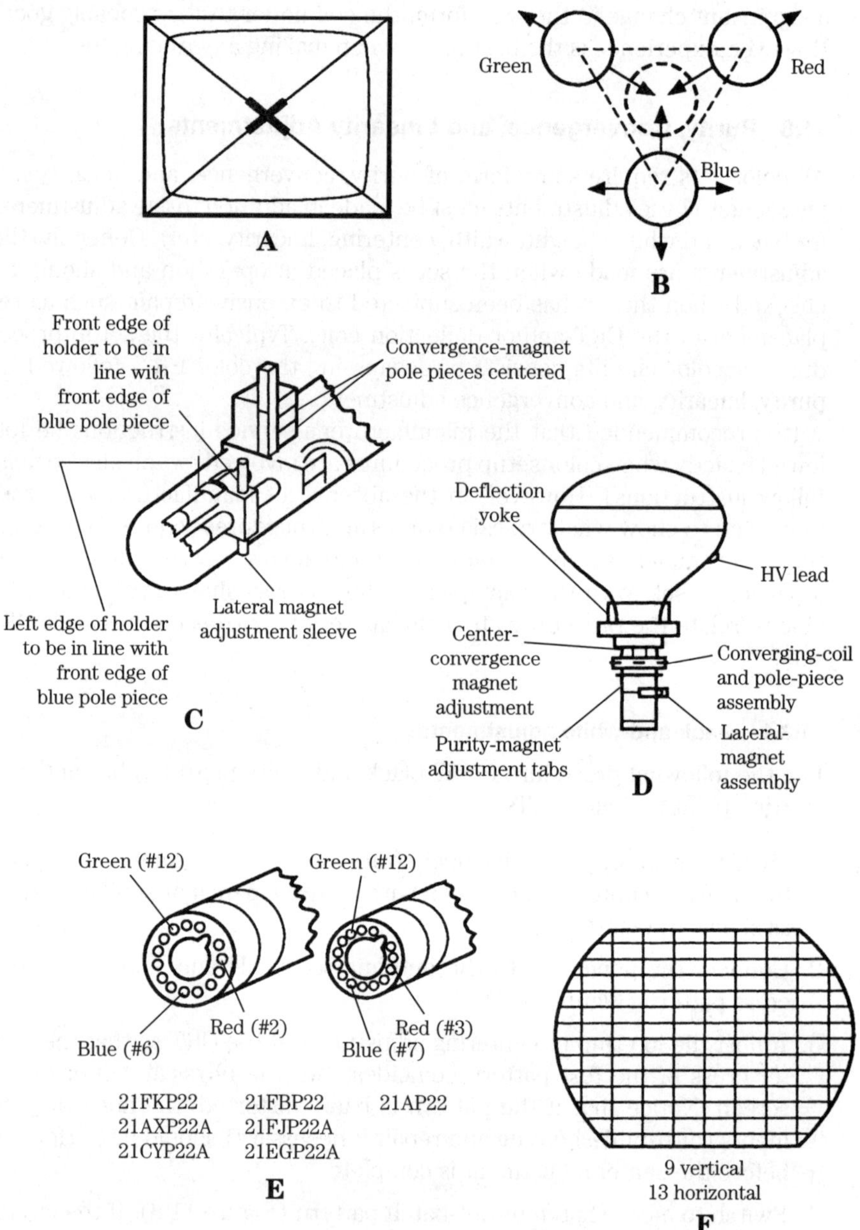

Figure 11.13 Center and convergence adjustments.

5. Adjust the height control (if any) so that the raster just fills the screen. (Again, check for any recommendations in the service literature and/or generator operating instructions as to a specific number of lines.)
6. Adjust the vertical and horizontal linearity controls (if any) so that the horizontal and vertical lines are straight. Remember that any bend in horizontal or vertical lines with a crosshatch pattern is likely to show up as a bend in the picture.
7. Wipe the grease pencil centering marks from the face of the CRT.

11.8.2 Preliminary convergence adjustments

Before making any convergence adjustments, check CRT focus. (Figures 4.7 and 4.12 show typical focus adjustment controls.)

1. Leave the generator connect to the antenna input, and select a dot pattern (Figure 11.7B).
2. Adjust the red, green, and blue magnets and the lateral magnet to get proper convergence of dots in the center of the CRT. The direction of dot movement using the magnets is shown in Figure 11.13B. Red and green movement is opposite that for blue. The red and green dots move diagonally, whereas the blue dot moves horizontally or vertically (on this particular CRT). Location of convergence and lateral-beam magnets on a typical CRT are shown in Figure 11.13C.
3. If a greater range of adjustment is necessary, the magnets of some CRTs (but not all) can be reversed (by rotating the magnet holder). If necessary, keep the CRT in focus when making this adjustment.

11.8.3 Color purity adjustments

The CRT and associated parts should be subjected to a strong demagnetizing field before any purity adjustments are made. All present-day color sets have some form of demagnetizing or degaussing system, usually a coil that is energized when power is applied (Figures 8.3 and 8.6). However, it is possible that the degaussing system might fail or the CRT might become contaminated. In either event, a commercial degaussing coil can be used to correct the problem.

Slowly move the coil around the CRT and around the sides and front of the set. Do not degauss the magnets around the CRT neck. If the CRT has a special built-in degaussing system, operate the system as described in the service literature, if not automatic.

Some sets (such as the larger-screen Hitachi models) have a geomagnetism-correction circuit. This circuit is used each time the set is moved to a different location. The landing (the point where the beam strikes the fluorescent material) of the CRT changes, depending on the influence of geomagnetism. With larger CRTs, color unevenness might appear because of this condition. The problem is aggravated when the direction of the set face is changed from north to south (or vice versa), causing the color unevenness to move around the edges of the screen. The geomagnetism coil reduces this effect by setting up a reverse magnetic field.

Once the CRT is properly degaussed, operate the generator to produce a blank raster (Figure 11.7). If the generator is not capable of producing a blank or solid-color raster, disable the output of the video detector (short the output to ground).

Loosen the screw on the yoke clamp and slide the yoke as far to the rear as possible. Figure 11.13D shows the location of the magnets and yoke on a typical CRT.

Shunt the blue and green guns to ground through individual resistors (typically 100-kV resistors). If the generator is provided with color-gun interrupters (gun killers), connect the leads from the generator to the corresponding leads or pins of the CRT socket. The location of color-gun pins for an assortment of color CRTs is shown in Figure 11.13E. Figures 4.7 and 4.12 show the color-gun pins (KRMKB,KG) for sets discussed in this book. Generally, it is not necessary to remove the socket from the color CRT or make any direct connection to the socket pins. Most color-gun interrupter leads have special alligator clips that pierce the insulation on the CRT lead and make contact with the wire.

Rotate the purity magnet around the neck of the CRT and at the same time adjust the tabs on the magnet to produce a uniform red screen area (solid red spot) at the center of the CRT.

Slide the yoke forward until the screen becomes completely red.

After checking the red field, it is helpful (but not absolutely necessary) to check the green and blue fields (individual for purity, and the white field for uniformity. The green field can be checked by shunting the red and blue guns (or use the color-gun interrupters). The blue field can be checked by shunting the red and green fields.

With all CRT guns shunted, there should be no color, and the screen should be a uniform white. Notice that if individual fields of red, blue, and green are present, and a white field is present with all guns off, you know that the CRT is good and is capable of producing all required colors.

Set the generator for a crosshatch pattern. Check that the height, width, centering, focus, overscan, etc. are properly adjusted (section 11.8.1) by comparing the crosshatch pattern with that specified for the CRT.

It is sometimes possible to get good purity only by pushing the yoke too far forward. This reduces width and height so that the set does not overscan properly. The crosshatch pattern provides a convenient method for adjusting the overscan (height and width) to ensure that the proper portion of the raster is extended beyond the range of the CRT mask.

The service literature for color sets often specifies a recommended amount of overscan at the left and right, and a different amount of overscan at the top and bottom. The recommended overscan varies in models. However, the crosshatch function provides a fixed number of vertical and horizontal lines, so it is relatively easy to judge the amount of overscan. The appearance of the crosshatch pattern with correct overscan (9 vertical and 13 horizontal lines in this case) in a representative set is shown in Figure 11.13F.

11.8.4 CRT temperature or white-balance adjustments

The compatible TV system is designed so that reception of black-and-white pictures is reproduced on color sets. The three CRT guns must be adjusted so that the telecast is reproduced as black and white within the normal usable range of the contrast and brightness controls.

The adjustments are called white balance or CRT temperature (or screen temperature) depending on the literature. No matter what they are called, the adjustments pertain to the reproduction of various luma values from black to white (gray-scale). Only a set correctly adjusted for gray-scale tracking can reproduce proper color when tuned to a color telecast.

1. Connect the generator to the antenna terminals and select a center-cross pattern. Set the CRT bias or static-convergence control (such as RV708 in Figure 4.7) and the screen or background controls (RV701, RV703, RF705 in Figure 4.7) to fully off. The pattern should be removed from the screen.
2. Advance the individual screen or background controls so that each control just produces a pattern on the CRT. When one or more of the controls fails to produce a pattern, advance the CRT bias or convergence control. After the convergence control is advanced to make the missing pattern appear, adjust the remaining screen or background controls to where the pattern just appears.

An alternate method for adjustment (that sometimes provides greater accuracy) involves extinguishing the pattern just after the pattern appears. That is, adjust the red, green, and blue screen or background controls to where each pattern is just cut off. If this method is used, increase the bias or convergence only as required to support the weakest screen or background adjustment.

11.8.5 Convergence checks

After purity and white-balance/temperature adjustments, check the convergence of dots on the CRT as described in section 11.8.2. If the temperature and purity adjustments have affected convergence, repeat the convergence procedure using a dot pattern. If the convergence magnets are touched, readjust the purity and white-balance/temperature controls.

11.9 Color Setup Using a Color Generator

This section describes how a typical color generator can be used to set up a color TV. Notice that the procedures cover essentially the same areas described in section 11.8, but they apply specifically to a color TV, using a particular color generator. With this generator, it is possible to vary the RF level as well as the chroma level of the signal applied to the antenna. Such features are not found on all color generators.

11.9.1 Preliminary procedure

Connect the generator output to the antenna input. Disconnect any antenna that might be connected to the set, including "rabbit ears". Let the generator RF-level and chroma-level controls to midrange. Select RF output will a full color-bar pattern (Figure 11.3). Disable any automatic fine-tuning (AFT) circuits. (This does not apply to the synthesized-tuning circuits with PLL described in chapter 2, but does apply to some older sets with manual and automatic fine tuning.) Adjust all TV set controls for best picture.

11.9.2 AGC adjustment

Use the following procedure on sets with any form of AGC:

1. Select a crosshatch pattern. Turn the AGC control (such as RV201 in Figure 3.4) until the pattern just begins to bend or distort. Then readjust the AGC to the point slightly below where distortion is eliminated. Notice that the AGC control has very little effect on some sets. If so, set AGC for best picture. Also, notice that the AGC control might be on the tuner module as shown in Figure 2.1.
2. Set the generator RF-level control to minimum. If the pattern distorts or loses sync, readjust the AGC as in step 1, setting AGC just below where the trace distorts.

3. Vary the RF-level control throughout the entire range. Readjust AGC as necessary to prevent the picture from distorting or losing sync through the widest possible range of RF levels.

11.9.3 Video peaking

Use the following procedure on sets with any form of video peaking adjustments. Select a gray-scale pattern. Adjust the video peaking coil (or coils) so that the edges of the bars are sharp. Notice that if you adjust the peaking controls too far in one direction, "ringing" (traces of the edge repeating across the pattern) might occur. If you adjust the peaking too far in the other direction, the edges might become soft or hazy.

11.9.4 Horizontal and vertical centering

Use the following procedure on sets with any form of horizontal or vertical centering. (Notice that vertical centering for the circuit of Figure 5.7 uses a switch S501, whereas the horizontal centering uses solder-bridge connections as shown in Figure 5.9.)

Select a center-cross pattern and adjust the centering controls so that the vertical and horizontal lines intersect at the center of the screen. Then select a crosshatch pattern and check that the outer lines are equally spaced from the edge of the CRT mask.

11.9.5 Purity

Use the following procedure on all sets:

1. Before adjusting purity, the set might need a rough check of static convergence as described in Sec. 11.9.10.
2. Select a blank raster. Turn off the blue and green guns or use gun-killer switches. This should produce a red screen.
3. Loosen the CRT yoke and move the yoke to the rear of the CRT neck (as far as possible without moving the purity-magnet assembly).
4. Adjust the purity magnets so that the area in the center of the screen is uniformly red with no dark or discolored areas.
5. Push the yoke forward until the entire screen is uniformly red. Tighten the yoke.
6. Turn on the blue and green guns. The raster should be uniformly white. If not, repeat the adjustment procedure.
7. If good purity adjustment cannot be obtained, part of the set might have become magnetized and require degaussing. Even sets with automatic degaussing can have a magnetized cabinet. Refer to Sec. 11.8.3.

11.9.6. Overscan

Use the following procedure on all sets. Select a crosshatch pattern, and adjust the width and height controls so that the raster extends beyond the edge of the CRT mask by the amount specified for the particular set. As discussed, the recommended amount of overscale varies but is usually specified in the service literature. As shown in Figure 5.5, there are two vertical height (or size) controls for this particular set, but no horizontal width control.

11.9.7 Linearity

Use the following procedure on all sets. Select a crosshatch pattern, and adjust the vertical-linearity controls (such as RV505 in Figure 5.6) to that the horizontal lines of the crosshatch are equally spaced (particularly at the extreme top and bottom of the screen). Adjust the horizontal-linearity controls (if any) so that the vertical lines are equally spaced.

11.9.8 Pincushion adjustment

Use the following procedures on sets with pincushion controls. Select a crosshatch pattern (or preferably a dot hatch) and adjust the pincushion controls (such as RV507 and RV508 in Figure 5.7) until all horizontal lines are straight.

11.9.9 White-balance and temperature (gray-scale tracking)

Correct white-balance and temperature applies to all sets, and is necessary to ensure that all three color guns are operating at a level to cause the red, blue, and green phosphors to glow at the same intensity. The beam currents of the three guns must maintain the ratio throughout the full range of the video signal and the brightness control. Proper balance must be maintained from black (low lights) through the grays to white (bright lights). Use the procedure in Sec. 11.8.4 to adjust white-balance and temperature, and to check gray-scale tracking.

11.9.10 Convergence adjustments

Misconvergence of a color set can be noted by observing the crosshatch or dot pattern. If the set is properly converged, the lines or dots are clear and sharp, with no color fringing (one or more colors appearing on the edges of the dots or crosshatch lines).

Notice that some sets require both static and dynamic convergence. Static convergence uses permanent magnets. Dynamic convergence requires a convergence yoke and corresponding circuits (such as the CY yoke, and circuits, of Figure 5.15). Most manufacturers use dynamic convergence for their larger-screen sets (such as the 25-inch Sony shown in Figure 5.15).

1. Before adjusting convergence, check other adjustments (especially purity), if specified in the service literature. These adjustments might include horizontal frequency, horizontal drive, high-voltage regulation, height, width, linearity, gray-scale tracking, focus, CRT screen grid, etc.
2. Select a dot pattern and perform the preliminary or static convergence adjustments described in Sec. 11.8.2. Adjust the permanent magnets on the CRT neck to converge the red, blue, and green beams at the screen center. A "perfectly converged" pattern has pure white dots with no color fringing.
3. Select a crosshatch (or dothatch) pattern, and perform dynamic convergence (if any) as described in the service literature.
4. Repeat static and dynamic convergence procedures as required until the complete crosshatch (or dothatch) pattern consists of crisp white lines (and dots) with the least amount of color fringing.

11.9.11 Check the color circuits

Use the following procedures on all color sets.

1. Select a full color-bar display. Switch out any special color controls (such as "accutint"). Adjust the hue or tint and color controls to midrange. Adjust the fine tuning (if any), brightness, picture, and contrast controls for best color-bar pattern (Figure 11.3).
2. Pay particular attention to the magenta and cyan (bluish-green) bars. Generally, if you can get good magenta and cyan with the hue or tint control, the color circuits are operating satisfactorily (and all of the remaining colors will be good). Ideally, the hue or tint control should be near the midrange position when magenta and cyan are off color. Typically, the magenta becomes red at one extreme of the hue or tint control, while the cyan goes from blue to green (or vice versa) at the other extreme. If it is impossible to get good magenta and cyan display, suspect the circuits that control the 3.58-MHz oscillator.
3. To check accuting or similar functions that enhance flesh tones, switch the function on, and check the color bars. The pattern should change, with several of the bars taking on a more reddish color. With an accuting or similar function on, the range of the hue or tint control is usually more restricted in most sets (only a slight shift of color at both extremes). The color control might also be restricted so that the color cannot be taken completely out of the pattern.
4. To check color-sync action, turn the generator chroma-level control slowly to minimum. The color should become pale and finally disappear. Because some sets have an automatic color control (ACC) circuit, the

rate of fading depends on the set. Most sets loose color-sync just before the color disappears (diagonal lines run through the colors). These conditions are normal. However, if a slight reduction in chroma level causes the colors to fall out of sync, suspect the color-sync circuits. As a further check, turn the RF-level control to reduce signal strength, and note the effect on color sync.

11.10 Using Vectorscopes

An NTSC vectorscope is used in TV broadcast work to troubleshoot and adjust color demodulators. The vectorscope also can be used to judge the general condition of color TV set circuits (chroma bandpass, burst amplifier, reference oscillator, phase detector, such as the circuits in Figures 4.5 and 4.6) when used with the NTSC color-bar output of a color generator. A vectorscope measurement is often more helpful for troubleshooting than merely observing the NTSC pattern on the CRT.

If an NTSC vectorscope is not available, a good lab-type scope can be substituted. The demodulated color signals (directly from the red and blue guns, such as KR and KB in Figure 4.12) can be used as X and Y inputs to the scope, as shown in Figure 11.14A. This connection should produce a scope display similar to that of Figure 11.14B. The scope can be set up for vectorscope operation as follows:

1. Connect the color generator to the antenna input and select the NTSC color-bar pattern (Figure 11.3).
2. Set up the scope for X-Y operation. Adjust the position controls to center the dot on the scope screen with no signal input to the scope.
3. Connect both the X and Y inputs of the scope to the red gun.
4. Adjust the horizontal and vertical gain of the scope to equal amounts. This should move the spot or dot to the 45° reference position (from the center), as shown in Figure 11.14B.
5. Now move the horizontal (X) input of the scope to the blue gun. Leave the vertical (Y) input connected to the red gun.
6. For a 90° picture tube, the scope display should be similar to that of Figure 11.14B. The typical 105° picture tube produces a more elliptical display.
7. If desired, the various vectors may be selected one at a time, by selecting the corresponding raster color display of a generator. Raster displays are discussed in section 11.4.6.

11.10.1 Basic vectorscope adjustment procedure

With the vectorscope connected as shown in Figure 11.14, rotate the hue or tint control through the full range, and note the effect on the vector pattern. The pattern should turn but should not change in size. If necessary (or possible) adjust the 3.58-MHz oscillator until the pattern is the same size throughout the range of the hue or tint control. In some cases, the reference oscillator is adjustable (such as the CV301 in Figure 4.6), but not in all color circuits.

As an alternate, set the hue or tint control to midrange and adjust the reference oscillator so that the color signals stays in sync as the chroma-level control on the generator is turned to minimum. The vector pattern (scope) should reduce in size, but should not rotate. Rapid spinning of the vector pattern indicates misadjustment of the oscillator (or possibly a defective crystal, such as X301 in Figure 4.6).

11.10.2 Troubleshooting with vectorscope patterns

It is possible to check operation of the color circuits by comparing the display in Figure 11.14 with the correct values shown in Figure1.6. That is, you can check each color for correct angle and amplitude. However, this is best done with an NTSC vectorscope, and is not too practical with a scope (even a good lab scope). However, any drastic deviations of the vectorscope pattern from that shown in Figure 11.14 indicate a major problem in the color circuits. Here are some examples.

A loss of R-Y signal causes a loss of vertical deflection, and the vectorscope pattern changes to a straight line. If the B-Y is good, but beam is deflected along the horizontal axis. This indicates that the trouble lies in the R-Y demodulator, matrix, or difference amplifier, depending on the circuit. (All of these circuits are in IC301 in Figure 4.6). If the R-Y signal is weak, some deflection of the vectorscope pattern occurs, but the pattern is extremely distorted.

A loss of B-Y signal results is no horizontal deflection, and the vectorscope pattern changes to a straight line (a vertical line if the R-Y signal is good). This indicates that the problem is in the B-Y difference amplifier, matrix, or demodulator, depending on circuits. Again, if B-Y is weak, some deflection occurs, but the vectorscope pattern is extremely distorted.

If there is a complete loss of color, the vectorscope pattern usually appears as a center dot or possibly a short line in the center of the vectorscope pattern. Remember that all color circuits do not produce identical vectorscope patterns. The patterns discussed here are for reference only, and must be considered as typical. Always consult the vectorscope instruction manual and all service literature.

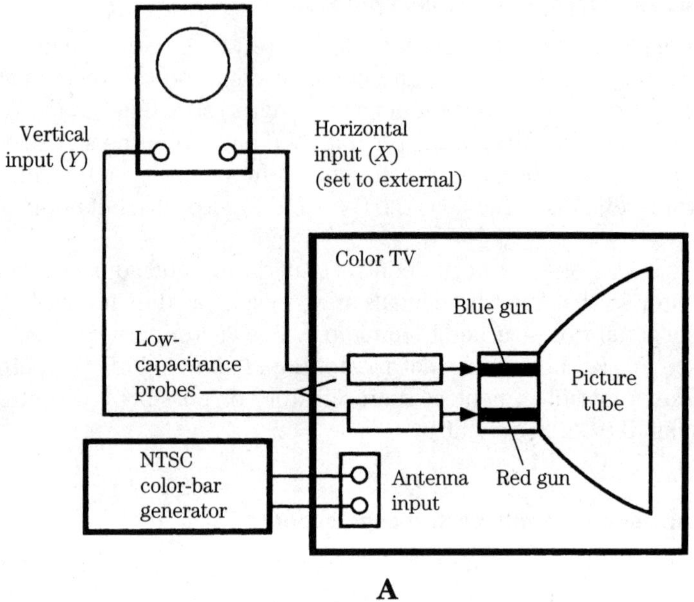

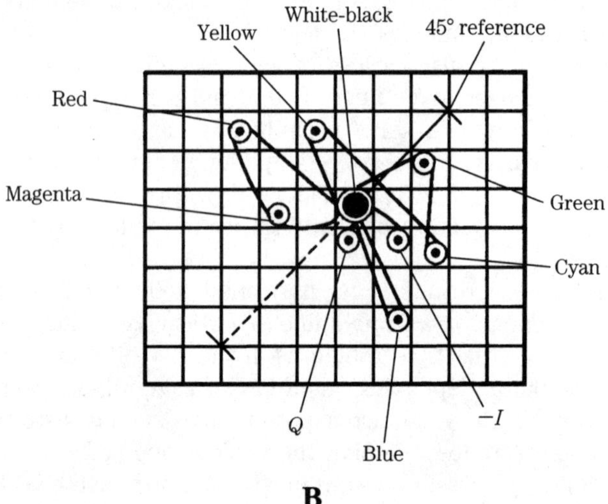

Figure 11.14 NTSC vectorscope connections and display.

11.11 Stereo-TV Generators

Although it is possible to align the stereo TV circuits of some sets using a conventional color TV or RF generator, it is generally simpler to use a special-purpose stereo TV generator. There are inexpensive stereo TV generators (such as the B&K Model 2009) that provide all of the signals necessary for service and troubleshooting. Such generators use spot modulation to simulate dbx encoding (chapter 10) at specific frequencies (rather than providing encoding across the entire TV frequency spectrum, as do expensive broadcast studio instruments).

The 2009 generates an FM-modulated RF carrier on TV channel 3 or 4 and at the standard TV IF carrier, as well as on a standard 4.5-MHz audio carrier. A mode selector permits four combinations of internal modulation: left-channel only (L), right-channel (R), baseband (L+R), and subband ($_{L+/-R}$). A 15,734-kHz pilot signal is generated and combined with the composite audio. The pilot can be switched on and off when desired, making it possible to test the pilot detector circuits in the decoder. A SAP mode is also selectable, providing a subband signal centered at 78,67 kHz, to test SAP operation.

11.11.1 Stereo TV generator operating notes

The following notes supplement the specific test and adjustment procedures described in section 11.12.

The L, L+R, and $L_{+/-}R$ modes should produce equal audio outputs so that the audio levels in the left and right channels should be equal for all modulating modes. The L+R or L–R mode should produce equal audio levels in each channel. Similarly, the L or R mode should produce the same level in the active channel as in the L+R or L–R modes. This can be used in troubleshooting.

For example, under a given set of conditions (same measurement points, same audio frequencies) the L and R channels should produce the same audio output voltage. If they do not, an unbalanced condition is indicated. (This unbalanced condition is not necessarily limited to the stereo-decoder circuit but could as likely be in the audio-amplifier circuits.)

The SAP function should produce equal SAP audio-output levels at all frequencies. The SAP function is monaural in this generator but has dbx encoding. This is not necessarily the case with a TV station signal, where both stereo and SAP can be broadcast simultaneously (even though you can be listed to only one at a time).

11.12 Stereo-TV Tests and Adjustments

Compare the following adjustment procedures with those shown in the service literature for the stereo TV being serviced. The test and adjustment points are shown in Figures 10.5 and 10.7.

11.12.1 L-R adjustment

The purpose of the L-R IC1 is to reinsert a carrier into the AM stereo $L_{+/-}R$ sidebands at pin 2 and produce corresponding audio output at pin 5. The missing L–R carrier is at 31.468 kHz and is produced by a VCO within the L-R decoder. Because there is no L–R carrier at the decoder output, the VCO is usually locked to the 15.734-kHz pilot (or a multiple).

In the circuit of Figure 10.5, the VCO can be set by adjustment of PLL-adjust VR1 at pin 16 and IC1, and can be monitored atg TP31. The VCO signal at TP31 is 15.734 kHz, even though the VCO operates at a different frequency (typically higher).

1. Ground the input to the stereo-decoder circuits at pin 2 of the SS connector. This removes all signals to the L–R decoder input, including the 15.734-kHz pilot. If the pilot is present, it is possible that the VCO will lock onto the incoming signal even though the VCO is not exactly on frequency when free-running. Check that stereo indicator D521 is off.
2. Connect a frequency counter to TP31, and adjust VR1 for a reading of 15,734 kHz. Remove the ground from pin 2 of the SS connector. Check that the TP31 reading remains at 15.734 kHz.

11.12.2 L–R Gain adjustment

The purpose of this adjustment is to set the level of the stereo L–R signal before dbx noise-reduction processing. This is not to be confused with the L–R separation adjustment that sets the L–R level after dbx processing (section 11.12.10).

1. Apply a modulated L–R signal to pin 2 of the SS connector. The stereo indicator D521 should turn on. Use a modulation frequency of 300 Hz (unless otherwise specified in the service literature).
2. Monitor TP37 for the 300-Hz signal, using a meter or scope. Adjust VR2 for the correct voltage level at TP37. Always use the values specified in service literature (for both input amplitude at pin 2 of the SS connector and output at the test point).

11.12.3 SAP level adjustment

The purpose of this adjustment is to set the level of the SAP signal (before dbx processing) in relation to the stereo L–R signal. If the SAP and L–R signals are not approximately equal at this stage in the audio path, the user audio output or volume control must be reset each time when switching between stereo and SAP.

1. Apply a modulated SAP signal to pin 2 of the SS connector. The SAP indicator D520 should turn on. Use a modulation frequency of 300 Hz unless otherwise specified in the service literature.
2. Monitor TP38 for the 300-Hz signal. Adjust VRF3 for the correct voltage level at TP38. Always use the values specified in the service literature (for both the input amplitude at pin 2 of the SS connector and the output at TP38). Notice that the SAP level adjustment is usually performed after adjustment of the SAP detector (but check the service literature for the preferred sequence).

11.12.4 SAP detector adjustment

The purpose of the SAP signal detector IC3 is to produce a low at output pin 7 when the 78.67-kHz SAP carrier is present at input pin 6. (Output pin 7 remains high when the SAP carrier is not present.) This is done by comparing the incoming SAP carrier with the IC3 decoder VCO. When both signals are present and locked in frequency and phase, pin 7 goes low. The VCO is set by adjustment of SAP PLL adjust VR4 at pin 1 of IC3.

1. Apply a 78.67-kHz signal to pin 2 of the SS connector. Monitor the voltage across TP35 and TP36. Set VR4 to full counterclockwise. The SAP indicator D520 should be off. The voltmeter should read about –1 V.
2. Adjust VR4 clockwise until the voltage at TP35 and TP36 changes to about +1 V. This indicates that pin 7 of IC3 is switched to low. You could measure at pin 7 of IC3 or at TP34, but the change in signal level is much more difficult to detect. Also, by checking at TP35 and TP36, you also confirm operation of Q13–Q15 simultaneously.
3. With the SAP carrier still applied at the circuit input (pin 2 of the SS connector), check that SAP indicator D520 has turned on when the TP35/TP36 voltage indication changes from –1 to +1V. This confirms operation of both D520 and Q33.

11.12.5 SAP detector signal output-level adjustment

The purpose of VR5 is to set the level of the SAP detect signal. (This signal appears when the SAP carrier is present, and pin 7 of IC3 goes low.)

1. Make certain that no SAP carrier signal is present. (On some circuits, it might be necessary to ground the SAP circuit input at pin 2 of the SS connector.) Monitor the voltage across TP35 and TP36. Adjust VR5 until the voltage at TP35/TP36 is –1V.

11.12.6 Stereo noise-reduction time-constant adjustment

The purpose of VR6 is to set one amount of control output from dbx noise-reduction IC4 for a given L–R signal amplitude. Connect a meter to TP39 and TP40. Adjust VR6 for the correct voltage across R5. Use the service literature values.

Notice that although you are measuring voltage across R5, VR6 is set for a given current through circuits within IC4 (the more current, the more control). Typically, R6 is set for a current of 15 mA at pin 13 of IC4. Because R5 is 1000 Ω, the voltage reading should be 15 mV.

11.12.7 Stereo-reduction VD adjustment

The purpose of VR7 is to set the L–R gain, after spectral processing by IC4 but before wideband processing, to produce the desired variable deemphasis (VD). (This is sometimes called the wideband or high-band VD adjustment.)

1. Make certain that no L–R signal is present. On some circuits, this can be done by grounding the input at pin 2 of the SS connector. In other circuits, it is necessary to disable the L–R audio path (such as connecting the emitter of Q18 to +9 V or +12 V).
2. Apply a 300-Hz signal to TP37, using the correct level specified in the service literature. A typical input level at TP37 is –24 dB.
3. Monitor the 300-Hz signal (after noise reduction) at TP43. Make certain that the level at TP43 is within limits specified by the service literature. The typical 300-Hz output level at TP43 should be between –23 and –35 dB. Notice the actual level at TP43.
4. Change the frequency of the audio signal applied at TP37 from 300 Hz to 8 kHz. Set the amplitude of the 8-kHz signal as specified in the service literature. A typical input level at TP37 with the 8-kHz audio is –17 dB.
5. Adjust VR7 so that the 8-kHz signal (after noise reduction) at TP43 is as specified in the service literature. A typical 8-kHz output level at TP43 is the actual 300-Hz level (as measured in step 3) less –11 dB. Remember that these noise-reduction adjustments are critical for proper operation of the stereo TV circuits. Also, the service literature generally recommends that the VD adjustment be performed before the separation adjustment described in section 11.12.10.

11.12.8 SAP noise-reduction time-constant adjustment

The purpose of VR8 is to set the amount of control output from dbx noise-reduction IC6 for a given SAP signal amplitude. Connect a meter to TP41 and TP42. Adjust VR8 for the correct voltage level across R26 (at TP41 and TP42). Use the service literature values.

Notice that although are measuring across R26, VR8 is set for a given current through the circuits within IC6 (the more current, the more control). Typically, VR8 is set for a current of 15 mA at pin 13 of IC6. Because R26 is 1000 Ω, the voltage reading should be 15 mV.

11.12.9 SAP noise-reduction VD adjustment

The purpose of VR9 is to set the SAP gain, after spectral processing by IC6 but before wideband processing, to produce the desired variable deemphasis. The procedure is the same as for adjustment of VR7 (section 11.12.7), except that TP38 and TP44 are used instead of TP37 and TP43.

11.12.10 L+/–R separation adjustment

The purpose of this adjustment is to set the level of the stereo L–R signal in relation to the mono L+R signal. Both signals are combined in the Q22 matrix (R59/R60). If the L–R signal is low in relation to L+R, you will only hear a mono signal. If the L–R signal is high in relation to L+R, you will hear both signals, but there will be poor separation between the left and right audio (the audio sounds like mono even though stereo is present).

1. Apply a modulated L–R signal to pin 2 of the SS connector. The stereo indicator D521 should turn on. Use a modulation frequency of 300 Hz unless otherwise specified in the service literature.
2. Monitor TP32 and TP33 for the 300-Hz signal. Adjust VR10 for the correct voltage level at TP32/TP33. Always use the values specified in the service literature (for both input amplitude and output at the test points). Remember that this adjustment can be critical in producing good stereo sound.
3. If you cannot find a separation adjustment procedure in the service literature, use the following as an emergency procedure only.
4. Set VR10 so that L–R is zero. (Generally, this means setting VR10 full counterclockwise.) Apply an L+R signal with 300-Hz modulation at pin 2 of the SS connector. Notice the voltmeter reading at TP32 and TP33. This is the mono L+R signal level.
5. Remove the L+R signal and apply L–R with 300-Hz modulation at pin 2 of the SS connector.
6. Adjust VR10 until the readings at TP32 and TP33 are the same as with the L+R (or just below L+R) in step 4.

11.12.11 Stereo separation tests

Virtually all stereo TV decoders require a stereo separation test and a stereo indicator (pilot) test. Compare the following procedures with those shown in the service literature.

1. A pilot signal must be present for all stereo tests. Operate the generator controls as necessary for a pilot signal, and check that the stereo indicator (such as D520) turn on.
2. Before making any stereo tests, make certain that the stereo function is selected on the TV set. For example, in the circuit of Figure 10.8, pin 17 of IC810 must be made low by system control (Figure 7.3) for the stereo function to be selected. IC810 responds by applying a signal from pin 19 to the STEREO indicator.
3. Select the desired audio-modulating frequency of 300 Hz, 1 kHz, or 8 kHz, using the appropriate generator controls.
4. Select the desired form of modulation (L, R, L–R, or L+R), using the corresponding generator controls. Notice that many stereo tests are conducted using L (left, followed by a repeat of the test with R (right) modulation. Often, this is followed with a test using L-R modulation.

On some generators, the controls are interlocked, so you can select only one of the four modulation signals at a time. No matter what controls are involved, remember that if you select L+R, there is no stereo operation at the decoder even though the pilot is present and the stereo indicator is on.

To measure stereo separation, select L (with the pilot on). The stereo indicator should turn on and sound should be heard from the left speaker. Monitor the left-channel speaker (or left-channel audio output, whichever is convenient). Measure the left-channel output voltage. Select R (with a pilot). The stereo indicator should remain on, and sound should be heard from the right speaker.

Compare the right- and left-channel voltages. The ratio of the left-channel output to the right channel is the stereo separation and is usually expressed in dB. Generally, the actual voltage are not critical. Instead, it is the voltage ratio that counts. Of course, both voltages (L and R) must be measured at the same corresponding point in the audio path (typically at the L and R speakers or outputs), and both must be measured using the same input voltage level and frequency.

Notice that when the pilot is present (stereo on), the channel outputs might be equal. However, with the pilot off (stereo off), the channel outputs should be substantially the same, no matter what modulating signal is used. Compare the measured stereo separation to that given in the set specification (if any).

Chapter

12

TV Troubleshooting Approach

This chapter summarizes my basic troubleshooting approach for TV. No matter what type of set is involved (color, stereo, etc.) a logical approach is needed to find and correct any fault. First, you must know how the circuits operate. Second, you must have an approach for troubleshooting each type of circuit. The circuit descriptions and recommended troubleshooting approaches in chapters 2 through 10 fill these two requirements. It is your job to compare the circuits of the set you are servicing with the circuits in this book.

There are service manuals, datasheets, or fact sheets available for most TV sets. Although these sheets do not include the elaborate descriptions found in training manuals, they do contain condensed data (schematics, waveforms, parts placement diagrams or photos, replacement parts lists, operating procedures, test and adjustment procedures, etc.) that are generally adequate.

No matter what information is available, study the data thoroughly before attempting to troubleshoot. As a bare minimum, you must have a good schematic, and know how to operate all controls for the set. (It makes a bad impression on the customer if you cannot find the on-off switch, especially on the second service call.) If there are no operating procedures in the service literature, try to get the Owner's Manual, which usually describes operation and any special precautions for the set. (Use the pretext of wanting to "look up the serial number" to lay hands on the manual.)

You must be able to use basic test equipment (meters, scopes, generators, counters, etc.). The procedures for the special-purpose generators (color, stereo TV) used for TV service are covered in chapter 11.

You must know how to use tools to repair a problem once it has been located. Most TV repairs can be made with basic tools (soldering and desoldering tools, pliers, screwdrivers, diagonal cutters, etc.). However, special techniques must be used for certain parts. Repair of PC boards, and removal and replacement of ICs and modules are typical examples.

Of most importance, you must be able to logically analyze the information about malfunctioning sets and apply a systematic, logical procedure to find the problem. In short, you must be able to think. The information analyzed might be the appearance of the picture on a TV screen, or indications taken from test equipment (waveforms, voltage and resistance measurements). Either way, it is the analysis of the information that makes for logical, efficient troubleshooting.

12.1 Troubleshooting Sequence

There are four basic steps in the troubleshooting sequence: determining the failure symptoms, localizing the trouble to a functional area, isolating the trouble to a circuit or module within the area, and locating the specific trouble. Before going into the details of the four steps, let us consider what is done by each step.

12.1.1 Determining failure symptoms

Determining symptoms means that you must know what the set is supposed to do when operating normally and more important, you must recognize when the normal job is not being done. All TV sets have operating controls and two built-in aids for evaluating performance (the picture tube and the loudspeaker). Using the normal and abnormal symptoms produced by these aids, you must analyze the symptoms to answer the questions: "How well is this set performing, and where in the set could there be trouble that produces these symptoms?"

12.1.2 Localizing trouble to a functional area

The term *function* is used in TV troubleshooting to denote an operation in a specific area of the set. For example, in any TV set, the functions can be divided in RF, IF, audio, video, picture tube, and power supply. To localize trouble systematically, you must have a knowledge of the functional units and you must be able to correlate all the symptoms.

As a classic (oversimplified) example, if both picture and sound are poor, the trouble might be in either the RF or IF stages, or possibly the PLL tuning system, because these functional areas are common to both picture and sound. On the other hand, if the picture is good, but the sound is poor, the trouble is probably in the audio stages that follow the IF, because these functional areas apply only to sound.

If the service literature has block diagrams, the blocks can be used to localize trouble with a technique known as *bracketing* (or good input/bad output). If the block diagram includes major test points, as it might in some well-prepared service literature, the block also permits you to use test equipment to help narrow down the probable trouble cause. However, test equipment is used more extensively during the isolation step of troubleshooting.

12.1.3 Locating the specific trouble

Although this step refers only to locating the specific trouble in the suspected functional area, the step also includes a final analysis or review of the complete procedure and use of repair to remedy the problem (probably to replace a part). This final analysis permits you to determine if some other malfunction produced the symptom, or whether the part located is the actual cause of trouble.

In sets where access to the circuits is relatively easy, perform a rapid visual inspection first. Among other things to look for during visual inspection are burned, charred, or overheated parts, arcing between PC traces, and cold-solder joints.

The next step in locating the problem is the use of a scope to observe waveforms and a meter to measure voltages. In most TV service literature the voltages (and possibly the resistances) are not given; you must be able to use test equipment to make the measurements. This is why test equipment and procedures are stressed throughout this book. After the trouble is located, you should make a final analysis of the complete troubleshooting procedure to verify the trouble. Then you can make repairs (probably replacement) and check out the set.

12.2 Practical TV Troubleshooting Sequence

Now that a basic troubleshooting sequence has been established, let us apply this approach to some practical situations. The following are examples of how the sequence can be used in troubleshooting a basic TV set.

Notice that these examples are very generalized and do not apply to any specific TV model. They are given here to establish how the sequence and basic techniques (bracketing, waveform measurement, voltage measurement, etc.) are used in practical TV service.

12.2.1 Some classic TV trouble symptoms

Table 12.1 lists some common troubles for a basic black-and-white TV set. The troubles are grouped into functional areas (or circuits) of the set. These areas, or circuit groups, correspond to those in the block diagram of Figure 1.3. Notice that the set is essentially a discrete component, with no

TABLE 12.1 Common Troubles in a Discrete-Component TV Set

RF tuner
No picture, no sound
Poor picture, poor sound
Hum bars or hum distortion
Picture smearing and sound separated from picture
Ghosts
Picture pulling
Intermittent problems
Problems on some (but not all) channels
IF amplifier and video detector
No picture, no sound
Poor picture, poor sound
Hum bars or hum distortion
Picture smearing, pulling, or overloading
Intermittent problems
Video amplifier and picture tube
Intermittent problems
Retrace lines in picture
Sound in picture
Poor picture quality
Contrast problems
No picture, sound normal
No picture, no sound
No raster (dark screen)
AGC
Poor picture
No picture, no sound
Sound IF and audio
Poor or weak sound
No sound

ICs. However, the following sections make direct reference to circuit diagrams that show corresponding IC circuit configurations.

Some of the symptoms listed in Table 12.1 point to only one area of the set as a probable cause of trouble. For example, if there is no vertical sweep (the CRT screen shows only a horizontal line) but other set functions are normal (good sound, for example), the trouble is probably in the vertical-sweep circuits (chapter 5). On the other hand, if both sound and picture are weak or poor, the trouble could be in the RF tuner (chapter 2), in the IF and video detector (chapter 3), or possibly in the drive of the video-amplifier and picture-tube circuits (in the set of Figure 1.3).

Here are examples of how the symptoms can be used as the first step in pinpointing trouble to an area, to a circuit within the area, and finally to a part within the circuit. Before going into these steps, let us follow the basic troubleshooting sequence already established.

12.3 Trouble Symptoms

It is not difficult to realize that there definitely is trouble when a TV set is plugged in and turned on and there is no sound, picture, or front-panel LED turn on. A different problem arises when the TV set is operating, but not properly. However, it is always possible for a customer to report a "problem" that is actually the result of improper operation. As a result, you must first determine the signs of failure (trust but verify).

12.3.1 TV symptom recognition

Symptom recognition is the art of identifying abnormal and normal signs of operation. For example, the normal TV picture is a clear, properly contrasted representation of an actual scene. The picture should be centered within the vertical and horizontal boundaries of the scene. If the picture suddenly begins to roll vertically, you would recognize this as a trouble symptom because the condition does not correspond to the normal performance that is expected.

Now assume that the picture is weak, caused perhaps by a poor broadcast signal in the area or a defective antenna (or a poor cable picture). If the RF/IF stages of the set do not have sufficient gain to produce a good picture under these conditions, you could mistake this for a trouble symptom. A poor picture (for this particular model of TV operating under these conditions) is not abnormal nor is it an undesired change. Thus, the condition is not a true trouble symptom and should be so recognized.

12.3.2 Equipment failure versus degraded performance

Generally, the terms *equipment failure* and *degraded performance* apply to the built-in indicators (TV screen, loudspeakers, etc.). However, in some cases it is necessary to use test equipment to distinguish between the two conditions. For example, assume that the circuit in question is a simple, single-stage video amplifier, such as shown in Figure 12.1. Such circuits are often found between ICs in TV sets (to provide both amplification and buffering of the signals passing from one IC to the other). Q221 in Figure 3.4 is an example.

Using the service literature, or your knowledge of similar circuits, you know that if a signal is applied to the base (input), an amplified signal

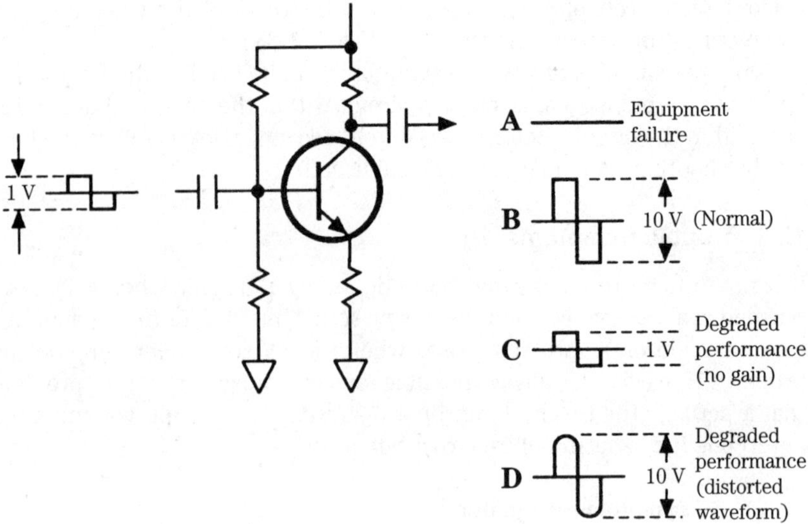

Figure 12.1 Single-stage video amplifier showing examples of degraded performance versus failure.

should appear at the collector (output). The amount of amplification depends on circuit gain. Assume that the input is 1 V and that the gain is 10. Let us analyze the four possible waveforms in Figure 12.1.

Waveform A is an example of equipment (or circuit) failure. With a normal signal at the base (input), there is no waveform at the collector (output), indicating that the circuit has failed completely. (The next troubleshooting step is substitution of parts and/or voltage and resistance checks, discussed in the following.)

Waveform B shows neither equipment failure nor degraded performance. That is, the collector signal (output) is substantially the same as the base (input) but about 10 times greater in amplitude. This is the normal indication, and no trouble exists.

Waveform C is an example of degraded performance. There is a signal at the collector (output), so the circuit is performing part of the normal function. However, the output waveform is not much greater in amplitude than the input waveform. Thus, the video amplifier is producing little or no gain (in any event, the gain is well below the desired factor of 10).

Waveform D is also an example of degraded performance. There is a video signal at the output, and the signal has the desired gain. However, the output waveform is distorted (the square-wave input is badly rounded at the output, always a problem for any TV video circuit).

12.3.3 Evaluation of trouble symptoms

The more recognition of trouble symptoms does not in itself provide enough information for you to decide on the probable cause because many faults produce a similar trouble symptom. The symptoms must be evaluated, usually by operating the controls and studying the results. The following is an example.

When there is no raster on the CRT, the trouble could be caused by a burned out CRT (filament or heater open), or by a failure of the high-voltage supply, among many other possible causes. However, the same symptoms can be produced by the brightness control being turned down (assuming that the power cord is plugged in and that the on-off button is pushed). Think of all the time you could save if you check the controls first, and evaluate the results.

12.4 Trouble Localization

Localizing trouble means that you must determine which of the functional circuit groups in the set are actually at fault. You do this by systematically checking each area selected until the faulty one is found. If none of the function circuit groups show improper performance, you must take a return path and recheck the symptom information (and observe more information, if possible). Several circuits could be causing the trouble, and the localize step narrows the list to those in one functional area.

12.4.1 Bracketing

Bracketing starts by placing brackets (at the good input and bad output) on the block diagram or schematic. Bracketing can be done mentally or physically by marking the brackets with a pencil, whichever is best for you. No matter what system is used, with the brackets properly positioned, you know that trouble exists somewhere between the two brackets.

The technique involves moving the brackets, one at a time (either the good input or the bad output), and then making tests to find if the trouble is within the newly bracketed area. The process continues until the brackets localize a circuit group. The most important factor in bracketing is to find where the brackets should be moved in the elimination process. This is determined from your deductions based on your knowledge of the set and on the symptoms. All moves of the brackets should be aimed at localizing the trouble with a minimum of tests.

12.4.2 Bracketing examples

Bracketing can be used with or without actual measurements of voltage or signals. That is, sometimes localization can be made on the basis of symptoms alone. In practical service, both symptom evaluation and tests are usually required, often simultaneously. The following examples show how bracketing is used in both cases.

Assume that you are servicing a set similar to the one shown in Figure 1.3, and there is no vertical sync. All functions are normal, but the picture rolls vertically. You could start by placing a good-input bracket at the input of the vertical-sweep circuits and a bad-output bracket at the vertical coils of the deflection yoke, as shown in Figure 12.2. However, from a practical standpoint, your first move should be adjustment of the vertical-hold control.

Notice that Figure 12.2 shows the vertical circuits in simple block form, and represents the basic set shown in Figure 1.3. Figure 5-5 shows the same circuits in a typical IC set. In Figure 5.5, the vertical oscillator and drive circuits are located within IC501, while the vertical output is in IC502. Also, the vertical hold, size, and linearity controls are connected to pins 19, 15, and 17 of IC501, respectively.

If the trouble is not cleared by adjustment of the vertical-hold control, confirm the good-input bracket by monitoring the input sync pulses. (This is at pin 19 of IC501 in Figure 5.5.) It is possible that the line between the sync separator Q506 and the vertical oscillator is open or partially shorted. (A complete short would probably cause failure of the sync-separator circuits.) If the sync pulses are absent or abnormal at pin 19, you must then move the good-input bracket to pins 20 and 22.

If the vertical-sync pulses are normal at the input of the vertical oscillator, but there is no vertical sync (even with adjustment of vertical hold), you have localized the trouble to a specific circuit in the vertical-sweep group, as discussed in section 12.5. With a good input at pin 19 of IC501, place a bad-output bracket at the vertical coils of deflection yoke L904.

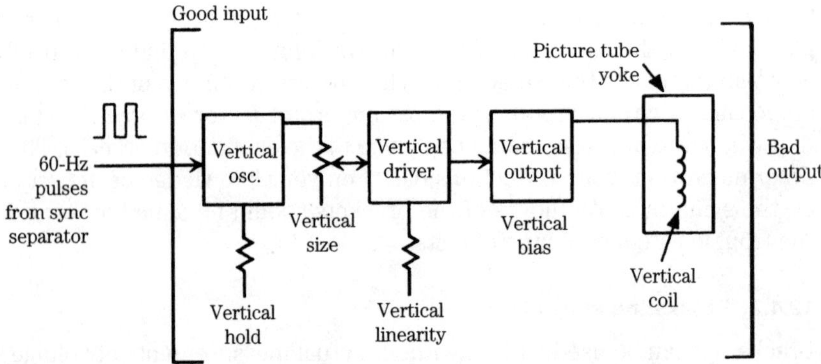

Figure 12.2 Vertical-sweep circuit showing example of bracketing.

12.4.3 Localization to more than one circuit group

Several factors should be considered when you have localized trouble to more than one circuit group. As a guide, if you can run a test that eliminates several circuits, or circuit groups, make that test first (before making a test that eliminates only one circuit). This requires an examination of the diagrams and a knowledge of how the set operates. The decision also requires logic.

Test-point accessibility is a prime factor to consider. A test point can be a special terminal located at an accessible spot (say on top of a PC board). The terminal is electrically connected to some important operating voltage or signal path, such as TP35 or TP82 in Figure 5.11. At the other extreme, a test point can be any point where wires join or where parts are soldered to terminals on PC traces.

Another factor (although definitely not the most important) is your past experience and history of repeated set failures. Past experience with identical or similar sets and related trouble symptoms should have some bearing on the choice of a first test point, but should not prevent you from testing at all related points in the circuit.

Those people with practical experience in troubleshooting know that the steps of a localization sequence rarely proceed in textbook fashion. Just as true is the fact that many troubles listed in the service literature might never occur in the set you are servicing. These troubles are included in the literature (in chart form or as troubleshooting trees) to guide you and arc not meant to be hard and fast rules. In many cases, it is necessary to modify your troubleshooting procedure. The physical arrangement of the set might pose special problems. Also, special knowledge gained from experience with similar sets might simplify the task of localizing the trouble.

12.4.4 Obvious and ambiguous trouble symptoms

Some trouble symptoms lead to an obvious localization process, while others are not quite so obvious. The following are some examples of this using the TV set in Figure 1.3, and the troubles listed in Table 12.1.

Obvious trouble symptoms

If the problem is one of sound only, with a good picture, suspect the sound IF and audio circuit group (not shown in Figure 1.3). In a set such as shown in Figure 3.3, localization for this symptom should start at pin 11 of IC201 and go to the loudspeaker through IC251. (Figures 3.4 and 3.5 show further circuit details.)

If there is good sound and a good raster but a poor picture, start localization at the video amplifier and CRT circuit group. In a set such as shown in Figure 4.4A, localization should start at Q351 (or possibly Q481) and go to the CRT (Figure 4.4B). If the video signal ahead of Q351 is bad, the sync would be affected, and you would probably not get a good raster. (Figures 4.5 through 4.7 show further circuit details of the video/CRT circuit group.)

If there are problems only on certain channels (in picture, sound, or both), suspect the RF tuner, although the problem could be in the tuning PLL, such as shown in Figures 2.5 and 2.6.

If there is sound and no raster, start localization by checking the input and output of the low-voltage power supply (not shown in Figure 1.3), as described in chapter 8.

If the horizontal sync is good but there is no vertical sync, or vertical sync is abnormal, start localization by checking the input to the sync-separator stages. In the circuit of Figure 1.3, this is at the video-driver output stage. In a set such as shown in Figure 5.5, start at pin 22 of IC501. You could also check at pin 1 but this is really not necessary because the horizontal sync is apparently good. If the sync is good at pin 22, go on through to the vertical coils of deflection yoke L904, as shown by the circuit details in Figures 5.6 and 5.7.

If the vertical sync is good but there is no horizontal sync, or horizontal sync is abnormal, again start at the input to the sync-separator stages as described for a vertical-sync problem, using Figures 5.5, 5.8, and 5.9.

If both horizontal and vertical sync appear to be abnormal, start localization at the video-detector output. In a set such as shown in Figure 5.6A, start at pin 20 of IC201. (Figure 3.6B shows typical waveforms.) Check all points from pin 20 of IC201 to pins 1 through 22 of IC501 in Figure 5.5.

If there is good horizontal and vertical sync but the picture is narrow or there is obvious distortion, start with the high-voltage supply and horizontal output circuit group, such as shown in Figure 5.9. You could also check the horizontal-oscillator (Figure 5.8) circuits, but any problem with the oscillator would probably show up as a horizontal-sync trouble.

Ambiguous trouble symptoms

AGC circuit problems often produce ambiguous symptoms and are therefore difficult to localize. Most present-day sets have AGC circuits in the IF/video-detector IC such as IC201 in Figure 3.4. The AGC signal developed in IC201 is applied to the IF circuits internally, and to the RF tuner through pin 28. The RF AGC is adjusted by RV201.

If the IF circuits in IC201 are defective, the AGC circuits do not receive proper IF signals, and do not produce a proper dc voltage to the AGC line (applied to the AGC input of the tuner) such as at the AGC input of TU101 in Figure 2.5. In turn, lack of proper AGC voltage can cause the RF tuner to operate improperly. Conversely, if the AGC circuits are defective, the IF amplifiers in IC201 and the RF tuner do not receive proper AGC voltages and do not deliver a proper IF signal to the AGC circuits.

To eliminate the AGC circuits as a trouble suspect, apply a fixed dc voltage to the AGC line and check operation. (This is known as clamping the AGC line.) If operation is normal with the AGC line clamped, but not when the clamp is removed, you have localized trouble to the AGC circuits. Of course, in a set such as shown in Figure 3.4, all of the circuits are in IC201, which must be replaced if any of the circuits (VIF, IF AGC, RF AGC, etc.) are bad. However, before you pull IC201, try correcting the problem by adjustment of TV AGC control RV201, as described in the service literature.

Here is a simple guide that can be applied to most AGC problems. If clamping the AGC line eliminates the problem, the trouble is probably in the AGC circuits (IC201). If the AGC line is clamped and the video-detector output (pin 20 of IC201) is not normal, check the signal at the IF input. This is at the base of Q221 in Figure 3.4, or at the IF terminal of TU101 in Figure 2.5. (To old time TV repair people, this was known as the "looker point" in the RF tuner.) If the signal at the looker point is abnormal, the trouble is localized to the RF tuner TU101. If the looker-point signal is normal, but the video-detector output is abnormal, you have localized trouble to the IF (IC201).

Horizontal circuits often produce ambiguous symptoms. If the horizontal oscillator or drive fail completely, there are no pulses to the high-voltage supply and horizontal-output circuits. As a result, there is no high voltage to the CRT and no raster. This same identical symptom can be caused by failure of the high-voltage transformer, the high-voltage output stage, or the CRT. In some sets, the high-voltage and output-circuit group also supplies voltages to the CRT accelerator grids. If these voltages are absent, there is no raster.

If you have good sound (indicating that the low-voltage supply is good), but there is no raster, there are two practical courses of localization for most sets. First, if convenient, check the high voltage for the CRT as well as the voltages to all CRT elements (Figures 4.12 and 5.13, for example). If it is not convenient to check the CRT, check the input to the high-voltage/horizontal-output group (which is also the output of the horizontal-oscillator/drive group). Figure 5.8 shows typical horizontal-drive output pulses at pin 10 of IC501.

If the drive output is normal, suspect the high-voltage horizontal-output group. If the drive signal is abnormal, the oscillator/drive is the likely problem area. Start by checking input pulses from the sync separator at pin 1 of IC501.

12.5 Trouble Isolation

As in the localization process, a block diagram is the most convenient tool for isolation. For example, consider how much easier it is to trace through the horizontal- and vertical-sweep circuits using the block of Figure 5.5

than it is through the detailed circuits of Figures 5.6 through 5.9. In turn, the actual schematic for the set is far more difficult to trace than Figures 5.6 through 5.9. Of course, if you do not have a block, you must use the schematic to narrow down the trouble area with tests and decisions.

The isolation process is much easier if you can recognize circuit groups as well as individual circuits. If you can subdivide the schematic into circuit groups, rather than individual circuits, you can isolate the group by a single test at the input or output of the group. For example, in Figure 1.3, all circuits after the video detector, and ahead of the vertical/horizontal circuits can be considered as part of the sync-separator group. This group has one input (from the video detector) and two outputs (60 Hz and 15,750 Hz). If either output is abnormal, with a good input, the trouble is isolated to the sync-separator group.

During the isolation process, you are interested in three bits of information: the signal path (or paths), the signal form (waveform, amplitude, frequency, etc.), and the operating and adjustment controls in the various circuits along the signal paths. If you know what signals are supposed to go where, and how the signals are affected by controls, you can isolate trouble quickly in virtually any set.

12.5.1 Comparison of signals or waveforms

Both signal tracing and signal injection (or substitution) are used in the isolation step. With either technique, the isolation step involves comparing actual signals or waveforms produced along the paths of the circuits against the signals/waveforms given in the service literature (if any!). If you have a signal/waveform that is absent or abnormal, you have isolated the problem to a circuit group (and possibly to a circuit in that group).

In the basic isolation process, you compare input and output of circuit groups (or circuits) in the signal paths. For a typical transistor circuit, the input is at the base (or gate in the case of a FET), while the output is at the collector or emitter (or drain end source for FETs). For a typical discrete transistor circuit group, the input is at the first base (or gate) in the signal path, whereas the output is at the last collector or emitter (or drain/source) in the same path.

Input-output relationships are not always easy to find when ICs are involved. For example, the IC is usually shown as a box, with the pins identified only by number. If you are lucky, there will be arrows pointing to inputs or away from outputs. If you are very fortunate, all of the pins will be identified by function (possibly with functional blocks inside the IC box, as is the case for the diagrams in this book).

Remember that the physical locations of circuit groups within the set almost never have any relation to the representation on diagrams (block and/or schematic). At best, the diagrams will show on which PC board the groups are located. For example, as shown in Figure 5.5, the vertical and horizontal circuit groups are located on the D board, while the adjustment controls for the same circuits are on the B board. You must consult part-placement diagrams (or photos) to find the physical location of the boards, and the parts on the boards.

12.5.2 Signal tracing and/or signal substitution

Both signal-tracing and signal-substitution (or signal-injection) techniques are used frequently in troubleshooting all types of TV sets.

Signal tracing is done by examining the signals at test points with a scope or meter. In signal tracing, the scope or meter probe is moved from point to point, with a signal applied at a fixed point. The applied signal can be from a generator or the TV-broadcast signal. For example, the waveforms shown as composite video at pin 20 of IC201 in Figure 3.6B are made with a color-bar generator at the antenna input.

Signal substitution is done by injecting a signal (from a color generator, sweep generator, etc.) into a circuit or circuit group (or to the complete set) to check performance. The injected signal is moved from point to point, with a device remaining fixed at one point. The monitoring can be done with a meter or scope, or with the CRT display.

12.5.3 Half-split technique

The half-split technique is based on the idea of simultaneous elimination of the maximum number of circuit groups or circuits with each test. This save both time and effort. When using half-split, place brackets at the good-input and bad-output points in the normal manner and study the symptoms. Unless the symptoms point definitely to one circuit or group that might be the trouble source, make the first test at a convenient point halfway between the brackets. Then move the brackets and make the next test halfway between the newly established brackets.

For example, assume that you are servicing a set such as shown in Figures 4.4A and 4.4B, and you have localized the problem to the video-chroma circuits. (There is good sound and a good raster, but obvious picture problems.)

Using half-split you place a good-input bracket at the base of Q351, and a bad-output bracket at the picture tube V901. The first tests in the group should be at pins 3 and 6 of IC301. If the signals are good, you have eliminated the comb-filter circuits, such as shown in Figure 4.5. If the signal is bad, you have isolated the problem to the same stages. If the signal is good at pin 3, but not at pin 6, the problem is isolated to chroma stages Q354, Q360, and T352. If the signal is good at pin 6, but not at pin 3, the problem is narrowed to the video stages Q356, Q355, Q359, and DL301. Of course, the problem could be in the comb-filter delay line DL351 and the associated splitters Q352 and Q353. Figure 4.5 shows typical splitter/delay-line waveforms.

Now assume that the signals at pins 3 and 6 of IC301 are both good. The next logical step is to monitor halfway between the known-good point and the bad output. Pins 31 through 34 of IC301 are at the approximate halfway point. If the signals are good, you have eliminated virtually all of the circuits shown in Figure 4.6. If the signals are bad, you have isolated the problem to the same stages.

Finally, assume that the signals are all good at pins 31 through 34 of IC301. The last step in isolation is to monitor at the circuits on the C board (Figure 4.7).

12.6 Locating a Specific Trouble

The last step of troubleshooting—locating the specified trouble—requires testing of the various branches of the faulty circuit to find the defective part. The proper performance of the locate step enables you to find the cause of trouble, repair it, and return the set to normal operation. A follow-up step is to record the trouble so that future troubles can be easier to locate. Such a history on a certain model can point out consistent failure patterns.

12.6.1 Inspection using the senses

After the trouble is isolated to a circuit, the first step in locating the trouble is to perform a preliminary inspection using the senses. For example, burned or charred parts can often be spotted by sight or smell.

Overheated parts, such as hot transistor cases, can be located quickly by touch. (If one case is much hotter than the others, give that transistor special attention.) Listen for high-voltage arcing, for "cooking" or overheated transformers, or for hum (or the lack of hum), whichever is the case. Although all of the senses are used, the procedure is referred to most frequently as a "visual inspection" in service literature.

12.6.2 Testing to locate faulty parts

Many transistors, ICs, and diodes are not easily replaced (unlike the ancient vacuum tube). Thus, the old-time procedure of replacing tubes at the first sign of trouble has not carried over into present-day TV service. Instead, the circuits are tested and analyzed to locate faulty parts.

Active devices. For testing and troubleshooting purposes, transistors, ICs, and diodes can be considered active devices in any set. Because of their key positions in the circuit, these parts are a convenient point from which to evaluate operation of the entire circuit (through waveform, voltage, and resistance tests). Making such tests at the terminals or pins of the active device often results in locating the trouble quickly.

Waveform testing. Usually, the first step in circuit testing is to analyze the output waveform of the circuit or active device. Of course in some circuits, there is no waveform (such as in a power supply), or there are no waveforms of significance. The waveform must be analyzed to check the amplitude, duration, phase, and/or shape. A careful study of waveforms can often pinpoint the most-likely branch of a circuit that is defective.

Transistor and diode testers. Testers are usually good for parts used at lower frequencies (audio range), but not necessarily at high frequencies (video range). However, if you are comfortable with testers (in-circuit or out-of-circuit), they might prove helpful to you in TV service.

Voltage testing. After waveform analysis and/or in-circuit tests, the next logical step is to measure voltages at the active-device terminals or leads. Always pay particular attention to those terminals that show an abnormal waveform. These are the terminals most likely to show abnormal voltage.

Resistance measurement. After waveform and voltage measurements are made, it is often helpful to make resistance measurements at the same point on the active device (or at other points in the circuit), particularly where abnormal waveforms and/or voltages are found. Suspected parts often can be checked by a resistance measurement, or a continuity check can be made to find point-to-point resistance of the suspected branch.

Current measurements. In rare cases, the current in a particular circuit branch can be measured directly with a meter. However, it is usually simpler and more practical to measure the voltage and resistance of a circuit and then calculate the current.

12.6.3 Waveform and signal measurements

When testing to locate trouble, the waveform measurements are made with the circuit in operation and usually with an input signal applied. The signals can be from a generator or you can use the TV broadcast signal (in some cases). If you use the waveform reproductions found in the service literature, follow all of the notes and precautions described in the literature. Usually, the position of operating controls, typical input signals, and so on are specified. Note that most TV waveforms are measured with the scope sweep setting at multiples or submultiples of 60 Hz and 15,750 Hz, so that two, three, or four waveforms are displayed.

Complete failure of a circuit usually results in the absence of a waveform. A poorly performing circuit usually produces an abnormal or distorted waveform. Also, exact waveforms are not always critical in all circuits. The representations of waveforms given in service literature are usually approximations of the actual waveform (unless photos are used). Also, the same waveform does not always appear exactly the same when measured with different scopes.

12.6.4 Voltage measurements

When testing to locate trouble, make the voltage measurements with the circuit in operation but (usually) with no special signals applied. If you use the voltage information found in service literature, follow all notes and precautions (position of operating controls, typical inputs, etc.).

In most service literature, the voltages are give on the schematic, together with the waveforms, as shown in Figure 12.3. Because of the safety practice of setting a voltmeter to the highest scale before making measurements, the terminals with the highest voltage should be checked first. Notice that in the circuits of Figure 12.3, the collector is grounded, and the emitter has the highest voltage. (The order in which you check voltages makes little difference when using an autoranging digital meter. However, it is good practice to check the highest voltages first, to establish the habit.)

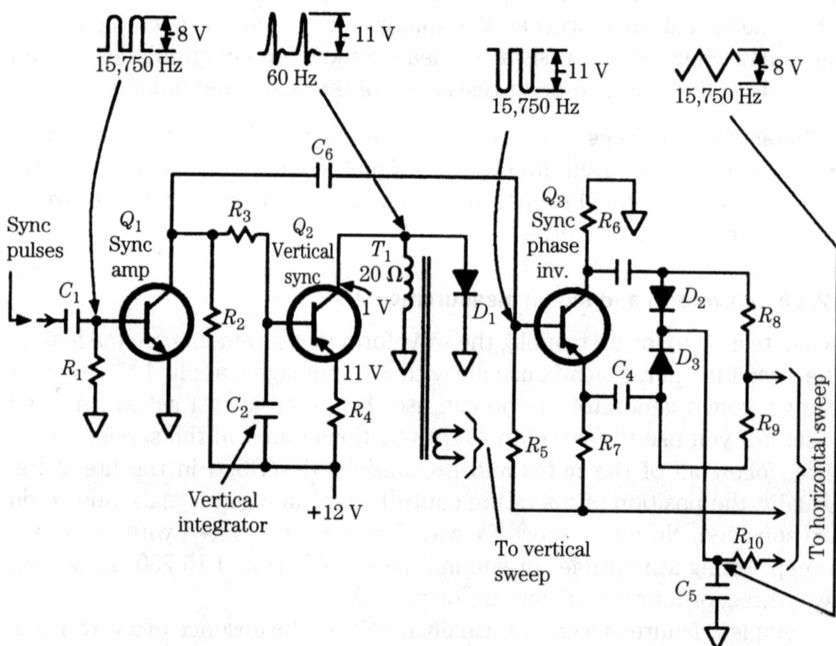

Figure 12.3 Sync-separator circuits showing typical waveforms.

12.6.5 Resistance measurements

Resistance measurements must be made with no power applied. Always observe any notes or precautions found in the service literature (position of operating controls, discharging of filter capacitors, etc.). In some cases, resistance information is given on the schematic as shown in Figure 12.3. However, do not be surprised at a great lack of resistance information in TV service literature.

12.6.6 Duplicating waveform, voltage, and resistance measurements

If you are responsible for service of one type or model of set, it might be helpful to duplicate all waveform, voltage, and resistance measurements found in the service literature with your own test equipment. This should be done only with sets known to be good and operating properly. With a good set of duplicate measurements, you can spot even slight variations during troubleshooting. If more than one set of test equipment is used, make the initial measurements with all available test equipment and record any variations.

12.6.7 Using schematics in TV troubleshooting

The following examples show how the schematic diagram is used to locate a fault within a circuit (the sync separator of a set such as shown in Figure 1.3). Assume that the circuit of Figure 12.3 is being serviced. The reasoning that led to this particular circuit is as follows. The symptom is a complete lack of vertical sync. That is, the TV picture rolls vertically and cannot be controlled by the vertical hold. All other functions, including horizontal sync, are normal. This localizes the trouble to either the vertical-sweep circuits or to the sync separator circuits (Table 12.1).

Isolating trouble to the sync separator. You can isolate the trouble to the sync separator with waveform measurements. Your first waveform is at the collector of Q2, which is (simultaneously) the vertical output of the sync separator, and the input to the vertical-sweep circuits. The waveform is absent (or abnormal), and you place a bad-output bracket at the collector of Q2.

All of the remaining waveforms shown in Figure 12.3 are normal. Thus, you could assume that Q1 is functioning normally (Q1 is passing the horizontal-sync pulses), and should pass the vertical-sync pulses. However, it is better not to assume anything. Instead, confirm that there are vertical-sync pulses at both the input (base) and output (collector) of Q1. Notice that neither of these pulses is shown in Figure 12.3 (which is quite typical for TV service literature).

Confirming vertical pulses. To overcome the lack of information, set the scope sweep to 30 Hz, and measure whatever waveform appears at the base and collector of Q1. If you find two pulses, they are the vertical-sync pulses. The amplitude of these pulses should be about equal to the horizontal-pulse amplitude (7 V base, 10 V collector). Assume that this is the case in making your measurement. Now, it is reasonable to assume that Q1 is probably good and that you have a good input to Q2.

Locating the fault in the Q2 circuit. With the trouble isolated to Q2, you must now locate the specific fault in the Q2 circuit, using voltage and resistance measurements. Notice that only two voltages are given for Q2 (1 V collector, 11 V emitter). This indicates a 1-V drop across T1, yet the normal waveform shows a pulse of about 10 V. This indicates that Q2 is normally biased to (or beyond) cutoff and is switched on by pulses from the junction of R3/C2 (the vertical integrator).

To bias a pnp transistor to cutoff, the base must be less negative (or more positive) than the emitter. Because a positive 12-V supply is used, the base of Q2 must be more positive than the emitter (11 V), but less than 12 V. Thus, if you find some voltage in the 11–12-V range from the base of Q2 to ground, the voltage is probably correct and R2 is probably good.

Clearing the vertical integrator. To clear R3 and C2 from suspicion, use 30-Hz and 7875-Hz sweeps to measure the waveforms to the collector of Q1 and the base of Q2. If R3 and C2 are good, two vertical waveforms should appear at the base of Q2 only when the 30-Hz sweep is used. If no waveforms appears, or if both the horizontal and vertical waveforms are found at the base of Q2, suspect R3 and C2.

Checking Q2. Now assume that there is a good pulse at the base of Q2. At this point, you could check Q2 by substitution, but it probably would be better to make voltage measurements at all terminals of Q2. This pinpoints any obvious part failures. For example, if R4 or T1 is open, all voltages are abnormal. If R4 is shorted (resistor leads shorted), the emitter is at the supply voltage (12 V) instead of 11 V.

Ambiguous measurements. Remember that voltage measurements alone might not solve the problem. For example, if D1 is shorted or leaking badly, it is still possible to get a near-normal voltage reading at the Q2 collector, but a poor pulse waveform. The same is true if D1 is open. If the primary winding of T1 has partially-shorted turns, there will be a near-normal dc voltage, but a drastically reduced pulse output to the vertical-sweep circuits.

Resistance measurements. Because there are no resistance-to-ground values give on Figure 12.3 (as there are in military-style service literature), you must check the resistance of each part on an individual basis. Notice that the primary of T1 should be 20 Ω.

Accurate resistance readings. To get accurate resistance readings of individual parts, you must disconnect one lead from the remainder of the circuit. If not, the effect of other parts in the circuit can further confuse the troubleshooting process. For example, assume that you measure the T1 primary resistance by connecting an ohmmeter across the winding. If the ohmmeter leads are connected so that the positive terminal of the ohmmeter battery is connected to the D1 anode, D1 is forward-biased, and the ohmmeter reads the combined D1 and T1 resistances.

This problem can be eliminated by reversing the ohmmeter leads and measuring the resistance both ways. If there is a difference with the leads reversed, check the schematic for possible forward-bias conditions in both diode and transistor junctions. Make it a routine procedure to reverse the leads whenever any abnormal resistance value is found.

12.6.8 Internal adjustments during troubleshooting

Do not make any internal adjustments during the troubleshooting process until the trouble has been isolated to a circuit and then only when the trouble symptom or test results indicate a possible maladjustment.

For example, assume that the vertical oscillator has an adjustment control (such as the vertical-hold control at pin 19 of IC501 in Figure 5.5) that sets the frequency of oscillation. If measurements show that the vertical oscillator is off frequency (not at 60 Hz), it is logical to adjust the vertical-hold control (as it would be if the picture rolls vertically). However, if waveform measurements show only a very low output (say at pin 14 of IC501) but that the output is on-frequency, adjustment of the vertical-hold control during troubleshooting could be confusing (and could cause further problems).

An exception to this rule is when the service literature recommends alignment or adjustment as part of the troubleshooting process. Generally, alignment and adjustment are checked after the repair is performed. This ensures that the repair has not upset circuit adjustments.

12.6.9 Repairing troubles

In a strict sense, repairing the trouble is not part of the troubleshooting process. However, repair is an important part of the total effort in getting the set back into operation. Repairs must be made before the set can be checked out and made ready for operation. Here are some important points to consider:

- Never replace a part if it fails a second time unless you are sure that the cause of trouble is eliminated.
- Always use exact replacements. Never install a replacement part that has characteristics or a rating inferior to those of the original.
- Always consult the service literature for any information on parts. Read all notes on the schematics. Many parts are critical, some to performance and some to safety.

- Install replacement parts in the same physical location as the original. In any circuit (but particularly in high-frequency video circuits) changing the location of parts (different lead lengths and so on) can cause the circuit to malfunction.

12.6.10 Operational checkout

Once the repairs are complete, make an operational check to verify that the set is free of faults and is performing properly again. Operate the set through all of the operating modes, not just the one where the failure occurred. When the check is complete and the set is "certified" (by you and/or the customer) to be operating normally, make a brief record of the symptoms, faulty parts, and remedy for future reference. This is particularly helpful if you must troubleshoot similar sets.

12.6.11 Safety precautions

It is assumed that you are familiar with all of the standard safety precautions for electronic troubleshooting. Such precautions include checking for leakage that can cause "hot" metal covers or surfaces, handling electrostatically sensitive (ES) devices, replacing leadless components, repairing PC boards, and so on. Always study the service literature (even if it is only a schematic) for any special safety precautions that apply to the particular set.

12.7 Troubleshooting Notes

The following notes summarize practical suggestions for troubleshooting all types of TV sets:

Transient voltages. Be sure that power to the set is turned off or that the line cord is removed when making in-circuit test or repairs (except for voltage measurements and/or signal tracing, of course). Parts can be damaged from the transient voltages developed when parts are changed (installing any kind of a plug-in, for example). Remember that certain circuits might be "live" even with the power switch or button set to off (such as with an instant-on TV set).

Disconnected parts. Do not operate the set with any parts disconnected, such as the loudspeakers or CRT yokes. This can result in damage to transistors and/or ICs.

Sparks and voltage arcs. Use test equipment (meter and high-voltage probe) to measure the CRT high voltage. Do not arc the high-voltage lead to ground for a spark test. This will result in damage to parts, and possibly to you!

Intermittent conditions. If you run into an intermittent condition and find no fault using routine checks, try tapping (not pounding) the parts. If this does not work, try rapid heating and cooling. (A small portable hair dryer and a spray-type cooler make good heating and cooling sources.) Apply heat first, then cool. The quick change in temperature normally causes an intermittently defective part to go bad permanently (the part opens or shorts). Never hold a heated soldering tool directly on a part or IC case!

Operating control settings. If transistor or IC pins appear to have a short, check the setting of any operating or adjustment controls associated with the circuit. For example, pin 4 of IC201 in Figure 3.8 is connected to the RF AGC control. If pin 4 shows what appears to be a short to ground, it is possible that the RF AGC control is set to minimum.

Test connections. Some transistors and ICs with metal cases have their cases tied to the collector (or to some point within the IC). Avoid using the case as a test point unless you are certain as to what point or circuit element is connected to the case. Avoid clipping onto subminiature resistors (which can break if handled roughly).

Shunting capacitors. Do not shunt suspected capacitors with known-good capacitors when troubleshooting. This technique is good only if the suspected capacitor is open, and could cause damage to other circuits (because of the voltage surge). If you suspect an open decoupling or bypass capacitor, monitor both sides (hot and ground) with a scope. There should be substantially no ac present at either side (ac should be bypassed through the capacitor to ground). If there is substantial ac at all decoupling capacitors, this usually indicates power-supply ripple. If there is ripple at only one decoupling capacitor, check that capacitor for a possible open. Of course, in a coupling capacitor, any signal at one side of the capacitor should appear on the other side, with no substantial change. Any drastic change in signals across a coupling capacitor indicates a problem.

Injecting signals. Make sure that there is a blocking capacitor in the generator output when injecting a signal into any circuit. If there is not, connect a capacitor between the generator output and point of signal injection (transistor base, IC pin, etc.).

12.8 Signal Injection versus Signal Tracing

With true signal tracing, you trace through all circuits of the set, checking that the waveforms, pulses, etc., are as they appear in the service literature. You compare the scope patterns obtained from the set with the standard patterns (found in literature) for amplitude, shape, frequency, etc. An absent or abnormal waveform indicates problems in the circuits being measured.

With true signal injection, you must have an analyst or NTSC generator to duplicate the broadcast signals, as well as the signals produced by the set. All analyst/NTSC generators are not the same, so it is difficult to generalize about the basic signal-injection technique. However, in the simplest of terms, you inject precise signals into all sections of the TV set and note the pattern produced on the CRT screen.

I do not promote either signal tracing or signal injection as the better method. This controversy has gone on for many years and eventually boils down to a matter of choice. However, it is generally conceded by most TV technicians that signal injection is the quickest (and thus the most profitable). Also, it is often most practical to use a combination of tracing and injection (with scope and generator) simultaneously.

12.9 Basic Analyst/NTSC-Generator Troubleshooting

This section is devoted to basic troubleshooting with an analyst/NTSC generator, where you inject test patterns (lines, dots, crosshatch, pulses, and color signals) at RF, IF, and video frequencies in a stage-by-stage troubleshooting sequence. Apply the signals at appropriate test points throughout the circuits and observe the displays on the CRT screen. You also can use a scope to view the test-pattern waveforms as the generator signal is processed through the set. With an analyst/NTSC generator, you have the advantage of being able to inject the signal directly into the stage being checked.

Any of the test patterns can be used. However, pulse and/or staircase waveforms are especially useful when checking IF and video stages. A pulse is a simple, sharply-defined trace, so distortion or loss of gain in the waveform is readily apparent. The staircase pattern is valuable for checking linearity in an amplifier (IF, video, audio, etc.). Nonequal steps monitored at the output of a circuit (say at the video detector) represent nonlinear distortion because all of the staircase steps are equal at the input.

The method described in this section starts by connecting the generator to the antenna terminals to evaluate the overall performance of the set, and to note the particular picture problem (no picture, smeary picture, etc.). Then you inject signals back to the video and chroma circuits, working forward through the stages until the picture problem appears. In this way, you isolate the defective stage. Of course, the defective stage can be found just as easily starting at the video detector and working back through the set until a normal picture is found. The technique you use is a matter of individual preference (and the capabilities of the generator).

12.9.1 Checking overall performance

1. Connect the generator to the antenna terminals as shown in Figure 12.4. Set the tuner and generator to the same channel (typically 3 or 4). Apply RF with a pattern to the set.

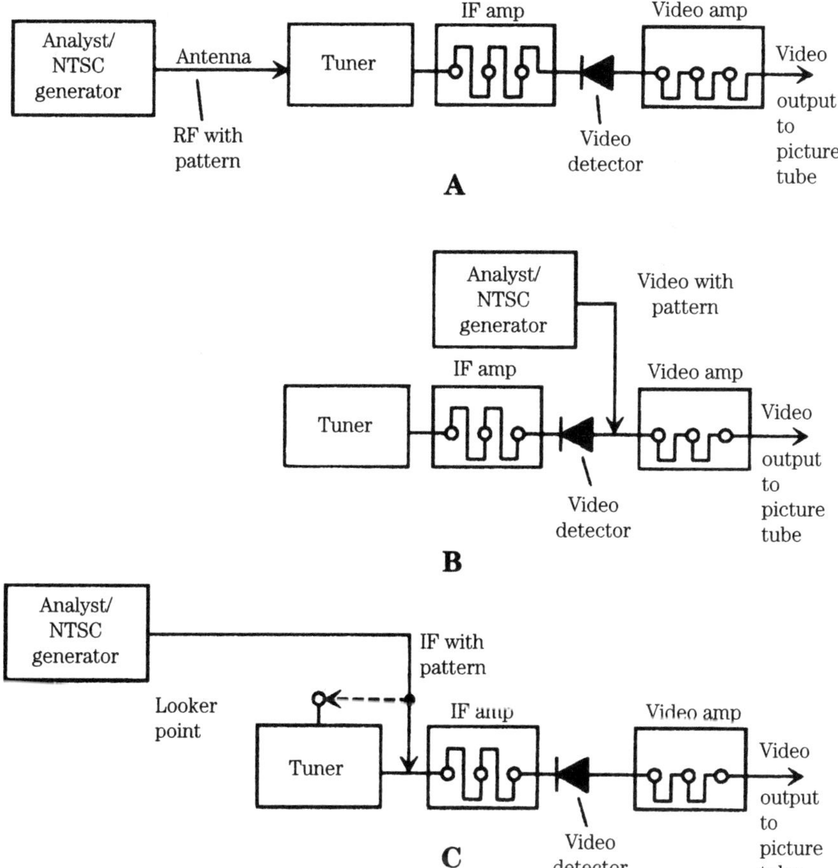

Figure 12.4 Checking overall TV performance with analyst/NTSC generator.

2. Adjust the set brightness, contrast, and other controls as necessary to get the best picture.
3. If the set is operating normally the pattern on the CRT should be good. If not, start the basic troubleshooting sequence.

12.9.2 Checking AGC

If the generator has a variable RF output, the overall ac function can be checked using the same connections shown in Figure 12.4A. Connect a meter to the AGC test point (such as pin 28 of IC201 in Figure 3.4) and vary the RF output of the generator. If the AGC system is operating, the voltage at the test point will vary as the generator RF-level control is adjusted. If the generator does not have a variable RF output, the AGC circuits can be checked by monitoring the AGC line voltage while applying and removing the RF signal. There should be an abrupt change in AGC voltage when the RF is applied and removed.

12.9.3 Checking sync lock-in range

If the generator has a variable sync, you can determine how well the set holds sync (horizontal and vertical) over a wide range of sync signals using the connections shown in Figure 12.4A. In this way, you can determine if the sync is overly sensitive to sync-level variations.

12.9.4 Checking video amplifier

Connect the generator to the video-amplifier input (video-detector output, such as pin 20 of IC201 in Figure 3.4). Do not reconnect the tuner to an antenna. Apply video with a pattern to the set.

If the generator has a variable video output or a video output where the polarity can be reversed, adjust the video control as necessary. For example, if the video control is set to the wrong polarity, the picture will be dark and there is no sync. Turn the control in the proper direction until the picture begins to "tear," and then back off the control just to the point where you get the best picture.

If the picture is about normal, the video circuits (from the video detector to the CRT) are good. Check the IF and tuner circuits as described in section 12.9.6. However, if the picture has the original problem as found during the overall-response check (section 12.9.1), you have isolated the trouble to the video circuits.

You can further isolate the defective circuit within the video circuits by additional signal injection into test points at each video stage (unless several stages are in a single IC). As the signal is applied to various test points, it might be necessary to readjust the video-level control. This is one of the advantages of a generator with variable video, because the video-sync polarity is often inverted at each stage.

As an alternate, you also can inject the signal at the video detector and then signal trace through the video circuits using a scope as described in section 12.9.7.

12.9.5 Checking video/chroma circuits

A rough check of the video/chroma circuits can be made by injecting a color-bar pattern at the video frequency into the video/chroma input (such as at Q351 in Figure 4.4A). The sync might not be too stable when the signal is injected at this point because you are beyond the sync takeoff point in some sets. However, the color pattern should be fairly good. If you find trouble in the video/chroma amplifiers, you can trace the color-bar signal using a scope.

12.9.6 Checking IF/tuner circuits

Connect the generator to the RF-tuner looker point or to a test point at the input of the IF amplifier as shown in Figure 12.4C. On some sets, you can unplug the cable connecting the tuner to the IF amplifier and apply the generator signal the cable. Apply IF (typically 45.75 MHz) with a pattern to the set.

Adjust the generator IF-level control (if any) for the best picture. Try different test patterns, including color bars. If you get normal patterns, the IF-amplifier stages are good, and you have isolated the problem to the RF tuner. If the picture still has the original problem, you have isolated the trouble as being in the IF amplifier or possibly the video detector. As shown in Figure 3.4, the IF circuits are in the same IC as the video detector (IC201). This is typical for most present-day sets.

12.9.7 Signal tracing with analyst/NTSC generators

When an analyst/NTSC generator is used in signal tracing, the generator serves as the signal source (in place of TV broadcast signals). The basic procedure consists of applying a test signal at the antenna and then using a scope to view the pattern as the signal is processed through the RF, IF, video, and chroma stages. When you reach a point where the waveform disappears, is badly distorted, or does not have sufficient gain, you have localized the trouble.

The scope must have a low-capacitance probe (for signal tracing of circuits after the video detector) and a demodulator probe (for signal tracing ahead of the video detector). Before you start, connect the probes directly to the generator output. In this way, you know what pattern is produced by each of the probes before the signals are processed.

It is usually recommended that the generator be set to produce a pulse or staircase pattern when tracing through the IF and video circuits (or any amplifier). If you are signal tracing in the video/chroma circuits, a color-bar pattern must be used to be of any real value.

When signal tracing a set where the picture will not lock in, inject the signal at the looker point or the input to the IF. If you connect the signal to the antenna, and there is a problem with RF-tuner to tuner PLL, the waveforms will be misleading.

If the generator has a variable sync control, vary the sync level while monitoring the waveform at each of the circuits. It is possible that a defective circuit will respond properly to a strong sync signal, but will fail on borderline sync signals.

Index

A

active-device testing, 292-293
AM frequencies for picture, 1-2
amplifiers, 4
analyst generator, 240-241, **240**, 300-303, **301**
  automatic gain control (AGC) testing, 301-302
  IF circuit testing, 303
  RF tuner circuits testing, 303
  signal tracing, 303
  sync lock-in range testing, 302
  video amplifier testing, 302
  video/chroma circuit testing, 302
antenna circuits, 37, **38**
arcing, 298
automatic fine tune (AFT), 23, 27, 29, 34
automatic frequency control (AFC), 6
automatic gain control (AGC), 6, 10
  adjustment, 266-267
  testing, 301-302

B

battery operations, low-voltage power supplies, 157
black-and-white adjustments, 261, **263**
black-and-white TV
  amplifier, 4
  automatic gain control (AGC), 10
  broadcast system, 1-3, **2**, **4**
  circuitry, 4, **5**, 7-10
  discrete components vs. ICs, 4
  electron beam (*see* picture tube)
  high-voltage power supplies, 6
  horizontal deflection, 3, 6
  horizontal driver, 6-7
  horizontal oscillator, 6-7
  IF amplifier, 8
  input-output points, 4
  intensity, 3, 9-10
  low-voltage power supplies, 6
  mixer, 4
  modulation, 3
  oscillator, 4
  picture tubes, 2-3, **2**, **3**, 9-10
  retrace, 3
  RF tuner, 4, 7-8
  sweep, 3
  sync separator, 7
  synchronization pulses, 3
  test points, 4
  vertical deflection, 3, 7
  video amplifier, 9-10
  video detector, 8
blanking pulse, 7
bracketing technique, troubleshooting, 281, 285-286, **286**
broadcasting TV, 1-3, **2**, **4**
  AM frequencies for picture, 1-2
  black-and-white, 1-3, **2**, **4**
  channels for TV, 1, 23, **25**
  color TV, 10, **11**, 12-17
  FM frequencies for sound, 1
  frequencies for broadcast, 1
burst-sync signal, 12

C

center cross convergence pattern, 249, **250**, 251
channels for TV, 1, 23, **25**
chroma (C) channel, 19
  video/chroma, 72-73, **74**
chrominance (*see also*video/chroma), 10, 12, 17, 59
clock, master clock signal, 142, 145
coils, 257-261
  ringing-test procedure, 261
color generators, 242-254, 266-270
  -IWQ patterns, 246-247
  amplifier linearity checks, 247
  audio or sound functions, 246
  automatic gain control (AGC) adjustment, 266-267
  center cross convergence pattern, 249, **250**, 251
  color adjustments, 244-245

Illustrations are indicated by **boldface** numbers.

color generators (*cont.*)
color setup steps, 266-270
color-circuit testing, 269-270
color-killer function, 246
comb filter adjustment, 247
convergence adjustments, 268-269
convergence patterns, 247-252, **248-249**, **250-251**
crosshatch convergence pattern, 251, **252**
CRT temperature adjustments, 268
differential gain test, 247
differential phase lock test, 247
dothatch convergence pattern, 251-252, **252**
dots convergence pattern, **250-251**, 251
gray-scale tracking adjustments, 268
horizontal centering adjustment, 267
IF performance test, 244
linearity adjustments, 268
NTSC color-bar generator, 242-244, **243**
NTSC color-bar pattern, 244-264, **245**
overall TV set performance test, 244
overscan adjustment, 268
phase lock of 3.59-MHz oscillator test, 245-246
pincushion adjustment, 268
purity adjustments, 267
raster patterns, 253-254, **253**
RF performance test, 244
staircase patterns, 247
sync signals, 254
vectorscope applications, 246
vertical centering adjustment, 267
video and chroma delay test, 244
video peaking adjustment, 267
white and black level adjustment, 247
white-balance adjustments, 268
color problems, 65-67, **66**
automatic gain control (AGC) adjustment, 266-267
black-and-white adjustments, 261, **263**
color-circuit testing, 269-270
color-sync out, 67
convergence adjustments, 261-266, **262**, 268-269
convergence checks, 266
CRT temperature adjustment, 265-266, 268
gray-scale tracking adjustments, 268
horizontal centering adjustment, 267
linearity adjustments, 268
linearity adjustments, 261-266, **262**, 268
no color, 65-66
overscan adjustment, 268
pincushion adjustment, 268
purity adjustments, 261-266, **262**, 267
vectorscope testing, 270-271, **272**
vertical centering adjustment, 267
video peaking adjustment, 267
weak color, 67
white-balance adjustments, 265-266, 268
wrong color, 67
color TV
1-Vp-p signal, 15-16
amplitude (NTSC), 15-16
black-and-white vs. color, 12
broadcast system, 10, **11**, 12-17
burst amplifier, 19
burst-sync signal, 12
chroma (C) channel, 19
chrominance, 10, 12, 17
circuitry, 17-22, **18**
color signal, 19-20
color signals (NTSC), 16-17, **16**
color-killer stage, 19
convergence, static vs. dynamic, 21-22
horizontal sync (NTSC), 13, **14**
hue or tint, 10, 16-17, **16**
interleave, 12
luma (Y) channel, 17, 19, 17
luminance, 10, 12, 10, 16-17, **16**
NTSC color video signal, 12
oscillator, 19
pedestal, horizontal blanking, 13, **14**
picture tubes, 19, 20-21, **21**
purity of color, 21
RGB color signal, 19-20
saturation, 10, 16-17, **16**
scanning rate, 12
sync pulses, 12
vertical sync (NTSC), 13-14, **15**
X and Z demodulators, 19
comb filter, 69, **70**, 71, **71**, 72
adjustment, 247
composite-video waveform analysis, 256-257, **257**, **258**
compressor/expander circuits, 189, **190**
contrast problems, 62
convergence, 21-22
adjustments, 261-266, **262**, 268-269
preliminary adjustments, 263
convergence circuits, 126-127, **128-129**
convergence patterns, 247-252, **248-249**, **250-251**
crosshatch convergence pattern, 251, **252**
CRT temperature adjustment, 265-266, 268
current measurement, 293
low-voltage power supplies, 157

D

dbx noise reduction (NR), 187, 189-191, **190**, 194-196, **195**, 199, **201**
 adjustments, 276
 compressor/expander circuits, 189, **190**
 deemphasis, 191, 196
 noise-reduction process, 189-191
 preemphasis, 191
 processing, 196
 spectral companding, 191
 wideband-amplitude companding, 191
deemphasis, dbx noise reduction (NR), 191, 196
deflection (*see* horizontal sweep circuits; vertical sweep circuits)
demodulator probes, 255-256
digital-TV circuits, 217-226, **219**
 checking ICs, 222
 circuitry, typical TV, 218-220, **219**
 clock signals, 222, **223**
 five-chip digital TV circuits, 218-220, **219**
 freeze mode, 218
 horizontal-drive circuits, 222, **225**
 input-output signals, 222
 layout of ICs on module, 219-220, **220**
 modular replacements, 221-222
 operation of digital-TV circuits, 217-218
 picture-in-picture, 218
 plug-in IC replacements, 221
 reset signals, 222, **223**
 special effects, 218
 strobe mode, 218
 troubleshooting approach, 221
  video circuits, 224, 226
 vertical sweep circuits, 224, **225**
 video circuits, 224, **225**, 226
diode testing, 293
discrete components vs. ICs, 4
Dolby Prologic surround-sound, 209-210, **209**
Dolby surround-sound, 208, **208**
dothatch convergence pattern, 251-252, **252**
dots convergence pattern, **250-251**, 251

E

electron beam (*see also* picture tube), 2-3, **2**, **3**
electron guns (*see also* picture tubes), 20
external-audio circuits, 210-211, 213, **213**

F

flyback transformer (FBT), 6, 89, 101
 ringing-test procedure, 259
FM frequencies for sound, 1
frequencies for broadcast, 1
freeze mode, digital-TV circuits, 218
frequency counters, 256
frequency synthesis (FS) tuning, 23, 24, 29-30
front end circuits, 23-39, **24**
 antenna circuits, 37, **38**
 automatic fine tune (AFT), 23, 27, 34
 channel allocation, 23, **25**
 components of typical front end circuits, 23, **24**
 frequency synthesis (FS) tuning, 23, 24, 29-30
 mixer, 23
 phase-locked loop (PLL), 23, 24, **26**, 27, **28**, 29-30
 pulse swallow control (PSC), 27
 RF amplifier, 23
 RF tuner, 28-29, **28**
  built-in PLL tuners, 35-38, **36**
  separate-PLL tuners, 30-35, **32-33**
 variable frequency oscillator (VFO), 23
 voltage controlled oscillator (VCO), 23, 24, **26**, 27

G

gray-scale tracking adjustments, 268

H

half-split technique, 291-292
Hall surround-sound, 205, **206**, 207, **207**
high-voltage power supplies (*see also* horizontal sweep circuits), 6, 89, 101-107, 113, **114-115**, 116
 APC/AFC signal, 6
 dark screen, 103, 104-105, 106
 flyback transformer (FBT), 6, 101, 104
 foldback or foldover picture, 106-107
 functions, 6
 high-voltage only absent, 104
 high-voltage or boost voltage absent, 104
 horizontal output troubles, 104
 horizontal sweep circuits, 101-107
  IC-type, 113, **114-115**
  jungle-clip, 119, 124
 measuring voltages, 104
 narrow picture, 106
 nonlinear horizontal display, 107

high-voltage power supplies (*cont.*)
operation of horizontal sweep circuits, 101, 103
overscanned picture, 106
pulse frequency, 6
signal path, 101, **102**
synchronization of pulses, 6
transistor Q1 measurements, 104
troubleshooting approach, 101, 103-110
yoke check, 104
high-voltage probe, 255
horizontal oscillator/driver, 6-7
feedback signal, 6
phase or frequency shift, 7
sync pulses, 6-7
horizontal sweep circuits (*see also* high-voltage power supplies), 3, 6, 89-129
convergence circuits, 126-127, **128-129**
dark screen, 99-100, 103, 104-105, 106
digital-TV circuits, 222, **225**
distortion, 100-101
flyback transformer (FBT), 89
foldback or foldover picture, 106-107
high-voltage power supplies (*see also* high-voltage power supplies), 89, 101-107, 113, **114-115**, 116
jungle-clip, 119, 124
IC-type horizontal sweep circuits, 107-116, **108**
jungle-clip type, 116-124
loss of sync, 100
narrow picture, 100, 106
no horizontal sync, 92-93
no sync, 91
no vertical sync, horizontal sync good, 91-92
nonlinear horizontal display, 107
operation of horizontal sweep circuits, 101, 103
oscillator, horizontal oscillator, 98-99
IC-type, 110, **112**, 113
jungle-clip, 120, **121**, **122-123**
overscanned picture, 106
phasing improper, 100
pie crust distortion, 100-101
pincushion correction circuits, 126-127, **128-129**
processing, single-IC, 126
pulled picture, 93, 100
signal path, 97, **98**
single-IC (25-inch set), 124-127, **125**
sync separator, 89-91, **90**
troubleshooting approach, 91-93, 99-101, 103-110
hue or tint, 10, 16-17, **16**

I

IF amplifier, 8
automatic fine tune (AFT), 8
automatic gain control (AGC), 8
circuit testing, 303
picture IF (PIF) module, 8
sound IF (SIF) amplifier, 8
video IF (VIF) module, 8
input-output points, 4
intensity, 3, 9-10
interleave, 12
intermittent conditions, 299
-IWQ patterns, color generators, 246-247
isolation transformer, low-voltage power supplies, 157

J

jungle clip, video/chroma, 76, **77**, 78-84, **78**, **80-81**

K

key encoder circuits, 152, **153**
keyboard service, 142-144

L

L+R circuits, 187, 189, 192, **193**, 194, 196
adjustments, 274, 277
Lenk's Digital Handbook, 139
linearity adjustments, 261-266, **262**
low-capacitance probe, 255
low-voltage power supplies, 6, 155-173, **156**
13-inch set, typical supply, 161, **162**, 163, 167
19-inch set, typical supply, 167, **168**
25-inch set, typical supply, 167, 169, **170-171**
auxiliary/standby power circuits, 163, **165**
battery operations, 157
brightness low, 160
circuitry, 161, **162**, 163, 167, **168**, 169, **170-171**
current measurement, 157

height and width insufficient, 160-161
isolation transformer, 157
line input circuits, 163, **164**
no sound, no raster, 158
no sound, no raster, transformer buzzing, 158-159
operation of low-voltage circuits, 155-156
oscilloscope testing, 157
pulled picture, 160
rectifier, 6
regulator, 6
resistance measurement, 157
snowy picture, 160-161
sound distorted, no raster, 159-160
sound weak, 160-161
substitute-supply, 157
switching supply, 163, **166**, 167, 169, **172-173**
troubleshooting approach, 157-161
vertical height excessive, 160
vertical height insufficient, 160
zener replacement, 157
luma (Y) channel, 17, 19
video/chroma, 72-73, **74**
luminance, 10, 12, 16-17, **16**

M

masking (*see also* picture tubes), 21
master clock signal, 142, 145
matrix surround-sound, 205, **206**, **207**
matrix-demodulator circuit, 64-65, **64**
mixer, 4, 23
modulation, 3
MPX decoder, 199, **200**

N

National Television Systems Committee (NTSC), color TV, 12
NTSC color-bar pattern, 244-264, **245**
NTSC-generators, 242-244, **243**, 300-303, **301**
automatic gain control (AGC) testing, 301-302
IF circuit testing, 303
RF tuner circuits testing, 303
signal tracing, 303
sync lock-in range testing, 302
video amplifier testing, 302
video/chroma circuit testing, 302

O

on-screen display (OSD) (*see also* remote-control circuits), 175-177, **176**
13-inch set, typical circuits, 180, **181**, **182**, 182-183
19-inch set, typical circuits, 184, **185**
blanking output, 183
circuitry, 184, **185**
reference inputs, 183
troubleshooting approach, 177
oscillator (*see* horizontal sweep circuits; vertical sweep circuits)
oscilloscopes, 241-242

P

pattern generator, 240-241, **240**
pedestal, horizontal blanking, color TV, 13, **14**
phase-locked loop (PLL), 23, 24, **26**, 27, **28**, 29-30
RF tuner with separate PLL, 30-35, **32-33**
phosphor screen (*see also* picture tubes), 21
picture IF (PIF) module, 8, 55, **57**, 58
picture problems (*see also* color problems)
brightness low, 160
color problems (*see* color problems)
contrast problems, 62-63
dark screen, 99-100, 103, 104-105, 106
distortion, 97-98, 100-101
foldback or foldover picture, 106-107
height and width insufficient, 160-161
height insufficient, 96
horizontal centering adjustment, 267
hum bars, 44
intermittent picture, 45
line splitting, 97
linearity adjustments, 268
narrow picture, 100, 106
no color, 65-66
no picture, 44, 62
no picture, no sound, 158
no picture, no sound, buzzing, 158-159
no picture, sound normal, 62
no raster, 61-62
no raster, sound distorted, 159-160
nonlinear horizontal display, 107
nonlinearity, 97-98
overloading, 44-45
overscanned picture, 106, 268

picture problems (*cont.*)
pie crust distortion, 100-101
poor picture, 44, 63
pulled picture, 44-45, 93, 100, 160
retrace lines in picture, 63-64
smeared picture, 44-45
snowy picture, 160-161
sound in picture, 63
vertical centering adjustment, 267
vertical height excessive, 160
vertical height insufficient, 160
weak color, 67
wrong color, 67
picture tubes, 2-3, **2**, **3**, 9-10
blanking pulse, 7
color TV, 19, 20-21, **21**
convergence, 21-22
CRT temperature adjustment, 265-266
electron beam action, 2-3, **2**, **3**
electron guns, 20
horizontal deflection, 3
intensity, 3, 9-10
masking, 21
modulation, 3
phosphor dots, 21
retrace, 3
ringing-test procedure, 261
sweep, 3
synchronization pulses, 3
testers/rejuvenators, 256
vertical deflection, 3
video amplifier, 9-10, 9
picture-in-picture, 218, 226-234
A/D converter, IC8, 233
chroma signal D/A converter, IC3, 233
circuitry, 231-234, **232**
clamper, IC 7, 232-233
clock signal generator, IC9, 234
controller, IC6, 233
field/display relationship in PinP, 227-228, **228**
main/subvideo input/output switch, 231
memory, IC10, 233
NTSC demodulator, sync separator, IC5, 232
NTSC demodulator, sync separator, Y/C multiplexer IC2, 231
processing basics, 226-228, **227**
signal-processing sequence, 228, **229**, 230-231, **230**
troubleshooting, 234
Y-signal D/A converter, IC4, 233
pincushion adjustment, 268
pincushion correction circuits, 126-127, **128-129**
power on/off, 144
preemphasis, dbx noise reduction (NR), 191
probes for meters and scopes, 254-256
demodulator probes, 255-256
high-voltage probe, 255
low-capacitance probe, 255
radio-frequency probes, 255
pulse swallow control (PSC), 27
purity adjustments, 261-266, **262**, 267

R

radio-frequency probes, 255
raster patterns, 253-254, **253**
rectifier, 6
regulators, 6
remote-control circuits (*see also* on-screen display (OSD)), 177-183, **178**
13-inch set, typical circuits, 180, **181**, **182**, 182-183
blanking output, 183
on-screen display (OSD) functions, 183
receiver circuits, 179, **180**, 182
reference inputs, 183
transmitter circuits, 177-179, **178**
troubleshooting approach
receiver, 179-180
transmitter, 178-179, 178
reset function, 142, 145
resistance measurement, 293, 295, 296-297
low-voltage power supplies, 157
retrace, 3
RF amplifier, 23
RF tuners, 4, 7-8, 28-29, **28**
automatic fine tune (AFT), 29, 34
built-in PLL tuners, 35-38, **36**
function chart, **37-38**
channel selection action, 29
frequency synthesis (FS) tuning (*see also* PLL), 29-30
manual fine-tune control, 8
phase-locked loop (PLL) (see also FS), 28, **28**, 29-30
separate-PLL tuners, 30-35, **32-33**
AFT up/down, 34
FM trap, 34
horizontal sync, 34-35
PLL tuning unit, 31
tuner, 31
tuning memory, 31, 34
tuning sequence, 35
testing, 303
RGB color signal, 19-20

video/chroma output, 73, **75**, 76
RGB interface circuits, 131-138, **132-133**
analog RGB mode, **136**, 137, **137**
digital RGB mode, 135, **136**, **137**
operation of circuit, 131, 134-135
RGB mode, 134-135
troubleshooting approach, 137-138
TV mode operation, 131, 134
video mode, 134
ringing-test procedure, 258-259
high-voltage flyback transformer, 259

S

safety precautions, 235-237, 298
disconnected parts, 298
shunting capacitors, 299
sparks and voltage arcs, 298
transient voltages, 298
saturation, 10, 16-17, **16**
schematics for troubleshooting, 295-297
separate audio program (SAP), 187, 189, 191-192, **192**, 197-198, **198**
adjustments, 274-275, 276-277
AP demodulation path, 197
signal-detect path, 197-198
troubleshooting approach, 197
shock hazards, 235-237, 298
shunting capacitors, 299
signal generators (*see also* analyst generators; NTSC-generators), 237-241, **238**
analyst generator, 240-241, **240**
pattern generator, 240-241, **240**
sweep-frequency alignment procedure, 239-240
sweep/marker generator, 237, **238**, 239
signal injection, 299-300
signal tracing, 291, 299-300
Sony Trinitone switching, video/chroma, 84, 86-88, **86-87**, **88**
sound IF (SIF) module (*see also* external audio circuits; sound problems; stereo-TV circuits), 8, 41-58
biasing, 41
hum bars, hum distortion, 44
intermittent picture and/or sound, 45
no picture or sound, 44
NTSC generator for troubleshooting, 43-44
output, 42-43
overloading, 44-45
picture IF (PIF) module, 55, **57**, 58
poor picture or sound, 44
signal analyzer for troubleshooting, 43-44
signal path, 41, **42**, **43**
smeared picture, 44-45
sweep generator for troubleshooting, 44
traps, 42
troubleshooting approach, 43-44
tuning, 41-42
sound problems (*see also* external-audit circuits; sound IF module (SIF); stereo-TV circuits)
hum, 44
intermittent sound, 45
no sound, 44, 62
no sound, no picture, 158
no sound, no picture, buzzing, 158-159
poor sound, 44
sound distorted, no raster, 159-160
sound in picture, 63
sound weak, 160-161
volume-control, electronic, 213-217, **214**
sparking, 298
spectral companding, dbx noise reduction (NR), 191
staircase patterns, color generators, 247
stereo-TV circuits, 187-210
adjustments, 273-278
audio output circuits, 202, **204**, 205
audio processing circuits, 202, **203**
basic system circuitry, 187, **188**, 189
circuitry, 198-205, **200**, **201**, **203**, **204**
compressor/expander circuits, 189, **190**
dbx noise reduction (NR), 187, 189-191, **190**, 194-196, **195**,199, **201**
adjustments, 276
generators, stereo-TV, 273
L+R circuits, 187, 189, 192, **193**, 194, 196
adjustments, 274, 277
MPX decoder, 199, **200**
MTS/MCS sound, 187, **188**
separate audio program (SAP), 187, 189, 191-192, **192**,197-198, **198**
adjustments, 274-275, 276-277
separation testing, 277-278
signal path, 191-192, **192**
stereo-reduction VD adjustment, 276
surround-sound circuits, 205-210, **206**, **211**, **212**
testing, 273-278
Zenith Transmission System, 187
strobe mode, digital-TV circuits, 218
surround-sound circuits, 205-210, **206**, **211**, **212**
circuitry, 210, **211**, **212**

surround-sound circuits (*cont.*)
Dolby Prologic surround, 209-210, **209**
Dolby surround, 208, **208**
Hall surround, 205, **206**, 207, **207**
matrix surround, 205, **206**, **207**
troubleshooting, 210
sweep (*see* horizontal sweep circuits; vertical sweep circuits)
sweep-frequency alignment procedure, 239-240
sweep/marker generator, 237, **238**, 239
symptom chart for troubleshooting, 281-283, **282**
sync lock-in range testing, 302
sync separator, 7, 89-91, **90**, **294**
troubleshooting, 295
synchronization pulses, 3, 12
system-control circuits, 139-153, **140**, **141**, **142**
19-inch set, typical controls, 145, **146-148**
25-inch set, typical controls, 148, **148-151**
basic operation, 139
ground connections test, 144-145
key encoder circuits, 152, **153**
keyboard service, 142-144
master clock signal, 142, 145
power connections test, 144-145
power on/off, 144
reset function, 142, 145
troubleshooting approach, 144-145

T

television basics, 1-22
AM frequencies for picture, 1-2
amplifier, 4
automatic gain control (AGC), 10
black-and-white TV
broadcast system, 1-3, **2**, **4**
circuitry, 4, **5**, 7-10
burst-sync signal, 12
channels for TV, 1, 23, **25**
chroma (C) channel, 19
chrominance, 10, 12
color TV
broadcast system, 10, **11**, 12-17
circuitry, 17-22, **18**
convergence, 21-22
discrete components vs. ICs, 4
electron beam (*see* picture tube)
FM frequencies for sound, 1
frequencies for broadcast, 1
high-voltage power supplies, 6
horizontal deflection, 3, 6
horizontal driver, 6-7
horizontal oscillator, 6-7
hue or tint, 10
IF amplifier, 8
input-output points, 4
intensity, 3, 9-10
interleave, 12
low-voltage power supplies, 6
luma (Y) channel, 17, 19
luminance, 10, 12
mixer, 4
modulation, 3
NTSC color video signal, 12
oscillator, 4
picture tubes, 2-3, **2**, **3**, 9-10
retrace, 3
RF tuner, 4, 7-8
saturation, 10
sweep, 3
sync separator, 7
synchronization pulses, 3, 12
test points, 4
vertical deflection, 3, 7
video amplifier, 9-10
video detector, 8
test connections, 299
test equipment/procedures (*see also* troubleshooting), 235-278
-IWQ patterns, 246-247
active-device testing, 292-293
amplifier linearity checks, 247
analyst generator, 240-241, **240**, 300-303, **301**
automatic gain control (AGC)
adjustment, 266-267
testing, 301-302
black-and-white adjustments, 261, **263**
center cross convergence pattern, 249, **250**, 251
coil testing, 257-261
color generators, 242-254, 266-270
color setup with color generator, 266-270
color-circuit testing, 269-270
comb filter adjustment, 247
composite-video waveform analysis, 256-257, **257**, **258**
convergence adjustments, 261-266, **262**, 268-269
preliminary, 263

convergence checks, 266
convergence patterns, 247-252, **248-249**, **250-251**
crosshatch convergence pattern, 251, **252**
CRT temperature adjustment, 265-266, 268
current measurement, 293
dbx noise reduction (NR), adjustments, 276
demodulator probes, 255-256
differential gain test, 247
differential phase lock test, 247
diode testing, 293
dothatch convergence pattern, 251-252, **252**
dots convergence pattern, **250-251**, 251
frequency counters, 256
gray-scale tracking adjustments, 268
half-split technique, 291-292
high-voltage probe, 255
horizontal centering adjustment, 267
IF circuit testing, 303
L+R circuits, adjustments, 274, 277
linearity adjustments, 261-266, **262**, 268
low-capacitance probe, 255
NTSC color-bar generator, 242-244, **243**, 300-303, **301**
NTSC color-bar pattern, 244-264, **245**
oscilloscopes, 241-242
overscan adjustment, 268
pattern generator, 240-241, **240**
picture tube (CRT) testers/rejuvenators, 256
pincushion adjustment, 268
probes for meters and scopes, 254-256
purity adjustments, 261-267
radio-frequency probes, 255
raster patterns, 253-254, **253**
resistance measurement, 293, 295, 296-297
RF tuner circuits testing, 303
ringing-test procedure, 258-259
coils, 261
high-voltage flyback transform, 259
yoke, 259, 261
safety precautions, 235-237, 298
SAP circuit adjustments, 274-275, 276-277
schematics for troubleshooting, 295-297
signal generators, 237-241
signal injection, 299-300
signal tracing, 291, 299-300, 303
staircase patterns, 247
stereo separation testing, 277-278
stereo-reduction VD adjustment, 276
stereo-TV circuit tests/adjustments, 273-278
sweep-frequency alignment procedure, 239-240
sync lock-in range testing, 302
sync signals, 254
test connections, 299
transistor testing, 293
vectorscopes, 242, 246, 270-271, **272**
vertical centering adjustment, 267
video amplifier testing, 302
video peaking adjustment, 267
video transformer tests, 257-261
video/chroma circuit testing, 302
voltage testing, 293, 294
waveform analysis or comparison, 290-291, 293-294, **294**
white and black level adjustment, 247
white-balance adjustments, 265-266, 268
yoke testing, 257-261
test points, 4
transient voltages, 298
transistor testing, 293
Trinitone switching (Sony), video/chroma, 84, 86-88, **86-87**, **88**
troubleshooting (*see also* test equipment/procedures), 279-303
active-device testing, 292-293
ambiguous symptoms, 287, 288-289
analyst generators, 300-303, **301**
automatic gain control (AGC) testing, 301-302
bracketing technique, 281, 285-286, **286**
current measurement, 293
diode testing, 293
disconnected parts, 298
equipment failure vs. degraded performance, 283-284, **284**
evaluating symptoms, 285
functional area localization of trouble, 280-281, 285-289
half-split technique, 291-292
IF circuit testing, 303
inspection procedures, 292
intermittent conditions, 299
internal adjustments during troubleshooting, 297
isolating trouble area, 289-292

troubleshooting (*cont.*)
localizing trouble to functional areas, 280-281, 285-289
locating specific trouble, 281
NTSC-generators, 300-303, **301**
operating-control settings, 299
operational checkout procedures, 298
repairs, 297-298
replacement parts , 297-298
resistance measurement, 293, 295, 296-297
RF tuner circuits testing, 303
safety precautions, 298
schematics for troubleshooting, 295-297
sequence for successful troubleshooting, 280-281
service manuals, datasheets, etc., 279
shunting capacitors, 299
signal injection, 299-300
signal tracing, 291, 299-300, 303
sparks and voltage arcs, 298
symptoms, 280, 281-283, **282**
evaluation, 285
obvious vs. ambiguous, 287-289
sync lock-in range testing, 302
test connections, 299
test equipment (*see also* test equipment/procedures), 279
testing to locate faulty parts, 292
tools, 280
transient voltages, 298
video amplifier testing, 302
video/chroma circuit testing, 302
voltage testing, 293, 294
waveform analysis or comparison, 290-291, 293-294, **294**

V

variable frequency oscillator (VFO), 23
vectorscopes, 242, 246, 270-271, **272**
vertical sweep circuits, 7, 89-129
adjustments, 96
blanking pulse, 7
convergence circuits, 126-127, **128-129**
deflection yoke, 95
digital-TV circuits, 224, **225**
distortion, nonlinearity, 97-98
flyback transformer (FBT), 89
frequency, 7
high voltage power supplies, 89
IC-type vertical sweep circuits, 107-116, **108**
insufficient height, 96
jungle-clip type, 116-124, **117**
line splitting, 97
no horizontal sync, 92-93
no sync, 91
no vertical sweep, 95
no vertical sync, horizontal sync good, 91-92
operation of vertical sweep circuits, 93, **94**, 95
oscillator
IC-type, 107, **109**, 110
jungle-clip, 116, **117**
vertical oscillator, 93, **94**, 95
pincushion correction circuits, 126-127, **128-129**
processing, single-IC, 124
pulled picture, 93
signal path, 93, **94**
single-IC (25-inch set), 124-127, **125**
sync pulse, 7
sync separator, 89-91, **90**
test points, 95
troubleshooting approach, 91-93, 95-97, 295-296
vertical output, 116, 118, **118**, **119**
IC-type, 110, **111**
vertical-sync, vertical hold adjustment, 96
video amplifier, 9-10
blanking pulses, 9
brilliance (brightness) control, 9-10
contrast control, 9-10
input signals, 9
output signals, 9
testing procedures, 302
trap, sound or SIF trap, 9
video detector, 8
video IF (VIF) module, 8, 41-58
audio outputs, 54-55
audio, 45, **46**, 49, **50**
biasing, 41
circuitry, 45, **46**, **47**, **48-49**
hum bars, hum distortion, 44
intermittent picture and/or sound, 45
no picture or sound, 44
NTSC generator for troubleshooting, 43-44
output, 42-43
overloading, 44-45
phase-lock VIF, 49, 51, **52-53**, 54-55, **54**
PLL detector, 51, **52-53**, 54, **54**, **55**
poor picture or sound, 44
pulling picture, 44-45
signal analyzer for troubleshooting, 43-44

signal path, 41, **42**, **43**
smeared picture, 44-45
sweep generator for troubleshooting, 44
traps, 42
troubleshooting approach, 43-44
tuning, 41-42
video outputs, 54-55, **56**

video peaking adjustment, 267
video transformer tests, 257-261
video/chroma processing circuits, 59-88
basic operation, 59-61
blanking pulses, 60-61
brightness control, 60
chroma processing, jungle clip, 79, **83**, 84
circuitry, 67, **68-69**
color problems, 65-67
color sync out, 67
color-circuit basics, 64-67
comb filter, 69, **70**, 71, **71**, 72
composite video signal, 59
contrast problems, 62-63
jungle clip, 76, **77**, 78-84, **78**, **80-81**
matrix-demodulator circuit, 64-65, **64**
no color, 65-66
no picture, 62
no picture, sound normal, 62
no raster, 61-62
no sound, 62
operation of video/chroma, 68
poor picture, 63
processing operations, 68
retrace lines in picture, 63-64
RGB drivers, jungle clip, 84, **85**
RGB output, 73, **75**, 76
signal path, 59, **60**
Sony Trinitone switching, 84, 86-88, **86-87**, **88**
sound in picture, 63
testing, 302
trap, sound trap, 60
troubleshooting approach, 61-64, 302
color problems, 65-67, **66**
video processing, jungle clip, 76, 78-79, **82**, **83**
weak color, 67
wrong color, 67
Y/C processing, 72-73, **74**

voltage controlled oscillator (VCO), 23, 24, **26**, 27
voltage testing, 293, 294
volume-control, electronic, 213-217, **214**

W

waveform analysis, 290-291, 293-294, **294**
composite-video waveform analysis, 256-257, **257**, **258**

white-balance adjustments, 265-266, 268
wideband-amplitude companding, dbx noise reduction (NR), 191

Y

yoke
ringing-test procedure, 259, 261
testing, 257-261

ABOUT THE AUTHOR

John D. Lenk has been a technical author specializing in practical electronic service/troubleshooting guides for more than 40 years. A long-time writer of international bestsellers in the electronics field, he is the author of 80 books on electronics, which together have sold well over one million copies in nine languages. Mr. Lenk's guides regularly become classics in their fields, such as Lenk's Video Handbook, Lenk's Audio Handbook, Lenk's Laser Handbook, Lenk's RF Handbook, Lenk's Digital Handbook, McGraw-Hill Electronic Testing Handbook, and McGraw-Hill Circuit Encyclopedia & Troubleshooting Guide, Volumes I and II.